D1356465

POST-IMPRESSIONISTS
IN ENGLAND

Edited by

J.B. BULLEN

Reader in English
University of Reading

ROUTLEDGE
LONDON AND NEW YORK

First published in 1988 by
Routledge
11 New Fetter Lane, London EC4P 4EE

Published in the USA by
Routledge
in association with Routledge, Chapman & Hall Inc.
29 West 35th Street, New York, NY 10001

Set in 11/12 Lasercomp Bembo
by Thomson Press (India) Limited, New Delhi
and printed in Great Britain
by T.J. Press (Padstow) Ltd
Padstow, Cornwall

Compilation, introduction, notes, chronology, bibliography and index
© J.B. Bullen 1988

Library of Congress Cataloging in Publication Data

Post-impressionists in England.
Bibliography: p.
Includes index.
1. Post-impressionism (Art)—France—Public
opinion. 2. Public opinion—England. 3. Art
critics—England—Attitudes. I. Bullen, J.B.
ND547.5.P6P674 1988 759.4 87-28505
ISBN 0-415-00216-8

British Library CIP Data also available

For my father
who was born in the year
of the first Post-Impressionist exhibition

Contents

CONTENTS

CONTENTS

Preface

The reception of Post-Impressionism in Britain was a very different affair from the reception of Impressionism. Not only was the impact of Post-Impressionist painting much more sudden and violent than that of Impressionist art which took the form of a gradual assimilation, but from the start the ideas and nuances contained in the term 'Impressionism' were quite distinct from the explosive associations of 'Post-Impressionism'.

The most important and obvious difference between the two terms lies in their respective origins. 'Impressionism', it will be remembered, is a translation from the French. It was invented in France and was readily adopted by the Impressionist painters themselves. The Impressionists looked upon themselves as a group, however loosely knit, and they shared many aesthetic aims and ideals. This was simply not the case with the so-called Post-Impressionist painters. In fact, the whole notion of Post-Impressionism was promulgated in England and none of the painters who were given the name would have recognised or understood it. It was formulated on the spur of the moment to form the title of Roger Fry's 1910 exhibition at the Grafton Galleries – 'Manet and the Post-Impressionists' – and it passed instantly into cultural history. It was convenient, potent, deeply misleading, yet in a strange way it admirably reflected the current state of English ignorance about modern French art. Many contemporary writers, artists and critics who were conversant with French painting tried, in vain, to resist it, pointing out that it did not do justice to the variety and complexity of French art. And as recently as 1979 the large exhibition at the Royal Academy entitled 'Post-Impressionism; Cross-currents in European Painting' offered a challenge to the exclusiveness of Post-Impressionism by demonstrating the enormous diversity and the persistent continuity in late-nineteenth- and early twentieth-century European art. Roger Fry and Clive Bell continued to use it, however, because for them it performed the useful function of separating a whole generation of painters of whom they approved from a generation – the

Impressionists – of whom they approved rather less.

The intrinsic difference between Impressionism and Post-Impressionism as art-historical concepts has led to the radically different organisation of *Post-Impressionists in England* from Kate Flint's volume, *Impressionists in England*. First, the time span of this volume is much shorter. Impressionism was absorbed slowly over a period of twenty years, whereas Post-Impressionism, as it was conceived by Roger Fry and his contemporaries, was assimilated between 1905 and 1914. Essentially this volume gives a cross-section of developments in taste in those years and concentrates especially on the brief, but complex period between 1910 and the First World War. Second, the terms of reference of this volume are substantially different from its companion. 'Impressionism' referred almost exclusively to painting, whereas from the start Post-Impressionism was perceived of as much as an ontology as a style in art. Though on one hand Post-Impressionism was wilfully exclusive, cutting off Cézanne, Gauguin and Van Gogh from their nineteenth-century predecessors, it was at the same time something of a portmanteau term which expanded to contain Fauvisim, Cubism, Futurism and Vorticism – indeed almost any modern style which did not conform to the British view of nineteenth-century art. Furthermore, Post-Impressionism was rapidly extended to the other arts. There were comic accounts of Post-Impressionist gastronomy and dress design, but on more than one occasion Stravinsky's *Rite of Spring* was seriously described as a Post-Impressionist work as was the music of Richard Strauss and Schoenberg. Granville Barker's avant-garde production of Shakespeare's *The Winter's Tale* in 1912 was called Post-Impressionist and one writer put forward the idea that D.H. Lawrence's first novel *The White Peacock* was an exercise in Post-Impressionist literature.

But the contemporary debate about the virtues or evils of Post-Impressionism was deeply imbedded in the social, political and philosophical climate of the period and though it would have been possible to give a broader account of the reception of painters like Cézanne, Gauguin and Van Gogh – an account which would extend into the 1920s and 1930s – it seemed more valuable to offer an insight into the tumultuous, rapidly changing and complex years of English cultural life between 1905 and 1914 – years which are still enigmatic and baffling for the historian. Consequently the arrangement of the material in this volume is strictly chronological and the reader can observe, month by month, the speed and violence with which English taste and English attitudes were transformed by successive waves of innovation from the Continent of Europe.

The introduction attempts to offer an overview of the period and provide a context within which the selection of items in the main text can be placed. Necessarily the critical items reproduced in that text represent only a tiny proportion of the published material but they have been chosen to demonstrate the momentum which built up around the subject of modern art in these years. The principle of selection has been two-fold. First, an attempt has been made to print the most intelligent, sensitive and enquiring literature on the subject and, with all his limitations, Roger Fry stands head and shoulders above most of his contemporaries in this respect. Second, an attempt has been made to represent the huge range of opinion on the subject. And though the hysteria of some of the writing has little critical merit, nevertheless, the apocalyptic letters to the press about the work of Cézanne, Matisse and Picasso possess considerable sociological interest. There are also many curiosities which could not be overlooked: few people will know that Arnold Bennett, G.K. Chesterton and Rupert Brooke were moved to write about Post-Impressionism in the popular press. The chronology at the end of the book offers another perspective on the subject. Because Post-Impressionism was essentially an English concept its influence was felt strongly in English painting, which in its turn deeply affected British critics. The chronology enables the reader to see how British painting and Continental art were being exhibited side by side and the way in which exhibitions of modern art in London after 1910 rapidly caught up with those in Paris.

Finally it will come as something of a surprise to many readers that there is so little material in this volume about the Neo-Impressionists – Seurat, Signac and their many followers in France. The fact is, however, that they were almost totally ignored in Britain at this time and their position in art history in English owes most to the efforts of American critics writing in America after the First World War.

The mis-spelling of many of the artists' names by the critics of this period is symptomatic of the unfamiliarity and newness of Post-Impressionism. It has been decided, therefore, to preserve the original orthography.

Acknowledgments

The author and publishers gratefully acknowledge the permission of the following to reproduce the copyright material in this book:

The Trustees of the Bateman Estate for the cartoon by H.M. Bateman from the *Bystander* of 23 November 1910 (no. 19); Birchem & Co. for 'The Cubist Room, Art of Pablo Picasso' by Wyndham Lewis (no. 99); the *Burlington Magazine* for an unsigned review of February 1908 (no. 2), Roger Fry's letter of March 1908 (no. 3), Roger Fry's translation of Maurice Denis of January 1910 (no. 7) and Clive Bell's article of January 1913 (no. 90); Chatto & Windus The Hogarth Press and the Estate of Leonard Woolf for an extract from *Beginning Again* by Leonard Woolf (no. 73); Constable & Co. Ltd for extracts from A.M. Ludovici, *The Letters of a Post-Impressionist* (no. 91); the *Contemporary Review* for Roger Fry's article in the *Fortnightly Review* of May 1911 (no. 30) and W.R. Sickert's article in the *Fortnightly Review* of January 1911 (no. 29); Mrs J. Finberg for A.J. Finberg's article in the *Star* of 14 December 1910 (no. 23); Mrs Nicolete Gray for Laurence Binyon's article in the *Saturday Review* of 12 November 1910 (no. 14); Chatto & Windus for quotations from Roger Fry's *Vision and Design* ed. J.B. Bullen (no. 71) and Clive Bell's *Art* ed. J.B. Bullen (no. 102); the Society of Authors for 'The Art of Pablo Picasso' by John Middleton Murry (no. 47); the *Spectator* for Desmond MacCarthy's review of 26 November 1910 (no. 20); Times Newspapers Ltd for three reviews by Frank Rutter in the *Sunday Times*: 1 October 1911 (no. 42), 10 March 1912 (no. 57) and 10 November 1912 (no. 83); the Wyndham Lewis Memorial Trust for an article in the *Egoist*, 1 January 1914 (no. 99).

Although every effort has been made to trace the holders of copyright, in some cases these have proved impossible to discover.

Introduction

Virginia Woolf's exaggerated sense of a change having taken place in human character 'in or about December 1910'[1] was related in part to her feeling about living in what she called 'a Post-Impressionist age'.[2] Human character may not have changed with the advent in Britain of Post-Impressionist painting, but the language and discourse of art certainly underwent substantial modification, and the vocabulary and grammar of painting which had been evolving in France seemed, from a British point of view, to overthrow established traditions and stable artistic values.

In 1910 many people in Britain thought they were witnessing a break in the artistic continuity of European art and the arrival of so-called Post-Impressionist painting in England served to polarise attitudes to both the function and status of art in contemporary life. To those writers and painters whose values derived from English realism and English idealism, the primitivism of Gauguin, the expressionism of Van Gogh, the subjectivism of Cézanne and the violent simplification of Matisse and the other Fauve painters posed a threat not just to artistic technique – to matters of 'correct' drawing and 'finish' – but they seemed to undermine the very ontology which had formed the basis of English art for so many years. Post-Impressionism was not just a new way of seeing the world, it offered a new way of understanding the world. It defined a new relationship between man and nature, and developed a new connection between the spectator and the work of art. It was not just that the subjects of Post-Impressionist art were new or different; the principles of mimesis itself – perspective, represent-ationalism, the articulation of the picture surface – all seemed to have been reorganised. To many commentators this shift in the ontology of art looked like regression, degeneracy, even lunacy; to others it was liberating – freeing art from traditions which had become stifling and constraining.[3]

The problem of accounting for Post-Impressionist painting was compounded by the inadequacy of contemporary critical language. The familiar vocabulary, diction and phrasing of art criticism was unable to cope with Post-Impressionist painting and the violent and inarticulate journalistic abuse directed at modern art in this period testifies both to the power of new and unfamiliar images and to the paucity of current critical terminology.

The violence of the British response to French Post-Impressionist art was born of unfamiliarity. Unlike Impressionism which, as Kate Flint points out,[4] was slowly absorbed into the British system over a period of twenty years, Post-Impressionism and its associated styles, Cubism, Futurism and Vorticism, burst upon the English public with startling rapidity. Within a period of three years – between 1910 and 1913 – Britain was forced to catch up with developments in Continental art which had taken twenty years to mature. Exactly why the British were so ignorant about French art, however, is not easy to explain, but self-satisfaction with English painting and a legacy of Victorian economic and cultural superiority were partly to blame. But the unstable and explosive state of French taste itself was also responsible. In the period 1890–1905 the French critical response to recent art makes it clear that the French themselves were deeply puzzled by the changes taking place in their own culture.

The history of art in this period focuses strongly on the two Post-Impressionist exhibitions organised by Roger Fry in 1910 and 1912. The significance of these shows remains unquestioned, but some adjustment is neccessary. The enormous prominence given to them in the history of art has tended to overshadow events both before and afterwards and to oversimplify a complex and tangled period of cultural history. Without diminishing the importance of these two shows it is necessary to see them in their context. They were neither unheralded, nor were they the only avenues by which modern art entered English life. After the Grafton Galleries exhibition of 1905 a number of Post-Impressionist works were seen in England before the first Post-Impressionist exhibition of 1910 and in any case many contemporary artistis and critics were familiar with what had been happening in Paris in the first years of the century. Furthermore between Roger Fry's two major shows, British taste and understanding underwent significant modification. There was a further exhibition of the painting of Cézanne and Gauguin, there were prints of cubist pictures in the press, there was an exhibition of the drawings of Picasso, and the British were treated to the work of the Italian Futurists. But perhaps most important, the British were strongly influenced by changes which took place in English art itself.

Most disturbing for the conservative critics in these years was the way in which British painting was infiltrated by Post-Impressionist techniques. The first Post-Impressionist exhibition seemed to encourage experimentation on a huge scale and Roger Fry's second Post-Impressionist exhibition in 1912 celebrated the British contribution to the modern movement by bringing together native talent and pictures from the Continent.

The acceptability of any new art form is intimately dependent upon the written word and during the period 1910–14 art criticism changed substantially; broadly speaking it evolved in the direction of formalism where concepts such as 'rhythm', 'movement', 'decorative power', or 'expressive colour' were used to translate visual experience into verbal experience. In this context, Clive Bell's book *Art* of 1914 is a considerable landmark. It is easy now to point to the shortcomings of terms like 'significant form' and 'asethetic emotion' but the advance they mark in the use of formalist language is considerable. Such a monograph as *Art* would have been impossible in 1910 and, though Bell's debt to Fry is enormous, their combined work in this field is a triumph of English empiricism.

FEBRUARY 1905–NOVEMBER 1910

The reception of Post-Impressionism in this country must begin with the Durand-Ruel exhibition of French art at the Grafton Galleries in 1905. It was a popular show which contained over three hundred paintings but amongst the brilliant and colourful examples of work by Boudin, Renoir, Monet, Manet, Degas and Pissarro were ten pictures which were largely neglected. They were by Cézanne. Frank Rutter remembered asking himself: 'What are these funny brown-and-olive landscapes doing in an impressionist exhibition?' (see no. 1) and later asked himself the question: 'Where was Mr Roger Fry in 1905?' Fry himself replied by admitting that, like most of his fellow countrymen, he was 'sceptical about Cézanne's genius' in 1905 since Cézanne was regarded by most authorities as a minor and peripheral Impressionist. By 1906, however, when Fry saw two more pictures by

Cézanne at the sixth annual exhibition of the International Society, he modified his view of the painter At this stage, however, Fry still had reservations about the central importance of Cézanne – reservations which he would rapidly abandon. In 1906 he claimed that Cézanne 'touches none of the finer issues of the imaginative life';[5] by 1908 he would be claiming that no other artist could more subtly touch the 'finer issues of life' than Cézanne.

It would be misleading to place too much stress on the weight of Fry's opinion at this early stage in the history of Post-Impressionism in England. Much more influential and widely read were the histories of French painting published in this country by Camille Mauclair, Wynford Dewhurst and George Moore. Mauclair's *French Impressionists* (translated by Paul Konody in 1904) referred to Cézanne's 'robust simplicity of vision' but condemned his figure painting as 'clumsy and brutal'.[6] Dewhurst, also writing in 1904, was more circumspect – realising that the tide was beginning to change in France and opinion was 'curiously divided' over Cézanne.[7] George Moore, however, in his *Reminiscences of the Impressionist Painters* (1906) was frankly dismissive of Cézanne. 'It would be untrue that he had no talent,' said Moore, 'but whereas the intention of Manet and of Monet and of Degas was always to paint, the intention of Cézanne was, I am afraid, never very clear to himself'.[8] Clearly, Moore had never given Cézanne's painting a moment's serious consideration, a fact which emerges amusingly in his confusion of Cézanne with Van Gogh. For George Moore the work of this composite personality represented 'the anarchy of painting...art in delirium' and he referred to paintings of 'crazy cornfields peopled with violent reapers, reapers from Bedlam.'

The year 1906 was the year of Cézanne's death, and even if George Moore did not know who Cézanne was, the anonymous writer of his obituary in the *Athenaeum* did. Not only was he conversant with the facts of Cézanne's life but he also pointed out that the ten pictures by Cézanne at the Paris Salon d'Automne of 1906 had gone unnoticed by British journalists.[9] In fact two years were to pass before the work of Cézanne or any other Post-Impressionist was to be seen in England, but an exhibition of the International Society in the early months of 1908 at the New

Gallery materially advanced the story of Post-Impressionism in this country.

The International Society had been founded in 1898 as a reaction against the provincialism of English art. It had never been wildly adventurous but a small section of the 1908 show contained a number of paintings unusual for the time. Along with some English work there were several pictures by Degas, Monet, Denis and Vuillard but there were also two pictures by Cézanne, a Van Gogh, a Gauguin, two Neo-Impressionist paintings by Signac and Henri Edmund Cross and a picture by Matisse (see headnote for no. 2).

The exhibition was largely ignored by the press, but one review was to have important consequences for Roger Fry. In what the anonymous writer in the *Burlington Magazine* called 'The Last Phase of Impressionism' he described the 'infantile' work of Matisse and his contemporaries as the last, fading and effete flickering of the earlier Impressionist movement. He saw a parallel between the developments of this new school out of Impressionism and the 'now forgotten Flemish and Italian eclectics' (no. 2). This review stimulated Fry to reply and in a letter to the *Burlington Magazine* he explained what he felt were the intrinsic strengths of the new movement. In doing so he employed a contrasting historical analogy – one that was to become an abiding principle in the Bloomsbury defence of Post-Impressionism. The change from Impressionism to 'Neo-Impressionism', Fry said, should be likened not to decadence, but to the changes which took place in art between the late Roman Empire and Byzantium. Cézanne and Gauguin, he asserted, 'are not really Impressionists at all. They are proto-Byzantines' (no. 3). In his defence of these artists Fry inverted the historical categories of the original review and transformed the role of modern artists from decadents into revivalists so that they became, in his terms, the initiators of a new, stronger and healthier tradition than that which they superseded.

Between 1906 and 1909 Fry published nothing further on modern art, but when he became editor of the *Burlington Magazine* he contributed a translation of an article by Maurice Denis on the work of Cézanne. The attraction for Fry of Denis's view of Cézanne was that Denis saw Cézanne in the role of both classical and contemporary, simultaneously ancient and modern. 'He is at once', says Denis in Fry's translation, 'the climax of the classic tradition and the result of the great crisis of liberty and illumination which has rejuvenated modern art' (no. 7).

Roger Fry was of course not the only critic who was aware that important developments were taking place across the Channel and the British public itself was treated to regular accounts of the major shows of modern art in Paris between 1905 and 1910. The *Burlington Magazine*, the *Connoisseur* and the *Studio* all published notices of contemporary activity in the French capital, but curiously it was the better quality journals and newspapers rather than the specialist art journals that directed their readers' attention to the exhibitions of modern art.

From 1900 both the *Athenaeum* and *The Times* devoted space to the Salon des Indépendants and the Salon d'Automne. The autumn salon – and that of 1905, in particular, which first showed the work of the Fauves – held no terrors for the critic of *The Times*. This exhibition, he said, 'is made up, it is true, of revolutionary painters, but not of anarchists'. 'There is no sensationalism here,' he added, 'but a quiet harmony of effort to record frankly and honestly sensations that neither the tyranny of a "school" nor the direct suggestion of a master have imposed'.[10] In the following year, 1906, the correspondent was equally indulgent to the spring exhibition of the Salon des Indépendants saying that 'to neglect this association and to pay attention solely to the more official societies of the Grand Palais or the Champs Elysées would be deliberately to warp one's judgement as to the artistic activities of Paris'.[11] Unfortunately this mood of open-minded tolerance did not prevail elsewhere, and Bernard Sickert (the brother of Walter Sickert) who reviewed the same exhibition for the *Burlington Magazine* poured scorn on pointillism. For him, it was 'an ugly heresy from the beginning' and one which by 1906 had become 'such a thing of terror that no man of any pretensions to taste can put up with.'[12]

In 1906, however, the most significant exhibition from the English point of view was a large retrospective collection of the works of Gauguin in the Salon d'Automne. Even *The Times's* critic found this impossible to digest and the strength of his feelings and the terms of his animus anticipate the response to Gauguin's pictures when they first appeared in England. The 'unnatural and obtrusive colour', the 'indifferent drawing' and the 'contorted attitudes' infuriated him, but what he found most offensive was the fact that in France, Gauguin was treated by an articulate and growing minority as a great painter and part-founder with Cézanne of a new movement in art.[13]

Between 1906 and 1908 it is possible to detect in English criticism the beginnings of a firmer and more entrenched attitude to French modernism. In the Salon d'Automne of 1908 the spectre of primitivism initiated by Gauguin was raised once again in the work of Vlaminck, Derain and Rouault. The vituperation which poured from the pen of the art critic of the *Nation* is characteristic of the English response and is a measure of the extent to which Fauvism was dominating the modern salons. These painters, he wrote, are merely 'fettered by the conventions of rock carving among savage people'.

> In trying to be new, to break loose from dark brown shadows, they now see everything in acid blue and black, fondly believing themselves original, and above all "new." Still lower in the scale are the perpetrators of the strident dissonances, which raise a smile from the indulgent, and despair from the more thoughtful; these apparently naif artists are indeed omniscient, and exasperated at their own inability to be original, seek refuge in the clumsy and would-be archaic.[14]

Even Robert Dell, who was later to be instrumental in bringing modern art to England (see headnote to no. 8), attacked 'the productions of ladies and gentlemen who can neither draw nor paint...and represent violet persons sitting under pink trees'.[15]

The year 1908, however, was the year of Matisse in France. He had thirty pictures on show at the Salon d'Automne; then in December of the same year his 'Notes d'un peintre' were published in *La Grande Revue* and, as we have seen, his first picture appeared abroad in the International Society's exhibition in London. Dell was impressed by Matisse's writing and he praised his pictures for their 'daringly original simplicity of colour and outline'.[16] But it was once again *The Times* which was most generous. If the readers of *The Times* thought that the New English Art Club represented revolutionary painting then, said the anonymous critic, they should go to Paris where they will see 'real revolutionary painting, compared to which the most extreme works of Mr. [Augustus] John are as timid as the opinions of a Fabian socialist compared with those of a bomb-throwing anarchist'. Modern artists in Paris, he continued, 'are sick of pictures that are the mere dull records of fact' and he sees Matisse as the new David whose 'main concern is to avoid rhetoric, the compromise and the dull imitation of the schools'. Matisse, said the critic, 'wants to create the art of painting anew . . . in his own fierce way' and the review concluded with a highly detailed and sympathetic account of Matisse's *Harmony in Blue* (no. 4).

By 1908 Continental support for Matisse's work was considerable. The French dealer Vollard displayed his work prominently; the Stein family had begun their large collection of works by Picasso and Matisse which was open to inspection by the curious and Bernard Berenson published a forceful letter in the pages of the American journal the *Nation* (no. 5) stressing the virtues of this new painter. Berenson who, like Fry, had created a reputation as a connoisseur of Italian art, may have acted as an influence on Fry at this point. Although relations were strained between them, Fry dined at Berenson's Italian home, I Tatti, in 1907 and the subject of modern art may well have been broached. Fry, however, was slower than Berenson to accept the vividness and vitality of Matisse's art. Even as late as 1909 he wrote to his wife after a visit to Matisse's studio that Matisse is 'one of those neo, neo Impressionists, quite interesting and lots of talent but very queer. He does things very much like Pamela's [Fry's seven year-old daughter]'.[17]

If English curiosity had been aroused by accounts from France of the painting and sculpture of Matisse, another event in England in 1908 served to stimulate that interest and perplexity still further. An English translation of a book appeared which focused for the first time in art history on the work of those painters that were later to be called 'Post-Impressionist'. Meier-Graefe's *Modern Art* (see no. 6) was unlike all other available accounts of nineteenth-century European painting in that it did not accept Impressionism as the height of the modern movement. Instead it stressed the importance of the work of Cézanne, Gauguin and Van Gogh, and went on to examine the part played in the modern tradition by the Nabis and the Neo-Impressionists. Meier-Graefe was a German who, when he lost his job as the editor of *Pan* in 1897, moved to Paris. There he became the editor of *Art decoratif* and published *Art* in German in 1904. The British found the work difficult, exasperating yet exciting. They had seen few or none of the pictures to which

Meier-Graefe referred, yet many writers, including Fry, were deeply impressed by his romantic and 'expressionist' account of Post-Impressionist art – an account which stressed the creative roles of artistic temperament in painting.

Back in England in 1908, the gap between London and Paris narrowed slightly through the efforts of Frank Rutter and what he called his 'Allied Artists' Association' – an exhibiting society modelled on the Salon des Indépendants. 'Independence' was not a quality very prominent amongst English artists at this period who were hard pressed by economic exigencies. Back in 1906 an exhibition entitled 'Some Examples of the Independent Art of Today' at Agnew's was much criticised for its lack of true imaginative independence and its narrow confinement to established British painterly conventions.

So Rutter had the idea of forming a body which was genuinely independent of official juries and where none of the exclusiveness of the academic tradition prevailed. He had long been familiar with the French salons and had taken every opportunity to visit Paris. During his career as art correspondent for the *Sunday Times* and the English correspondent for *L'Art et les artistes* he had witnessed the outstanding success of the Salon des Indépendants and the Salon d'Automne, so like them his 'London Salon', as it became known, threw open its doors to all and sundry.

The response was enormous – so enormous that in 1908 the Royal Albert Hall was hired for the first show and many painters who in the next few years would be labelled as English Post-Impressionists – Alfred Wolmark, Roderic O'Conor, J.D. Fergusson and Harold Gilman – exhibited their work. The Allied Artists' exhibition provided a platform for all kinds of styles and techniques and its catholicism and openness prepared the way for Roger Fry's much more concentrated, explosive and dramatic exhibition of French art in 1910.

It is a fact, however, that many of the successful artists who showed at the Allied Artists' exhibition were familiar with the work of the first generation Post-Impressionists and with more recent French art long before either were seen in London. The Scottish painter J.D. Fergusson, for example, had lived in Paris for many years and had frequently exhibited at the modern salons. In 1909 Frank Rutter invited him to review the Salon d'Automne in the journal of the Allied Artists' Association, *Art News*. Of the pictures by Manguin, Puy, Friesz, Van Dongen, Anne Estelle Rice and Sickert which Fergusson mentioned, it was, for Fergusson, the work of Matisse that was most impressive.[18]

Like Fergusson, his friend Walter Sickert was no stranger to French art. When in 1905 he returned from his self-imposed exile on the Continent he actually confessed that he wished to 'create a Salon d'Automne milieu in London'[19] and his Fitzroy Street Group of 1907 grew, in part, out of that desire. Like Fergusson, too, Sickert had shown his pictures regularly at the Salon d'Automne and had even put on a one-man show in Paris in 1907. Unlike Fergusson, however, Sickert did not like the work of Matisse, and when he reviewed the English translation of Théodore Duret's *Manet and the French Impressionists* (1910) in *Art News* he poked fun at Berenson's defence of Matisse, making it clear that he felt that Berenson had made a fool of himself about modern art.[20] Nevertheless Sickert, in the same review, bemoaned the insular state of English art, and his own 'Salon d'Automne milieu' did something to change this by bringing together in one group many painters familiar with Continental art.

Spencer Gore, for example, enjoyed the works by Cézanne at the Durand-Ruel exhibition in 1905, and was deeply moved by the Gauguins at the Salon d'Automne of 1906 which he visited in Sickert's company. Unlike Sickert he also liked the work by Matisse in that same exhibition. Charles Ginner, too, joined the Fitzroy Street Group with a good knowledge of French art. He had been born in France and educated at the Académie Vitti. His love of Van Gogh caused the displeasure of his teachers at Vitti's, and when Rutter persuaded him to review the International Society's exhibition for *Art News* in 1910 he expressed his admiration also for Vallotton, Vuillard and Matisse.[21]

Other painters less directly involved with the Fitzroy Street Group had also encountered something of the ferment in French painting. Robert Bevan had met Gauguin as early as 1894 at Pont Aven; in 1905 Clive Bell and Vanessa Stephen visited Roderic O'Conor at his Paris studio and both expressed admiration for his collection of works by Cézanne and Gauguin; Duncan Grant met Matisse and Picasso on his visits to Paris and C.R.W. Nevinson said that he had heard of 'the "mad" painting of Van Gogh some five years before their "discovery" by Roger Fry and the dealers'.[22] The list of artists familiar with the paintings at the independent salons before 1910 could be extended to include J.B. Manson, Eric Gill, Wyndham Lewis, Phelan Gibb, Augustus John and the Americans Nan Hudson, Ethel Sands and Anne Estelle Rice. Roger Fry was a relative newcomer to the field, and it is a testimony to his energy and intuition that in spite of this disadvantage it was he who was primarily responsible for bringing modern art to England.

The main inspiration for Fry's first Post-Impressionist exhibition in 1910, however, came neither directly from France nor from Frank Rutter's Allied Artists' Association but from an exhibition held earlier in the year on the south coast. The mayor of Brighton felt that one way of employing the resources of the recently opened gallery and at the same time of establishing an *entente cordiale* with the annual influx of Continental visitors would be to put on a show of modern French art. The exhibition was divided into three sections, two of which posed no problems for visitors. The first contained work by late nineteenth-century realists, *plein-airistes* and genre painters; the second was devoted to Impressionist work; but the third comprised a collection of pictures of a kind never before seen in Britain. The earliest work in Room Three was a portrait of Albin Valabrègue by Cézanne; the most recent were two still-life studies by Matisse. There were also three pictures by Gauguin, Neo-Impressionist work by Signac, Cross and Luce and paintings by Vuillard, Bonnard, Vallotton and Denis. There were works by Marquet, Friesz, Sérusier, Puy, Valtat and Laprade together with 'Fauve' paintings by Derain and Vlaminck. This was by no means a complete cross-section of advanced painting in France – there were no pictures by Van Gogh or Seurat, for example – and no one knew this better than Robert Dell, who was responsible for organising the contents of this section. Dell, who lived in Paris, explained in the preface to the catalogue why he had had so much difficulty in obtaining representative examples (no. 8). Dealers in France were simply not interested in expanding into the British market. By 1910 they were fully employed supplying the demands of private individuals and the organisers of exhibitions in Europe and Russia. Their intuitions were correct. Almost nothing was sold from the third section, but it was this section which received the most critical attention.

The principal objections to the new art were poor technique, absence of naturalism and archaism. 'What is one to think of Paul

Gauguin's ideas of oxen – "Les Boeufs"?' asked one critic. 'They are wooden looking beasts akin to those of the nursery Noah's ark variety, and their landscape environment is innocent of any attempt at perspective.'[23] 'Must one', asked another, 'accept a picture, in which oxen are represented as flat things with giraffe-like necks standing before a house, all open in front so that one can see the contorted inhabitants, as real art?'[24] The attention of almost every critic of the exhibition, however, was captured by two views of London (see headnote to no. 8). They were both painted by André Derain on a visit to the capital in 1905 and they fascinated critics because not only was the subject well known, but so, too, was its treatment at the hands of the Impressionists. The problem was mainly one of colour. Valtat could be forgiven for turning his picture of a Provençal garden into a riot of colour, but, as Walter Higgins asked the readers of the *Art Chronicle*, 'can an artist who paints the Thames Embankment with yellow sky, pink trees and pavements, yellow water, blue cabs and green houses, by any means be serious in his art?'[25]

In their attacks on what they saw as the childishness, self-consciousness, affected primitivism and offensive crudeness of modern French art all the hostile critics shared one quality – bewilderment. The puzzlement and confusion was, of course, much greater at the first Post-Impressionist exhibition, but the Sussex local press provided an example in miniature of what was to happen on a much larger scale in November 1910.

Not all the notices of the Brighton exhibition, however, were unfavourable and those that praised the exhibition can be divided into two groups. One group led by Robert Dell stressed the historical continuity of modern French painting. In the introduction to the catalogue he reminded his readers about the controversy over Impressionism and how the Impressionists were once labelled 'anarchists' because they appeared to sever all links with the art of the past. He urged his readers to look closely at what he called 'Neo-Impressionism' and there they would see that many of the painters were operating within a set of values whose origins lay deep in the nineteenth century (no. 8). This theme was taken up by a long and anonymous review in *The Times*. The writing has all the hallmarks of Roger Fry's style as he argues that Maurice Denis 'tries to achieve the grandeur of Piero della Francesca' and how Matisse is 'more primitive than the early Sienese in his reaction against photographic realism'. He concludes with the hope that out of this art 'a new Giotto will arise and once more unite representation with expression, so that European painting may start again upon a steady course of progress' (no. 9). The other group is best represented by the writing of Frank Rutter. In three substantial reviews he emphasised the revolutionary newness of the paintings in Room Three and praised André Derain in particular for '[taking] his courage in his hands and, throwing overboard the whole cargo of art history, ancient and modern [trying] to forget that a picture was ever painted, and with eyes freed from traditional vision . . . seeks to recreate the barbaric art of infancy.'[26].

The uncertainty amongst critics and the confusion amongst the public about French art of the period was summed up by Lewis Hind, art critic of the *Art Journal* and the *Daily Chronicle*. In 1910 he went to Paris where he saw the large Stein collection of Matisses. He was aware that something important was happening in the French capital and on the eve of the first Post-Impressionist exhibition he rhetorically asked his readers whether they had heard of Matisse: 'Do you know his works?' Hind asked:

> Ninety-five out of a hundred laugh when they see them. Five think. Do you know the 'cube' paintings of Pecasso [sic], the improvisations of Wassily Kandinsky, or the chaotic interpretations of London by André Derain? Pause before you embark on that voyage. You will find yourself without a chart, without a compass. Beauty, say these protagonists, quite rightly, is not final. To understand us you must break through your conventional ideas of beauty.[27]

NOVEMBER 1910–JANUARY 1911

Roger Fry held the private view for his first Post-Impressionist exhibition on Saturday 5 November 1910 and the exhibition opened to the public on the following Monday. The story of the arbitrary way in which it was put together has often been told and perhaps the best account came from the pen of the exhibition Secretary, Desmond MacCarthy, in his book *Memories*.[28] The exhibition itself was an instant *succès de scandale* and Frank Rutter gave an eye-witness account of the audience in those early days. 'Every day people flock to the galleries,' he said, and most of them give vent to their feelings in language more audible than polite. Angry old gentlemen shake their fists in their impotence and cry aloud that all this is just done for advertisement.... Scandalised ladies murmur their disgust and wonder how anybody dared to exhibit such disgraceful daubs.... Fashionably dressed young men pry closely into the canvases in the hope of discovering some immorality to explain the uproar, and find nothing there so shocking as their own prurient imagination. What an audience! the painters, if still alive, might well retort to those who cry upon them, 'What an exhibition'.[29]

The press response was overwhelming and in November alone there were more than fifty articles, reviews, letters, cartoons and parodies in the daily papers and the journals. The Saturday edition of the *Sphere* was the first off the mark with a full-page spread of reproductions under the heading 'The Latest Revolt in Art'[30] but by Monday hostile reviews began to appear in large numbers. Above all the 'Post-Impressionists' – and the label was instantly taken up[31] – were interpreted as an anarchist group bound in an unholy alliance and bent upon the destruction of the civilised values of the west. Their work, said the critic in *The Times*, was 'like anarchism in politics' and represented 'the rejection of all that civilization had done, the good with the bad'.[32] But it was Robert Ross's long article in the *Morning Post* which stated most clearly the special fears generated by Post-Impressionism. He wittily pointed out that a more appropriate date than 5 November could not have been chosen on which to reveal 'a widespread plot to destroy the whole fabric of European painting' but what disturbed him most was the apparent subversion of reason, sanity and decorum in the painting. Van Gogh's work was the 'visualised ravings of an adult maniac' and Matisse followed 'the Broadmoor tradition' in his use of discordant colouring. Ross likened the taste for Post-Impressionist work in Europe to a terrible disease or infection and suggested that it should be treated 'like the rat plague in Suffolk'. The source of the infection', he said, referring to the pictures at the Grafton Galleries, 'ought to be destroyed' (no. 11).

The attacks on Post-Impressionist art in the following weeks took three main forms and adopted three main themes. The first was connected with the political and cultural threat posed by the new movement; the second was the subjective 'insanity' and 'egoism' of Post-Impressionism and the third was an attack on the self-conscious primitivism of modern art. All the adverse comments, however, were based upon a conspiracy theory which quite inaccurately assumed that the Post-Impressionists were a closely linked fraternity with common aims and common ideals.

'Anarchy in High Art' – a headline in the *Tatler* on 23 November – expressed a fear which was widespread amongst the critics of Post-Impressionism. For many writers what they saw in the Grafton Galleries represented an attack on the values of western culture from the ground of art. Though this now seems bizarre, it must be remembered that in 1910 political and social disturbance was a constant topic of public debate. The suffragette movement was strong and gaining strength; rumblings of unrest were developing amongst the miners of South Wales – discontent which culminated in the violent strikes of the following year; Kropotkin was living in London and publishing works on anarchist theory, and *The Times* of 1910 was filled with accounts of the activities of anarchist groups in Italy, Japan, Russia, Spain, Switzerland and, above all, France, the home of Post-Impressionism. The worst fears of the British about anarchism were confirmed in December 1910 when policemen were killed in a gun battle with anarchists in Houndsditch. 'Anarchism' was a word which occurs again and again in British reviews of French painting in the earlier years of the century[33] and the fear of anarchism plainly heightened the hysteria of traditional painters like Philip Burne-Jones and William Blake Richmond and conservative critics like P.G. Konody and Robert Ross. The panic was infectious. The critic of the *Spectator* pointed out that 'in France a reform movement always has its section who are for barricades, the guillotine, and the Anarchist's bomb'[34] and in an article entitled 'Pop Goes the Past' Holbrook Jackson explicitly connected what he saw at the Grafton Galleries with a 'deep-rooted revolt against the past' and the 'dark and forbidding names' of Bakounin and Max Stirner (no. 26).

But it was Ebenezer Wake Cook whose letters to the *Morning Post* were the most violent and extreme. Though Cook was primarily a painter (see headnote to no. 17), in 1904 he published *Anarchism in Art and Chaos in Criticism* which was much influenced by Max Nordau's book *Degeneration* (1895) with its prophecies of western decline. Now, in Post-Impressionism, Cook found the confirmation of his worst fears. The Post-Impressionist painters, he said, 'are the analogue of the anarchical movements in the political world, the aim being to reduce all institutions to chaos; to invert all accepted ideas on all subjects' and they turned the Grafton Galleries 'into a Morgue for "Modernity" art' (no. 17). Cook's views were based on Nordau's theory of psychological decadence and the notion that mental disease had become endemic in the west. This, according to Nordau, was most clearly traceable in the styles of modern art – a view which found echoes in many of the early reviews of the Post-Impressionist exhibition. Matisse's *Femme aux yeux verts*, for example, represented 'the imbecility of an intentionally childish daub' for P.G. Konody and Van Gogh's pictures were 'merely the ravings of a maniac' or 'the expression of an unhinged mind'.[35] His painting was, claimed Robert Ross, 'of no interest except to the student of pathology and the specialist in abnormality' (no. 11). It was, however, the academicians who were the most active opponents of Post-Impressionism. Blake Richmond, Philip Burne-Jones the painter son of Sir Edward Burne-Jones, Cook himself and Charles Ricketts (no. 13) were all trained in a nineteenth-century British tradition which seemed totally at odds with the techniques of Post-Impressionist art. Blake Richmond's letter to the *Morning Post* about the 'daubs by living French experimentalists' betrayed a fear itself which tottered on the brink of mania. 'One cannot reason', he wrote, 'where there is no reason to start with. The marvel is that these hysterical daubs should have been perpetrated at all, but various forms of disordered mentality are common to these times' (no. 16). Back in the 1880s Richmond had inveighed against Turner's late pictures; in 1893 he had spoken out against Degas's picture *L'Absinthe*,[36] and now, nearly thirty years later, he was castigating the 'egoism' of the Post-Impressionists for failing to follow the path dictated by nature and for abandoning themselves to self-indulgence. 'Health' and 'virility' are threatened with contamination by Post-Impressionism, and curses are the only recourse of the old painter. 'There is no regeneration for deluded egoists', he wrote; 'They are lost morally in the inferno where Dante places the unfaithful to God and to his enemies' (no. 16).

If the attitude of academicians to Post-Impressionism was unintentionally comic, by the end of the first month many journals had begun to discover the consciously comic aspect of this exhibition. The *Bystander* published the first set of cartoons showing the uncontrollable laughter on the faces of visitors complete with parodies of pictures by Van Gogh (no. 19). *Punch*, for whom aesthetic controversies had always been a fruitful source of amusement, published comic dialogues set in the Grafton Galleries and the *Tatler* obtained a cheap laugh by printing a reproduction of a work by Herbin in a number of positions claiming that it mattered little which way up it was seen.[37] As for the supporters of the new art, they wisely kept their powder dry and spoke up only in the second month of the show. Lewis Hind, Frank Rutter and Cunninghame Graham all published tempered responses to the barrage of criticism and Roger Fry sent an article to the *Nation* in which he stressed what he saw to be the strengths of French art. 'They are in revolt', he admitted, 'but against an effete and dead nineteenth-century tradition; they *are* anarchists, he said, though, 'they are not destructive and negative, but intensely constructive' (no. 18). The second month of the show was marked by a substantial change in the tone of the critical response. The large number of items in the press dedicated to the subject was almost as great as in the previous month, but discussion of Post-Impressionism shifted from the review columns to the letters. What is most striking, however, is the way in which legions of protesters in November disappeared and gave way to an almost equally large number of champions of the new movement. It is true that Ebenezer Wake Cook kept up his clumsy attack and was joined by A.J. Finberg in the *Star*[38] while Robert Morley spoke of 'the debasement of the lives of the painters living in the Gay City' and their symptoms of 'disease and pestilence'.[39] But even after a mere thirty days such views were beginning to appear uncritical and absurd and even writers who were out of sympathy with the new 'movement' began to take a more reasoned and conciliatory approach.

Almost immediately a host of letters appeared in print accusing Wake Cook and Henry Holiday of indulging in mere mindless abuse, while Arnold Bennett reminded the readers of the *New Age* of Berenson's support for the painting of Matisse in the American journal *Nation* (no. 22). In fact the readers of the *English Review* of 3 December had already had an opportunity to read Berenson's words in a long article by Lewis Hind – an article which, though slightly naïve in its approach, set the tone for the apologists in the month of December. In it Hind went step by step through the stages of his own conversion to modernism, quoting liberally from Matisse's 'Notes of a Painter' and Meier-Graefe's *Modern Art*.[40] The case for the defence in this month rested on three main points. The first was that there already existed a large group of witnesses all willing to testify on behalf of Post-Impressionism. To the name of Berenson were added those of Rodin and Degas in France, von Tschudi in Germany, and Claude Philips, Herbert Horne, Lionel Cust and C.J. Holmes in England. The second point rested on the issue of historical continuity. One writer to the *Nation* accurately detected the link between Cézanne's peasant figures, Van Gogh's *La Berceuse* and the depiction of peasant life in the work of Millet and suggested that there were connections between the drawing of Picasso and Matisse and that of Augustus John. Another correspondent went so far as to liken the much-reviled painting by Matisse, *La femme aux yeux verts*, to the *Mona Lisa*. 'There are', he said in Paterian prose, 'strange secrets in her green eyes, deep tragic eyes that will fascinate the dreamers of future centuries, her destined votaries, as the eyes of the Mona Lisa have fascinated men from Leonardo's day to ours.'[41]

The third argument was concerned with the formal and technical achievements of the Post-Impressionists and it was this argument which constituted the substance of Fry's two articles for the *Nation* (nos 18 and 21). His second, published on 3 December, disarmed opposition by the candid admission that in many cases the examples of painting at the Grafton Galleries were highly inadequate or unrepresentative. Similarly candid is his description of his growing realisation of the importance of Cézanne to the movement as a whole. Even though he was a newcomer to the subject, Fry's writing possesses an authority and subtlety which none of his contemporaries could match. Speaking of Cézanne he says:

> As I understand his art, and I admit it is exceedingly subtle and difficult to analyse – what happened was that Cézanne, inheriting from the Impressionists the general notion of accepting the purely visual patchwork of appearance, concentrated his imagination so intensely upon certain oppositions of tone and color that he became able to build up and, as it were, re-create form from within; and at the same time that he re-created form he re-created it clothed with color, light, and atmosphere all at once (no. 21).

JANUARY–OCTOBER 1911

'The exhibition of the so-called "Post-Impressionists" was the most disturbing feature in the history of British Art in 1910', said the critic of the *Pall Mall Gazette* (no. 31) and the third month of critical reaction to that 'disturbance' was significantly different from the previous two. When the exhibition of Post-Impressionist painting closed in January 1911 the attitude of the public had changed significantly. Derision had given way to admiration and the *Daily Graphic*, in an article entitled 'An Art Victory: Triumphal Exit of the Post-Impressionists', was not the only newspaper to notice how 'public taste in pictures is advancing faster than the critics' (no. 32). The contrast between English painting and Post-Impressionist work was admirably pointed up by the exhibition of the National Portrait Society which replaced 'Manet and the Post-Impressionists'. Many writers were quick to point out that the colour, vibrancy and vitality of French art had instantly given way to English conventionality and dullness when the works of Cézanne, Van Gogh, Gauguin, and Matisse were replaced with pictures by Charles Ricketts, William Blake Richmond, Philip Burne-Jones, and Ebenezer Wake Cook.

By the early months of 1911 the supporters of Post-Impressionism had had time to gather their forces and three extended monographs on the new art appeared in quick succession. Frank Rutter's *Revolution in Art* was the first, quickly followed by Charles Holmes's *Notes on the Post-Impressionist Painters* and Lewis Hind's *The Post Impressionists* (nos 33, 34 and 35). Each took a very different view of the subject, but each offered a view which was broadly positive. At the same time substantial and explanatory articles began to appear in the journals. The scholarly *Burlington Magazine*, which in November 1910 had published a long piece by Meier Riefsthal on Van Gogh, now printed a substantial piece by Clutton-Brock who likened the creation of a new language of painting amongst the Post-Impressionists to the efforts of Wordsworth and Blake in creating a new language of poetry (no. 36) and even the popular and conservative *Art Journal* carried a long and well-illustrated article by Frederick Lawton on the life and work of Cézanne who also compared Cézanne with William Blake.[42]

Much of the writing about Post-Impressionism now began to be more discriminating, and it became clear to even the most enthusiastic that all the painters included in this so-called group were far from alike. In a long, witty and penetrating article in the *Fortnightly Review*, Walter Sickert took Roger Fry to task for devising an exhibition which so poorly reflected the state of late nineteenth-century and contemporary French art. Sickert was careful to distinguish his own views from the strident opposition of the academicians in England, and made it clear that he had a long-standing familiarity with the work of Cézanne and Gauguin, as well as a personal knowledge of many of the artists themselves. But he wondered what Cézanne had in common with the Fauves and wondered too, with justification, why Maurice Denis had been overlooked and why Vuillard and Bonnard had been excluded. Sickert's discrimination, however, was coloured by deep personal prejudice: 'my teeth are set on edge' by much of the work of Van Gogh, he said, and he made it quite clear that he had no time for what he called Matisse's 'art-school tricks' (no. 29).

In the early months of 1911 Post-Impressionism continued to supply material for comedy. 'Well if this is Art,' says a painter pointing to an 'infantile' Post-Impressionist work in a cartoon in *Black and White*, then 'I-You-and Michael Angelo are *all* wrong.'[43] In *Punch* a man gets his revenge on a Post-Impressionist portrait painter by paying for it in an illegible Post-Impressionist cheque,[44] and, again in *Punch*, the primitive Post-Impressionist 'Novel of the Future' is projected as an extended nursery rhyme complete with full-scale baby noises.[45] The Chelsea Arts Club staged a Post-Impressionist evening in the form of a fancy dress ball for a group they entitled 'Racinists' where the participants were invited to adopt some of the more extreme manifestations of Post-Impressionist art.

During this period Wake Cook kept up his barrage of letters to the press about the 'decadence' of Post-Impressionism; John Singer Sargent wrote saying that he was 'absolutely skeptical as to their having any claim whatever to being works of art';[46] Sir Alfred East at a dinner of the Authors' Club gave his opinions on 'the morbid art of the Post-Impressionists' and W.B. Richmond continued to fulminate at the Royal Academy, warning the students against the 'intellectual, emotional and technical degeneracy' of the Post-Impressionists with their 'wilful anarchy and notoriety hunting' which, he said, 'verged on criminality'.[47]

The reception of Post-Impressionism in England is intimately bound up with its influence on British art and from the moment that the first Post-Impressionist exhibition closed it became apparent that British artists had already been experimenting in the new forms. The exhibitions at Vanessa Bell's Friday Club became a platform for Post-Impressionist techniques where, for example, Duncan Grant's *Lemon Gatherers* (1910) was a testimony to the fact that he had already entered the 'Post-Impressionist age'. The work of J.D. Innes, said one critic, early in 1911, was imitative of Maurice Denis[48] and the influence of Picasso was recognised in both Ethel Wright's *The Arbour* and Anne Estelle Rice's *Egyptian Dancers* at the Women's National Art Club exhibition.[49] Phelan Gibb was declared to be 'an English artist who had debauched his by no means inconsiderable talent in the new cult'[50] and at the exhibition of the New English Art Club, Ethel Wright, according to P.G. Konody, proclaimed 'her sympathy with the new movement'.[51] Less surprisingly, Konody also detected the influence of Post-Impressionism at the Allied Artists' exhibition in July 1911. Robert Bevan's *The Courtyard* betrayed 'the unbroken primaries of the Matisse school' while amongst the newly formed Camden Town Group, Charles Ginner, Konody correctly remarked, 'appears to worship at the shrine of Van Gogh'.[52]

Just as the arguments in favour of the new art began to take on a more reasoned and discriminating tone, so the attitudes of the detractors hardened into something more calculated and sinister. The idea of a connection between modern art and insanity mooted in the earliest reviews of the Grafton Galleries exhibition was taken seriously by a number of artists. Consequently Dr T.B. Hyslop, Physician Superintendent to the Royal Hospitals of Bridewell and Bedlam, was invited to give an illustrated talk at the Art Workers' Guild in January. His subject was the art of the insane and his aim was to demonstrate the connection between the diseased minds of the 'lunatics in one of the metropolitan asylums' and the work of the Post-Impressionists. Roger Fry was invited to reply, but his eloquence did nothing to convince many members of the artistic fraternity. The substance of Hyslop's extraordinary lecture, entitled 'Post-Illusionism and Art of the Insane', was published in the *Nineteenth Century* (no. 38) and in it Hyslop accused modern artists and their admirers of being mental 'degenerates'.

'Degenerates', he wrote, 'often turn their unhealthy impulses towards art, and not only do they sometimes attain to an extraordinary degree of prominence but they may also be followed by enthusiastic admirers who herald them as creators of new eras in art'. The paper, which says more about T.B. Hyslop than about either art or madness, attempts to cast a slur on Post-Impressionism by invoking the authority of medical science. It addresses itself to all the prominent features of Post-Impressionist painting as they appeared to the public in 1911 and systematically identifies them with retrogression, cerebral degeneration and '"coprographia" – i.e. pertaining to lust, filth, or obscenity'. Sadly this was not an isolated incident. Hyslop's lecture was followed by a series of articles in *Art Chronicle*, entitled 'Health and Disease in Art', also written by an 'eminent physician' and which repeated Hyslop's arguments at greater length and illustrated them with drawing and painting by the insane.[53]

This initial period of ferment about Post-Impressionism concluded, however, on a saner note with the publication in a May issue of *Fortnightly Review* of a lecture given by Roger Fry at the end of the first Post-Impressionist exhibition. In contrast to so much of what had preceded it, it is urbane and full of lightly worn scholarship characteristic of Fry at his best. He put his audience at their ease by placing modern French art in its historical context and by pointing out that 'distortion' in art works is nothing new and is a common feature of early Italian painting. He also developed his ideas about the relation between the practical life and the imaginative life – ideas which he had first put forward in 1909 – arguing that the Post-Impressionist painters 'speak directly to the imagination through images created, not because of their likeness to external nature, but because of their fitness to appeal to the imaginative and contemplative life' (no. 30).

Yet even in these early months of 1911 it is possible to detect the beginnings of the next wave of disturbance which was to break over Britain from the Continent. This time it was a style of much more recent origin – a style which had been christened in 1908 by Louis Vauxcelles as 'cubism'.

In France examples of cubist painting could be seen in private collections and small galleries, but it was most prominently displayed in the annual modern salons. As we have seen, the Salon d'Automne and the Salon des Indépendants, had long been noticed by the English press, but one of the effects of the first Post-Impressionist exhibition was to create a much more lively interest in England about what was happening in Paris. Frank Rutter, who was a conscientious watcher of events in the French art world, welcomed what he frankly called the new 'anarchy' in art. 'The innocent abroad', he wrote in *Revolution in Art*, 'wandering through the Grand Palais, views with increasing concern and bewilderment room after room filled with what (to him) appear to be horrible monstrosities'. But this, for Rutter, was a sign of energy and life. He applauded the 'veritable passion for simplicity' in the work of Matisse and the Fauves, and he realised that Picasso was what he called 'the chief of post-Matisseism'. Nevertheless he was deeply puzzled by Picasso's 'new vision of form, building up his paintings with a series of cubes, greyish to yellow green in colour, about three inches square as a rule, cubes some square to the spectator, others at angles, and all ingeniously fitted together to express his feeling for form'.[54]

When Roger Fry visited the Salon des Indépendants in June 1911 he, too, was confused and puzzled by the huge array of different styles. He praised the work of Maurice Denis, Othon Friesz and André Lhote for their 'classic feeling for pure beauty' but Fry, like Rutter, was bafffled by the cubist work. 'There are those', he wrote, 'who, like Herbin, are following Picasso in his search for an artistic philosopher's store [sic], endeavoring to get at the intellectual abstract of form, whereby they can recreate a world of pure significance; and there are those who', he added, making a contrast which was later to have great significance in English criticism, 'following Matisse, search for an intenser unity in the balance of directions and volumes, and the just disposition of intervals' (no. 40). Huntly Carter's review of the same exhibition in *New Age* gives a particularly valuable insight into developments in French art at this period. He, too, realises that 'Picasso is the father of the new extremists' and commented on 'the latest development of Picassoism or Cubism' in the work of Herbin and Braque as being perhaps 'the most remarkable feature of the exhibition' (no. 39). Carter also mentioned the work of a group of Anglo-Saxon Fauves whose influence would very soon be felt in England – J.D. Fergusson, Anne Estelle Rice and Jessie Dismorr. One of Fergusson's paintings exhibited at the 1910 Salon d'Automne entitled *Rhythm* attracted the notice of John

Middleton Murry and Michael Sadleir. According to Fergusson, whom they sought out in Paris in 1911, these two had decided to start a magazine and wanted to call it *Rhythm* and use his 'picture as a cover'.[55] This magazine, which was to contain 'all the latest information about modern painting from Paris', appeared for the first time in the summer of 1911. 'Nobody', said the *Art News*, 'who wishes to keep pace with the movements and ideals of modern artists can afford to ignore *Rhythm*'[56] and even P.G. Konody was forced to admit that 'Post-Impressionism has evidently come to stay. It now has its official organ in the shape of the new shilling quarterly "Rhythm"'.[57]

NOVEMBER 1911–FEBRUARY 1912

By November 1911 the critical confraternity was able to distinguish readily between four main groups within the Post-Impressionist movement. Cézanne, Van Gogh and Gauguin were already beginning to look like established masters; the Fauves, with their intense colouring and their insistence on 'rhythm', were still outlandish in the eyes of many critics, though the modified form of Fauvism in the work of Fergusson, Rice, Peploe and Dismorr was beginning to be acceptable. Cubism, however, was quite another matter and, in spite of the fact that neither Braque nor Picasso exhibited, it was cubism which attracted most critical attention in the French Salon d'Automne of 1911.

Three events late in 1911 demonstrate very clearly how critical responses in England had changed. The first was an exhibition in November at the Stafford Gallery organised by John Neville at the suggestion of Michael Sadleir. Sadleir, who was an energetic collector, had been converted to Post-Impressionism earlier in the year, and had sent Neville to Paris to secure for him work by Gauguin and Cézanne (see headnote to no. 45). Neville's exhibition was well received. Cézanne, according to *The Times*, was the 'Wordsworth of painting'[58] and P.G. Konody said that Gauguin's *L'Esprit Veille* was so 'beautiful in design as well as in colour, and intense in expression, that it well deserves to be called a masterpiece' (see no. 45). 'Post-Impressionism', he wrote, 'has taken firm root among us' and its implantation was celebrated in a painting by Spencer Gore entitled *Gauguins and Connoisseurs at the*

Stafford Gallery (private collection) illustrating the pictures and containing small portraits of Augustus John, Wilson Steer and Neville himself.

The second event was an exhibition at the Goupil Gallery, also in November, which featured the British Post-Impressionism of Alfred Wolmark, S.J. Peploe and Augustus John. The reception was mixed. The critic of the *Outlook* thought that Wolmark's *Decorative Panel* drew attention to itself 'simply by its over-emphasis of crude, noisy colour', while Peploe's *Tulips* 'has certain unrestful decorative qualities which quite fail to charm'.[59] Huntly Carter, on the other hand, who used every opportunity to further the cause of French Post-Impressionism, had plenty of advice for Wolmark to intensify further his Post-Impressionist techniques.[60]

The item which caused the greatest controversy in the latter part of 1911, however, was a picture by Picasso. His *Mandolin, Wine Glass and Table* (now called *La Mandoline et le Pernod*) was reproduced in a poor monochrome illustration in the *New Age* of 23 November. Even in this form, however, it created a storm of protest which was remarkable in its ferocity and its duration. All the old arguments about 'decadence' were resurrected, and many of the earlier antagonists of Post-Impressionism reached once again for their pens. A huge correspondence was generated in the *New Age* and many other journals, and letters, articles and comments continued to pour in well into 1912. Huntly Carter was quick to defend Picasso and published an article in the same issue as the illustration (no. 46). Middleton Murry tried to support him but not before Ebenezer Wake Cook was in print speaking of 'the depths of degradation, inanity... [and] sheer lunacy' of 'Picasso-ism'.[61] G.K. Chesterton was also quick to condemn the 'sodden blotting paper' which Picasso had created, and in an article in the *Daily News* he elaborated his argument by condemning what he saw to be the paucity, shallowness and vacuity of modern art criticism which could not 'explain' Picasso to him (no. 48). Lewis Hind, who had championed Matisse so vocally, came out strongly against cubism and Roger Fry was uncharacteristically silent. It was left, then, to Carter and Middleton Murry to wage a war in favour of Picasso and to welcome in the 'new age' of non-mimetic art.

The debate over Picasso had hardly died down when a one-man exhibition of his drawings opened at the Stafford Gallery in April

1912. 'M. Picasso,' said *The Times*, 'the leader of the most advanced school of French painting, has been called an incompetent charlatan.' 'The exhibition of drawings by him,' the reviewer went on, '. . . proves that he is not that' (no. 64). Much to the dismay of Picasso's critics, all but one of the drawings in this exhibition were figurative studies. One of them, a donkey's head which *The Times* described as 'extraordinarily delicate and precise in chararacter' where, 'the animal is treated . . . with the imaginative seriousness of the great Chinese artists' was reproduced in the *Art Chronicle* for all to see. P.G. Konody, however, remained unconvinced by anything that Picasso did. 'Scraps that should never have been rescued from the waste-paper basket' was his verdict for the readers of the *Observer* and 'to exhibit them as works of art is simply *fumisterie*'.[62]

The idea that the whole modern movement was *fumisterie* or a spoof was strongly endorsed by two authorities who, in earlier years, had been staunch supporters of modernism. Walter Sickert adopted the kind of derogatory terms which one had become used to reading from the pen of Wake Cook when he said that:

> The conspiracy of semi-unconscious 'spoof,' which is looked upon by some as an alarming symptom of the artistic health of the present day, is in reality a very small and unimportant manifestation. In the story of the 'Emperor's New Clothes,' it was the whole nation that affected not to see that his Majesty was naked. The modern cult of post-impressionism is localised mainly in the pockets of one or two dealers holding large remainders of incompetent work. They have conceived the genial idea that if the values of criticism could only be reversed – if efficiency could be considered a fault, and incompetence alone sublime – a roaring and easy trade could be driven. Sweating would certainly become easier with a post-impressionist *personnel* than with competent hands, since efficient artists are limited in number; whereas Piccassos and Matisses could be painted by all the coachmen that the rise of the motor traffic has thrown out of employment. It is, after all, an extremely small circle of very unoccupied ladies who find amusement and excitement in going one better than the other in ecstasy at the incomprehensible (no. 49).

Sickert's tone is recognisable as that of the wounded professional at bay – the highly trained professional who feels jealous of those he considers amateurs. Perhaps even more remarkable is the attitude of D.S. MacColl who in the 1880s and 1890s had been an articulate champion of Impressionism. In a long article in the

Nineteenth Century he attacked Fry's interpretation of Cézanne's work and that of the first generation Post-Impressionists. Cézanne, says MacColl, 'was not a great classic: he was an artist, often clumsy, always in difficulties, very limited in his range, [and] absurdly so in his most numerous productions'; Gauguin's 'fine period was short,' he said, and for MacColl it was 'a drop from *L'Esprit Veille* to fantastic rubbish like *Christ in the Garden of Olives*'. Though his response to Van Gogh's work was more tempered, he had no time for Picasso and Matisse. He dismissed Picasso's work as 'geometrical mania' and Matisse's *La Femme aux yeux verts* was merely 'a silly doll' (no. 52).

But by the latter part of 1911 both Sickert and MacColl were beginining to look like two Canutes resisting the tide of modern art. Commenting on the change which had taken place in public opinion, the critic of the *Pall Mall Gazette* in January 1912 pointed out how in little less than a year 'scoffers have turned into admirers' and 'many of those who were most abusive and bitter in their denunciation have come to look with indulgence upon what only twelve months ago appeared to them intolerable!' 'It is no good,' he continued, 'closing one's eyes to the fact that the principles of Post-Impressionism in its milder form have permeated British art and renewed its vitality. Scores of exhibitions have familiarised us with the new rhythm, reconciled us to the synthetic, in place of the imitative, rendering of the facts of nature'.[63] Yet even now a new development was taking place in the language and scope of criticism. Ideas of 'rhythm' imported from France by J.D. Fergusson, enshrined in the title of Middleton Murry's journal, were being developed in different ways by Huntly Carter and Michael Sadleir. Carter's stress on the significance of rhythm as an important element of the new art led him to extend his interpretation of modernism to include music and drama. His championship of Picasso late in 1911 was developed in a full-length study entitled *The New Spirit in Drama and Art* where he explained that his contact with Post-Impressionist painting drew his attention to the fact that

> there was a vital connection between the advanced movement in painting and the movement in the theatre which, once established, would bring about a union of the two, set them mutually acting and reacting upon one another, and tend to remove all difficulties to the proposal to lift the theatre into the region of art. My subsequent

observations in Paris led me to the conclusion that rhythm is the connecting link between plastic forms of art and the 'scene,' and the continuous and consistent search for this is hourly bringing them closer together (no. 67).

For Carter it was the painting of Fergusson, Peploe and De Segonzac which had such strong links with the work of the directors of the *Ballet Russe* but for Michael Sadleir the synthesis between sight and sound was to be discovered in quite a different area. In an important article for *Rhythm* entitled 'After Gauguin' Sadleir examined Kandinsky's *Über das Geistige in der Kunst* where he paraphrased Kandinsky's ideas that 'Music, poetry, painting, architecture are all able in their different way to reach the essential soul, and the coming era will see them brought together, mutually striving to the great attainment' (no. 53). The route would be, of course, through abstraction and the 'psychological effects on the observer of various colours' – a development out of Mallarmé's theories about 'synthetic word painting'.

MARCH–OCTOBER 1912

The feeling that the principles of visual art were being enlarged by reference to the dynamics of physical movement, and that in their turn music, theatre and ballet were adopting the expressive potential derived from painting was enhanced by the appearance of the Futurists at the Sackville Gallery in March 1912. 'What must be rendered is "dynamic sensation – that is to say, the peculiar rhythm of each object, its intention, its movement, its interior force"' – so the *Spectator* quoted from the catalogue, commenting that 'the Futurist idea claims a certain kinship with the modern movement of French music – sensation rather than emotion'.[64] In spite of the declaration in the catalogue that Futurists were 'absolutely opposed' to the art of the Post-Impressionists and the cubists of France, nevertheless in 1912 Futurism seemed to be more closely related to Post-Impressionism than it appears in retrospect. Severini's pictures looked very much like the cubism of Picasso to a writer in *The Times* and many of the critics thought that they were attempting to outdo the Post-Impressionists in experimental terms.

Though P.G. Konody hailed the appearance of Futurism in

London with the headline 'Nightmare Exhibition at the Sackville Gallery' (no. 54), generally speaking most critics treated the incident as an *opéra bouffe*. Roger Fry in the *Nation* and C.H. Collins Baker in the *Saturday Review* gave a cautious welcome to Marinetti, Severini, Boccioni and Russolo, but they both felt that their experiments in the psychology of perception promised more than they achieved (no. 55). Even Sickert had some words of praise for Futurism. He was delighted to be able to report that 'Both Severini and Boccioni...are competent workmen' (no. 60) and the painting which received most critical acclaim was Severini's picture *Pan Pan Dance at the Monico* (destroyed in the second World War).

OCTOBER–DECEMBER 1912

The central months of 1912 were uneventful when compared with the same period in the previous year, but on 5 October the Second Post-Impressionist exhibition opened at the Grafton Galleries. In spite of Fry's contention that the critical response to this exhibition showed British philistinism 'as strong and self-confident and as unwilling to learn by past experience as ever' (no. 82), the fact is that the tone of most journalists was more guarded and more tolerant than it had been in 1910. There were, of course, the derogatory remarks that had become commonplace in British journalism. The critic of the *Morning Post* called it a 'deplorable and degrading show'; the *Star* warned its readers that the whole thing was 'vapid, empty, stupid and above everything, dull'; Collins Baker called it a 'mysterious, rather boring conspiracy to fool the public' and Anthony Ludovici thought that most pictures were 'pot-boilers', describing Marchand, Derain and Vlaminck as heralds of 'the decay and dissolution of art'.[65] But they were in a minority and most critics found complimentary things to say especially about the small section of works by Cézanne or the work of Duncan Grant, Frederick Etchells, Vanessa Bell or the sculpture of Eric Gill.

There were many reasons for this. One, of course, was familiarity. The period between the two exhibitions had been a highly educative one for the British. The Second Post-Impressionist exhibition was also much better organised and better thought out in its choice of pictures. Furthermore, with British work in the show, Post-Impressionism could no longer be accused of being a French conspiracy. The exhibition was divided into three parts; Fry chose those for the French section, Clive Bell those for the English group, and Boris Anrep was responsible for the Russian section. Post-Impressionist works from other countries were excluded because Fry felt that they had 'not yet added any positive element to the general stock of ideas' and in the case of the Italian Futurists they had merely developed a system based on 'a misapprehension of some of Picasso's recondite and difficult works' (no. 69). But, as in the case of the first Post-Impressionist exhibition, the introduction to the catalogue played an important part in the general response to the exhibition.

Bell's introduction to the French section was historically important because it contains one of the earliest formulations of the notions of 'significant form'. How, Bell asks, does the Post-Impressionist regard a simple household object such as a coal-scuttle? 'He regards it', says Bell, 'as an end in itself, as a significant form related on terms of equality with other significant forms. Thus have all great artists regarded objects' (no. 70). But the great critical problem of the show was presented by the cubist pictures, and both Fry and Bell adopted musical analogies to account for these. In Post-Impressionism, said Bell, 'We expect a work of plastic art to have more in common with a piece of music than with a coloured photograph' and Fry suggested that the logical extreme of Post-Impressionist art would 'be the attempt to give up all resemblance to natural form, and to create a purely abstract language of form – a visual music' (nos 70 and 71). In point of fact these analogies would have been better adapted to the work of Kandinsky, yet surprisingly Kandinsky's painting, which had been seen in London on several previous occasions, was not shown.

If Cézanne, Gauguin and Van Gogh had been the *enfants terribles* of the First Post-Impressionist exhibition, then Matisse and Picasso fulfilled a similar role in the Second. Responses to the forty-two works by Matisse ranged from the gushing enthusiasm of Lewis Hind to the condemnation of his 'indescribable outrages' by P.G. Konody (no. 76). His large design, for Prince Shchukin's *La Danse*, received most attention, closely followed by *Le Luxe* – a picture which appears prominently in Vanessa Bell's study *A Room at the Second Post-Impressionist Exhibition* (1912). Picasso, however, caused universal consternation. His cubist *Tête d'Homme, Tête de Femme* and *Buffalo Bill* were 'unintelligible to the eye and the mind' (*The Times*), as 'unsympathetic as they are unintelligible' (Lewis Hind), an 'utterly unintelligible tangle of...lines' (P.G. Konody) or just 'mad' (*Architect*).[66] The critic of the *Queen*, Martin Hardie, unintentionally but amusingly summed up the general dilemma when he wrote: 'most bewildering and confusing' is the *'cubisine* [sic] of Pisasco [sic].'[67] Even Fry betrayed considerable uncertainty about cubism. 'They may or may not be successful...', he wrote, adding that 'It is too early to be dogmatic on the point, which can only be decided when our sensibilities to such abstract form have been more practised than they are at present' (no. 71).

Some of the most constructive criticism came from Frank Rutter who in his articles for the *Sunday Times* had long been questioning the usefulness of the general term 'Post-Impressionism'. His first-hand familiarity with French art in Paris made it clear to him that the term tended to induce 'utter confusion' in England since it represented 'some half-a-dozen distinct and separate art movements which in France are given separate names' (no. 83). He offered a far more logical, chronological arrangement for the exhibition and when he organised the Post-Impressionist and Futurist exhibitions at the Doré Gallery in the autumn of 1913 he put his ideas into practice.

JANUARY–OCTOBER 1913

The year 1913 was a year of both consolidation and diversification for the Post-Impressionist movement. The British public had become more or less shock-proof to new forms and new movements and at least two events in 1913 which a few years previously would have caused outrage and alarm – the exhibition of Post-Impressionist furniture at the Omega Workshops and the lectures at the Doré gallery by Filippo Marinetti – were greeted with amusement and good nature by the press. The sense of bewilderment about cubism persisted but gone was the talk of 'insanity' and 'decadence'. In fact in August 1913 Sir George Savage (one of the doctors who treated Virginia Woolf) organised an exhibition of art works from lunatic asylums around the country, but now no comparison was made between the art of the insane and the modern art of France. On the contrary, *The Times* went out of its way to make any such suggestions inadmissable.[68]

But this year was marked by open hostility between those who had previously appeared united in the cause of modernism. Even in 1910 it was possible to detect the seeds of division in the differences between the conciliatory historicism of Roger Fry and the more revolutionary stance of Frank Rutter, and in June 1911 Fry himself noticed a radical distinction between those who followed Matisse and those who painted in the wake of Picasso at the Salon des Indépendants. Until 1913, however, unity had prevailed in the face of the common philistine enemy but the growing tolerance of the British public made internal disagreements less fatal. Consequently in 1913 there was real division about the relative status of the work of Cézanne and Matisse on the one hand and that of Picasso and the Futurists on the other. An article by Clive Bell in the *Burlington Magazine* in January 1913 marked the opening salvo in a war which developed into the secession of Wyndham Lewis and others from Roger Fry's Omega Workshops later the same year. Bell admitted that Cézanne was the 'one giant who moves me supremely,' whereas the 'latest works of Picasso' left him 'cold' (no. 90). For Bell formal beauty had strongly Platonic overtones; it was, he said, 'the boat in which artists ferry us to the shores of another world' – a view which received considerable support from that other Bloomsbury writer, Desmond MacCarthy. MacCarthy claimed that Post-Impressionism was 'the desire for pure form and colour ridden home to the last extreme', while 'Cubism', he said, 'certainly fails in this respect'.[69] MacCarthy's remarks were made in March 1913 in a review of the first exhibition of the Grafton Group at the Alpine Club – an exhibition which represents one of the last moments of unity in the British avant-garde of this period. It was dominated by Bloomsbury and included Vanessa Bell, Duncan Grant together with Wyndham Lewis and Frederick Etchells plus Kandinsky and the American Max Weber who were invited to join them. All the pictures were exhibited anonymously.

The idea of anonymity was undoubtedly Fry's, for whom the notion of a *botega* held great appeal. It played a significant part in the formation of the Omega Workshops very soon afterwards and one of the functions of the Grafton Group exhibition was to show how Post-Impressionist ideas were being exploited for decorative purposes. When the Workshops opened in April 1913 Fry gave an interview to the *Pall Mall Gazette*, under the title 'Post-Impressionism in the Home', saying, 'We should cease to insist on the extreme individuality of the artist' and he spoke nostalgically of the 'common effort' which went into the design and production of the Borough Polytechnic murals in 1911.[70] Several newspapers ran articles on the Workshops on the lines of 'How Mr Fry is trying to bring a "spirit of fun" into our sedate homes' and in his catalogue Fry emphasised the spirit of primitive joy with which the various objects of furniture had been constructed or painted. Wyndham Lewis did not share Fry's 'primitive joy' nor did he fit into the domestication of Post-Impressionism. Consequently by the time Frank Rutter's important exhibition of Post-Impressionists and Futurists opened at the Doré Gallery in October 1913 it was well known that 'some of Roger Fry's crew are in open revolt'.[71]

NOVEMBER 1913–MARCH 1914

The Doré exhibition was in many ways the summing up of much that had taken place since Roger Fry's first Post-Impressionist exhibition exactly three years before. It was deliberately historicist in its organisation and was a conscious effort on Frank Rutter's part 'to set forth in a coherent and so far as possible in a chronological order examples of various schools of painting which have made some noise in the world during the last quarter of a century'. 'The loose way', he added, 'in which the term "post-impressionist" has been used to cover a number of varying, and in some respects contradictory movements, has naturally confused a public seldom inclined to push very far its analysis of modern painting' (no. 98).

The scope of the exhibition ran from Pissarro to Cézanne, Gauguin and Van Gogh, through Neo-Impressionism, Fauvism and Intimism to the Cubism of Picasso and Herbin and the Futurism of Severini. English paintings which were effectively variants on these movements, were well represented and there was a section devoted to the recently established Camden Town School. Of the English work, however, it was probably Wyndham Lewis's *Kermesse* which attracted most attention. It had already been exhibited at the Allied Artists' exhibition at the Royal Albert Hall in the summer of 1912 where Roger Fry and Clive Bell had noticed it favourably.[72] Now even P.G. Konody, who had no taste for the new geometric art, praised Lewis's efforts. 'The dancers, it is true,' he wrote, 'look like some gigantic fantastic insects descended upon earth from some other planet; but his picture is a true piece of decoration and has real nobility of arrangement'.[73]

In 1913 Lewis had clearly been more at home decorating Madame Strinberg's café night club The Cave of the Golden Calf than making lampshades for Roger Fry, and when he finally broke with Omega he took with him Etchells, Hamilton and Wadsworth. The division in the ranks of modernism became fully public at an exhibition held in Brighton in 1913 under the auspices of the Camden Town Group and entitled 'English Post-Impressionists, Cubists and Others'. For this Lewis wrote the section in the catalogue for the so-called 'Cubist Room' which he published in the Egoist on 1 January 1914 (no. 99). Lewis was eager to distinguish his work from that of Camden Town on the one hand and Bloomsbury on the other, so he stressed – contra Fry, Bell and MacCarthy – the contemporary spirit of the new art and its debt to Marinetti and Picasso. 'Post impressionism,' he claimed, 'is an insipid and pointless name invented by a journalist' which, he said, 'has been naturally ousted by the better word "Futurism" in public debate on modern art'. Lewis was not the only artist who objected to the labelling of all modern movements with the term Post-Impressionism. On the same day that Lewis published his article on the 'Cubist Room' Charles Ginner expressed his belief in the power of what he called 'Neo-Realism' in British art, which, together with Spencer Gore, he had been practising since early 1913. Ginner's 'neo-realism' had nothing in common with Lewis's Cubism but was instead an appeal for a return to nature and to natural non-geometric forms in art. Nevertheless Ginner, like Lewis, condemned Post-Impressionism as an academic movement 'based,' as he put it, 'on a formula' (no. 100).

The debate was joined by T.E. Hulme who took up the cause of 'geometrical art' from a much more theoretical standpoint based in part on ideas derived from German aesthetics and the writings of Bergson. Like Lewis he, too, was set against the liberal humanism of Bloomsbury and saw in the new art the 'Break up of the Renaissance' (no. 101). In practical terms the new movement in English art found expression in the first exhibition of the London

Group and the establishment of the Rebel Arts Centre – both in March 1914 – from which, significantly, the work of Bloomsbury was excluded.

In spite of internal rivalry and dissent about the significance of Post-Impressionism, the Bloomsbury view of French art was popularised and widely publicised with the publication of Clive Bell's *Art* in the early months of 1914. This book, for all its lack of logic and its unfounded assertions, served to put the case for the primacy of form in a humanist context simply, readably and comprehensibly. *Art* could not have come at a more opportune moment. Since 1910 the British public had been coming to terms with ideas about rhythm, plasticity and decoration in modern art; it had learned the lesson of Post-Impressionism that form was more important than content and *Art* not only repeated that lesson but extended it to the arts of different periods and different cultures. 'There is no mystery about Post-Impressionism', Bell wrote, 'a good Post-Impressionist picture is good for precisely the same reasons as any other picture is good. The essential quality in art is permanent' (no. 102).

Back in 1910, Post-Impressionism had been associated with anarchy, revolution, social and psychological disturbance. Now in 1914 Bell's stress on permanence, continuity and universality was music to the ears of the British who were about to be plunged into the turmoil of the First World War.

NOTES

1 Virginia Woolf, 'Mr Bennett and Mrs Brown', in *Collected Essays by Virginia Woolf* (Hogarth Press, 1975), i, 320.

2 Virginia Woolf to Vanessa Bell, 21 July 1911, in *The Flight of the Mind: The Letters of Virginia Woolf 1888–1912*, ed. Nigel Nicolson (Hogarth Press, 1975), 476.

3 Katherine Mansfield, for example, confessed to Dorothy Brett that when she saw Van Gogh's paintings at the first Post-Impressionist exhibition they 'taught [her] something about writing...a kind of freedom': A. Alpers, *Katherine Mansfield* (Jonathan Cape, 1954), 151–2.

4 *Impressionists in England*, ed. Kate Flint (Routledge & Kegan Paul, 1984), 26.

5 Roger Fry, 'The New Gallery', *Athenaeum* 13 January 1906, 56.

Frances Spalding identifies the two pictures by Cézanne in this exhibition as *Still Life* (Venturi no. 70) and *Winter Landscape* later illustrated as fig. 27 in Fry's book on Cézanne. See Frances Spalding, *Roger Fry, Art and Life* (Granada, 1980), 116.

6 Camille Mauclair, *The French Impressionists 1860–1900*, trans. Paul G. Konody (Popular Library of Art, 1904), 141.

7 Wynford Dewhurst, *Impressionist Painting* (1904), 16.

8 George Moore, *Reminiscences of the Impressionist Painters* (Maunsel, 1906), 35.

9 *Athenaeum*, 3 November 1906, 557. The reviewer's remarks were quite accurate and the pictures by Cézanne were not noticed in England in spite of enormous coverge in the French press. In *La Revue bleu* (21 October 1905, 522) Camille Mauclair suggested that though Cézanne was once ostracised he was, by 1905, the 'most copied' artist in the salon.

10 *The Times*, 19 October 1905, 5.

11 *The Times*, 2 April 1906, 4.

12 Bernard Sickert, *Burlington Magazine*, July 1906, ix, 222.

13 *The Times*, 9 October 1906, 10.

14 'Art: The Autumn Salon', *Nation*, 7 November 1908, 221.

15 R[obert] E. D[ell], 'Art in France', *Burlington Magazine*, November 1908, xiv, 118.

16 Ibid.

17 Letter of 17 May 1909, quoted in Frances Spalding, *Roger Fry*, 118–19.

18 *Art News*, 21 October 1909, 7.

19 See Wendy Baron, *The Camden Town Group* (Scholar Press, 1979), 9.

20 Sickert expressed his relief as one who 'had hitherto cherished modest and chilly doubts at the sufficient length and weight of [his] kilt of culture, when [he] saw Berenson prone and bare in the field of modern art, revealing deficiencies [he] had long suspected, but dared not hint at!' 'Manet and the Impressionists', *Art News*, 17 February 1910, 120.

21 'Neo-Impressionist' [Charles Ginner], 'Note on the International Society', *Art News*, 21 April 1910, 194.

22 C.R.W. Nevinson *Paint and Prejudice*, i (Methuen, 1937), 9.

23 'Modern French Art at Brighton: Some Nightmare Impressionists', *Brighton Standard and Fashionable Visitors' List*, 11 June 1910, 2.

24 Walter Higgins, 'Modern French Painting', *Art Chronicle*, 25 June 1910, 118.

25 Ibid.

26 Frank Rutter, 'Round the Galleries: Rebels at Brighton', *Sunday Times*, 28 August 1910, 6. This was the third article which Rutter

wrote on the Brighton exhibition; the others appeared in the two previous weeks.

27 Lewis Hind, 'The Consolations of an Injured Critic – VII', *Art Journal*, October 1910, n.s. xxx, 294.

28 Desmond MacCarthy, *Memories* (MacGibbon & Kee, 1953), 178–83.

29 Frank Rutter, 'Round the Galleries', *Sunday Times*, 13 November 1910, 14.

30 *Sphere*, 5 November 1910, 130. Three pictures by Van Gogh and three by Gauguin were reproduced in this issue.

31 Frank Rutter seems to have been the first writer to use the term 'Post-Impressionists' in print. In a review of the Salon d'Automne in *Art News*, 15 October 1910, 4, he described Othon Friesz as 'a post-impressionist leader', and the same issue of *Art News* carried an advertisement for 'The Post-Impressionists of France' (5). This was some three weeks before 'Manet and the Post-Impressionists' opened at the Grafton Galleries.

32 *The Times*, 7 November 1910, 12.

33 *Tatler*, 23 November 1910, 228.

34 'Art: The Grafton Gallery', *Spectator*, 12 November 1910, 798.

35 P.G. Konody, 'Art Notes: Post-Impressionism at the Grafton Galleries', *Observer*, 13 November 1910, 9.

36 *Impressionists in England*, 291–2.

37 *Punch*, 30 November 1910, 386 and *Tatler*, 23 November 1910, 229.

38 A.J. Finberg: 'The Latest Thing from Paris', *Star*, 8 November 1910, 2.

39 Robert Morley, letter in the *Nation*, 8 December 1910, 406.

40 C. Lewis Hind, 'The New Impressionism', *English Review*, December 1910, vii, 180–92.

41 F. McLean Stowell, letter in the *Nation*, 17 December 1910, 503 and John. J. Adams, letter in the *New Age*, 15 December 1910, 167.

42 Frederick Lawton, 'Paul Cézanne', *Art Journal*, February 1911, n.s. xxxi, 55–60. The article was illustrated with seven works by Cézanne.

43 'The New Art by Tony Sarg', *Black and White*, 21 January 1911.

44 *Punch*, 25 January 1911, 60.

45 *Punch*, 1 February 1911, 78.

46 J.S. Sargent, letter in the *Nation*, 7 June 1911, 10.

47 'Sir William Richmond on Post-Impressionism', *The Times*, 10 January 1911, 11.

48 *Daily Graphic*, 28 January 1911, 469.

49 P.G. Konody, 'The Women's International Art Club', *Observer*, 5 March 1911, 9.

50 P.G. Konody, *Observer*, 9 April 1911, 5.

51 P.G. Konody, 'The New English Art Club', *Observer*, 28 May 1911, 16.

52 P.G. Konody, 'The London Salon', *Observer*, 9 July 1911, 4.

53 'Health and Disease in Art', *Art Chronicle*, 1 May 1911, 163–6; 15 March 1911, 179–86; 1 April 1911, 199–202; 15 April 1911, 211–12.

54 Frank Rutter, *Revolution in Art* (Art News Press, 1910), 46; 53; 54.

55 Margaret Morris, *The Art of J.D. Fergusson* (Blackie, 1974), 64.

56 *Art News*, 15 August 1911, 85.

57 P.G. Konody, *Observer*, 16 July 1911, 7.

58 *The Times*, 28 November 1911, 11.

59 *Outlook*, 25 November 1911, 738–9.

60 *New Age*, 9 November 1911, 36.

61 E. Wake Cook, letter in the *New Age*, 30 November 1911, 119.

62 P.G. Konody, 'The Stafford Gallery', *Observer*, 28 April 1912, 6.

63 *Pall Mall Gazette*, 15 January 1912, 439.

64 J.B., 'The Italian Futurists', *Spectator*, 16 March 1912, 439.

65 *Morning Post*, 4 October 1912; A.J. Finberg, *Star*, 5 October 1912; C.H. Collins Baker, *Saturday Review*, 9 November 1912, 577; Anthony M. Ludovici, *New Age*, 21 November 1912, 66–7.

66 *The Times*, 4 October 1912, 9; C. Lewis Hind, *Daily Chronicle*, 5 October 1912, 6; P.G. Konody, *Observer*, 6 October 1912, 6; *Architect*, 8 November 1912.

67 Martin Hardie, *Queen*, 12 October 1912, 646.

68 *The Times*, 8 August 1913, 4.

69 Desmond MacCarthy, 'Abstract and Elementary', *New Witness*, 27 March 1913, 661.

70 Basil Williams, the chairman of the House Committee Borough Polytechnic, invited Fry to decorate the Polytechnic's dining hall. So Fry brought together Duncan Grant, Frederick Etchells, Bernard Adency, MacDonald Gill and Albert Rutherson to make designs which are now in the Tate Gallery, London.

71 P.G. Konody, 'Post-Impressionism at the Doré Gallery', *Observer*, 26 October 1913, 10.

72 Fry said that Lewis had 'built up a design which is tense and compact' and went on to praise the geometry and the 'rhythm' of the picture (*Nation*, 20 July 1912, 583). Bell said that *Kermesse* 'holds together in the way that a sonata by Beethoven holds' (*Athenaeum*, 27 July 1912, 98).

73 P.G. Konody, 'Post-Impressionism at the Doré Gallery', *Observer*, 26 October 1913, 10.

The history of Impressionism in England and the history of Post-Impressionism touch at many points in the first decade of the twentieth century, but perhaps the most important landmark is the Durand-Ruel exhibition of French painting at the Grafton Galleries in 1905. This show, in terms of both the number of paintings displayed and the impact which they made, was dominated by Impressionist work and the largest group of paintings by Cézanne to be seen so far in Britain – ten in all – were ignored or misrepresented in the press. Surprisingly, Cézanne was not to have a one-man exhibition in England until 1925 but the Durand-Ruel exhibition marks the beginning of a change in his reputation and the beginning, too, of the entry into Britain of works later known as 'Post-Impressionist', a development which culminated in Roger Fry's first Post-Impressionist exhibition of November 1910.

1. Frank Rutter, *Art in My Time*

1933, pp. 111-14

Frank Rutter (1876–1937) became a professional art critic in 1901. He was Curator of Leeds Art Gallery (1912–17). As the art critic for the *Sunday Times* (from 1903) and *Financial Times* and the editor of *Art News* he was one of the most eloquent champions of modernism. For some years the Allied Artists' Association, which he set up in 1908, provided exhibition space for modern British and Continental artists.

In 1933 he recalled seeing the pictures by Cézanne at the Grafton Galleries in 1905, but admits that at the time he was able to make very little of what he saw.

Of the ten pictures by Cézanne, one was a portrait of Choquet now in the Gallery of Fine Arts, Columbus, Ohio.

Two others are identifiable from the photographs which M. Koechlin gave to the library of the Musée des arts decoratifs, Paris: they are *Un Dessert* (1873–7: Philadelphia Museum of Art) and *Sous Bois* (1882–5: Fitzwilliam Museum, Cambridge).

'Hullo! What's this? What are these funny brown-and-olive landscapes doing in an impressionist exhibition? Brown! I ask you? Isn't it absurd for a man to go on using brown and call himself an impressionist painter? Who are they by? Oh, Cézanne. That's the man who paints still life. Now, I like those better. Those apples over there are really very good. And this other thing, *Dessert*. That's not so bad. The apple's quite good, isn't it? and the knife. But that right-hand side of the flask is pretty wobbly, and that glass hasn't quite come off. He's not very strong on drawing, is he? But I like his draperies, that curtain and the table-cloth; and the table too, that's really quite good. Yes, there's something in it, but it's rather dark and brown. I don't like his colour. Let's go back and look at the Monets.'

That is how the 'fans' of impressionist painting talked about Cézanne in 1905. The general opinion of those who were sympathetic to the impressionist movement was very fairly stated by the critic of the *Daily Mail* when, after commenting on the work of Manet, Monet, Degas and Renoir, he wrote: – 'Cézanne, after these masters, strikes one as an exceptionally able amateur.' That was putting it very kindly, even generously, for few would admit at the time that Cézanne was 'exceptionally able,' though many would agree that he looked like an amateur.

Where, oh where was Mr. Roger Fry in 1905, and why was not his voice heard in the land? How could he allow anybody to call Cézanne an 'amateur' with impunity?

One of these days no doubt Mr. Fry will explain, but for me there was no excuse. In the Salon d'Automne of 1904 there had been a special retrospective Cézanne Exhibition, and I had seen it. At least, I had walked through it on the occasion of a hurried visit to Paris, and I fear I was not as impressed as I ought to have been. My opinion in 1904–5 was very much that of M. Durand-Ruel Senior; Cézanne was the painter of some quite good still-life pictures who had bungled nearly everything else he attempted. It took me fully two years to revise this opinion and learn that Cézanne had painted some really beautiful landscapes, notably the *Avenue in Provence* (*c.* 1885), *The Marne Bridge* (1888), and *Lac d'Annecy* – and had occasionally, but much more rarely, done figure subjects – notably *Woman with a Rosary* (1896) – which had great moving power despite their evident imperfections. But I still think, as I did then, that Cézanne was most completely successful in the painting of still life and that this will ultimately be recognised as his most important and lasting contribution to the art of his time.

But there was really no time to think much about Cézanne in 1905. We were far too busily occupied trying to persuade the pundits of British art to accept Manet, Monet, Degas and Renoir.

---

2. Unsigned review, 'The Last Phase of Impressionism'

Burlington Magazine, February 1908, xii, 272–3

The 8th annual exhibition of the International Society in January and February 1908 contained one of Van Gogh's watercolour studies of a weaver, a picture entitled *Pommes et Pains,* possibly *Un Dessert* (1873–7: Philadelphia Museum of Art) and another still life by Cézanne, Gauguin's *Haere-Pape* (1892: Barnes Foundation, Merion), together with pictures by Signac, Cross and, paradoxically, the first work of Matisse to be shown outside France – an oil entitled *La Jetée à Collioure.* This review acted as a stimulus to Roger Fry's first public pronouncement of his views on Post-Impressionist art (see no. 3).

The annual exhibitions of the International Society of Sculptors, Painters and Engravers at the New Gallery, in addition to their material success, have an interest for English artists and English art patrons such as no other modern show in London can claim. The mere variety of the exhibits, coupled with the fact that they aim at representing the arts of all Europe, ought in itself to be no small stimulus to the inhabitants of an island who still remain, in many respects, insular. But these exhibitions have a special value, since in them year after year the pioneers of the great modern movements exhibit side by side with their artistic heirs. We can thus trace more than one remarkable school back to its source, and in the process of the survey estimate what it has done and what yet remains for it to do.

This year's admirable exhibition, for example, gives us the chance of examining the movement which is commonly described by the nickname of Impressionism. The three works by Claude Monet, and Renoir's portrait, which hang together in the West Room, will serve as a point of departure. In the charming *Printemps* (No. 144) Monet is hardly attempting more than the other French naturalistic painters of his time, and his delicate vision of a spring morning might hang with as little incongruity by the side of pictures by Corot and Daubigny as the portrait by Renoir (No. 142) might show among pictures by Alfred Stevens. In the *Vue de Hollande* (No. 140) Monet's art has become more bright, more vibrant, more vaporous, but he might claim that he had gained in freshness and quality of colour at least as much as he had sacrificed of pictorial coherence. The difference, in fact, between the early work and the late one is no more than the difference between the middle and late periods of Turner. In France a similar method of work has been applied with success to an entirely new range of subjects by Degas, and with perhaps rather less discretion by Besnard and many others. In England it has attracted the talent of Mr. Wilson Steer, Mr. Clausen and Mr. Mark Fisher and the like, who have employed it with originality and success. With the younger generation of French artists, however, it would seem as if the movement is in the last stage of decay. Of this the two pictures by Maurice Denis in the North Room, and the collection of works grouped at the end of the balcony, afford ample illustration. The two large panels by M. Denis are, apparently, a burlesque of a certain type of archaistic religious painting, and had they been executed in water-colour on a scale suitable to the pages of *Le Rire* or *L'Assiette au Beurre* they might have deserved a smile. At present the impression created is one of sadness that so much labour should have been expended to produce so trivial a result.

It is, however, in the balcony that Impressionism reaches the stage of positive disintegration. In the pictures of M. Joaquim Mir there is effectiveness of a kind, though it can hardly be called pictorial. In the picture of M. Paul Signac the method is reduced to the level of a mosaic of neat briquettes of colour, with a result once more absurdly out of proportion to the labour spent upon it, while in that of M. Matisse the movement reaches its second childhood. With M. Gauguin, who is placed hard by, some trace of design and some feeling for the decorative arrangement of colour may still be found; with M. Matisse motive and treatment alike are infantile.

It is usually unwise for those who are not themselves extremely young to condemn forthwith work which they do not understand, since the history of the fine arts for the last two hundred years is one long record of the misunderstanding of youthful talent by established authority. Hogarth, Reynolds, Turner, Constable, the Preraphaelites and Whistler were each in turn attacked by their seniors, but in one or two cases at least there was just a modicum of truth in what their seniors had to say. Hogarth is. undeniably unequal in his work; Turner and Constable are neither of them faultless; the first works exhibited by the Preraphaelites are still open to the charge of mannerism or inexperience; Whistler's art is not greatest when it is most slight. Thus if the balance of available evidence is in favour of youth, there is a little to be said on the other side.

But in the case of this modern French work are we really dealing with youth at all? Have we not rather to deal with the extreme old age of Impressionism, young though the painters may be. Now if history proves that revolting youth is generally right and conservative old age is generally wrong, it proves with equal certainty that the best work done by any movement is done by its pioneers. In the hands of their immediate followers the revolt loses its freshness: in the hands of the next generation it sinks into callous imitation or empty caprice.

If we could for a moment imagine ourselves in the twenty-first century, should we not in looking back at the Impressionist movement regard it as something analogous to the naturalistic movement in fifteenth-century Florence? Impressionism has done a similar service to art by proving that a far wider range of tone and colour and luminosity was possible in oil painting than had been previously admitted except by Turner; the pioneer Impressionists have thus a scientific value as well as an artistic one, although in comparison with more complex manifestations of human genius they may appear just a little barbaric. But if this be the position of the leaders of the movement, their followers and imitators must be placed on a very different level, and if we attempt to judge their future by analogy, it is with the now forgotten Flemish and Italian eclectics that they must be classed. Impressionism in France has run its course, and salvation for the next outburst of original talent must be expected from some entirely different quarter. Our national art appears to conspicuous advantage at the New Gallery, chiefly because, whether consciously or not, it has recognized this vital fact, and thereby has avoided the senile puerilites to which we have referred. France may still dominate the world of sculpture, but the immediate future of painting seems to be with Great Britain or, in her default, with Germany.

3. Roger Fry,
letter to the *Burlington Magazine*

March 1908, xii, 374–6

In 1908, Fry (1866–1934), who had a wide reputation as an authority on Italian Renaissance art, was still educating himself in French art of the late nineteenth century. In this response to the criticism levelled at the exhibition of the International Society in 1908 (see no. 2) he devised a historical analogue for what he called 'neo-Impressionism' by likening it to proto-Byzantine work – a theory which was to play an important part in later defence of what came to be called 'Post-Impressionism'.

THE LAST PHASE OF IMPRESSIONISM

Sir,

As a constant reader and frequent admirer of your editorials on matters of art I should like to enter a protest against a tendency, which I have noticed, to treat modern art in a less serious and sympathetic spirit than that which you adopt towards the work of the older masters. This tendency I find particularly marked in an article with the above heading. The movement which is there condemned, not without a certain complacency which to me savours of Pharisaism, is one that surely merits more sympathetic study. Whatever we may think of its aims, it is the work of perfectly serious and capable artists. There is, so far as I can see, no reason to doubt the genuineness of their conviction, nor their technical efficiency. Moreover in your condemnation you have, I think, hit upon an unfortunate parallel. You liken the pure Impressionists, of whom we may take Monet as a type, to the naturalists of the fifteenth century in Italy, and these neo-Impressionists to the 'now-forgotten Flemish and Italian eclectics'. Now the eclectic school did not follow on the school of naturalism; there intervened first the great classic masters who used the materials of naturalism for the production of works marked by an intense feeling for style, and second, the Mannerists, in whom the styles of particular masters were exaggerated and caricatured. The eclectics set themselves the task of modifying this exaggeration by imbibing doses of all the different manners.

Now these neo-Impressionists follow straight upon the heels of the true Impressionists. There has intervened no period of great and then of exaggerated stylistic art. Nor has Impressionism any true analogy with naturalism, since the naturalism of the fifteenth century was concerned with form, and Impressionism with that aspect of appearance in which separate forms are lost in the whole continuum of sensation.

There is, I believe, a much truer analogy which might lead to a different judgement. Impressionism has existed before, in the Roman art of the Empire, and it too was followed, as I believe inevitably, by a movement similar to that observable in the neo-Impressionists — we may call it for convenience Byzantinism. In the mosaics of Sta Maria Maggiore as elucidated by Richter and Taylor (*The Golden Age of Classic Christian Art*) one can see something of this transformation from Impressionism in the original work to Byzantinism in subsequent restorations. It is probably a mistake to suppose, as is usually done, that Byzantinism was due to a loss of the technical ability to be realistic, consequent upon barbarian invasions. In the Eastern Empire there was never any loss of technical skill; indeed, nothing could surpass the perfection of some Byzantine craftsmanship. Byzantinism was the necessary outcome of Impressionism, a necessary and inevitable reaction from it.

Impressionism accepts the totality of appearances and shows how to render that; but thus to say everything amounts to saying nothing – there is left no power to express the personal attitude and emotional conviction. The organs of expression – line, mass, colour – have become so fused together, so lost in the flux of appearance, that they cease to deliver any intelligible message, and the next step that is taken must be to re-assert these. The first thing the neo-Impressionist must do is to recover the long obliterated contour and to fill it with simple undifferentiated masses.

I should like to consider in this light some of the most characteristic painters of this movement. Of these M. Signac is the only one to whom the title neo-Impressionist properly applies. Here is a man feeling in a vague, unconscious way a dissatisfaction at the total licence of Impressionism and he deliberately invents for himself a restraining formula – that of rectangular blobs of paint. He puts himself deliberately where more fortunate circumstances placed the mosaic artist, and then he lets himself go as far in the direction of realistic Impressionism as his formula will allow. I do not defend this, in spite of the subtle powers of observation and the ingenuity which M. Signac displays, because I do not think it is ever worth-while to imitate in one medium the effects of another, but his case is interesting as a tribute to the need of the artist to recover some constraint: to escape, at whatever cost, from the anarchic licence of Impressionism.

Two other artists, MM. Cézanne and Paul Gauguin, are not really Impressionists at all. They are proto-Byzantines rather than neo-Impressionists. They have already attained to the contour, and assert its value with keen emphasis. They fill the contour with wilfully simplified and unmodulated masses, and rely for their whole effect upon a well-considered co-ordination of the simplest elements. There is no need for me to praise Cézanne – his position is already assured – but if one compares his still-life in the International Exhibition with Monet's, I think it will be admitted that it marks a great advance in intellectual content. It leaves far less to the casual dictation of natural appearance. The relations of every tone and colour are deliberately chosen and stated in unmistakable terms. In the placing of objects, in the relation of one form to another, in the values of colour which indicate mass, and in the purely decorative elements of design, Cézanne's work seems to me to betray a finer, more scrupulous artistic sense.

In Gauguin's work you admit that 'some trace of design and some feeling for the decorative arrangement of colour may still be found', but I cannot think that the author of so severely grandiose, so strict a design as the *Femmes Maories* or of so splendidly symbolic a decoration as the *Te Arti Vahiné* deserves the fate of so contemptuous a recognition. Here is an artist of striking talent who, in spite of occasional boutades, has seriously set himself to rediscover some of the essential elements of design without throwing away what his immediate predecessors had taught him.

And herein lies a great distinction between French and English art (I am speaking only of the serious art in either case), namely, that the French artist never quite loses hold of the thread of tradition. However vehement his pursuit of new aims, he takes over what his predecessors have handed to him as part of the material of his new formula, whereas we in England, with our ingrained habits of Protestantism and non-conformity, the moment we find ourselves out of sympathy with our immediate past, go off at a tangent, or revert to some imagined pristine purity.

The difference is one upon which we need not altogether flatter our selves; for the result is that French art has a certain continuity and that at each point the artist is working with some surely ascertained and clearly grasped principles. Thus Cézanne and Gauguin, even though they have disentangled the simplest elements of design from the complex of Impressionism, are not archaizers; and the flow in all archaism is, I take it, that it endeavours to attain results by methods which it can only guess at, and of which it has no practical and immediate experience.

Two other artists seen at the International deserve consideration in this connexion: Maurice Denis and Simon Bussy. Against the former it might be possible to bring the charge of archaism, but he, too, has taken over the colour-schemes of the Impressionists, and in his design shows how much he has learned from Puvis de Chavannes. His pictures here are not perhaps the most satisfactory examples of his art, but any one who has observed his work during the past five years will recognize how spontaneous is his sense of the significance of gesture; how fresh and genuine his decorative invention.

M. Bussy is well known already in England for his singularly poetical interpretation of landscape, and though at first sight his picture at the International may strike one as a wilful caprice, a little consideration shows, I think, that he has endeavoured to express, by odd means perhaps, but those which appeal to him, a sincerely felt poetical mood, and that the painting shows throughout a perfectly conscientious and deliberate artistic purpose. Here again the discoveries of Impressionism are taken over, but applied with quite a new feeling for their imaginative appeal.

I do not wish for a moment to make out that the works I have named are great masterpieces, or that the artists who executed them are possessed of great genius. What I do want to protest against is the facile assumption that an attitude to art which is strange, as all new attitudes are at first, is the result of wilful mystification and caprice on the artists' part. It was thus that we greeted the now classic Whistler; it was thus that we expressed ourselves towards Monet, who is already canonized in order to damn the 'neo-Impressionists'. Much as I admire Monet's directness and honesty of purpose, I confess that I see greater possibilities of the expression of imaginative truth in the tradition which his successors are creating.

4. From an unsigned review, 'The Autumn Salon'

The Times, 2 October 1908, 8

This early review of the Salon d'Automne is remarkable for its sympathetic treatment of the work of Matisse. It deals in some detail with a picture Matisse painted for Sergei I. Shchukin entitled *Harmony in Blue* and almost immediately repainted as *Harmoney in Red* (1908–9: Hermitage Museum, Leningrad).

Those who may fancy themselves shocked by occasional examples of what they take to be revolutionary painting at the New English Art Club or elsewhere in London should pay a visit to the Autumn Salon in Paris. There they will see real revolutionary painting, compared to which the most extreme works of Mr. John are as timid as the opinions of a Fabian Socialist compared with those of a bomb-throwing anarchist. It would be easy to break into eloquent denunciation of this revolutionary painting and to thank Heaven that we have nothing like it in England. It will be more useful to attempt to discover why it exists, and why in Paris it has become so common that no one notices or resents it any more than the motors that incessantly trumpet and clatter by in the streets.

And first one must insist that these revolutionary pictures are not inferior, as art, to the great mass of pictures in the ordinary Academy Exhibition, and are certainly more amusing. True, the ordinary Academy picture may convey more information; but, if pictures express nothing and have no beauty, they are of no account; and it matters not whether they shock us or bore us. Nor must we assume that these revolutionary pictures are all produced by painters who wish to conceal with mere effrontery their ignorance of the rudiments of their art. Many of them have been regularly trained in the schools; and that is the very reason why they have broken out in blind revolt. They are sick of the pictures that are mere dull records of fact, and determined not to produce any more of them. The tyranny of fact in modern art has provoked them to a rage of mere destruction, and they have not yet emerged from that rage. Their pictures express a determination to do nothing that modern academic taste demands; but, unfortunately, art cannot be made of negatives, as academic painters have proved long ago. Art which is more intent upon avoiding than upon achieving slips at once into a convention, for there is no individuality in mere avoidance. Thus a revolutionary convention has grown up with great rapidity, and nearly all these revolutionary painters are the slaves of it. You would expect them to differ violently; but they are nearly all violently alike, with the same staring toyshop figures, the same absence of composition, the same streaks of primary colours.

M. Matisse is one of their leaders; and crowds of them follow him as obsequiously as David was ever followed by his pupils. His drawings prove him to be a strong and accomplished draughtsman; but in his paintings his main concern is to avoid the rhetoric, the compromise, and the dull imitation of the schools. He wants to create the art of painting anew. He will be a primitive at all costs, not like the mild English and German imitators of Italian primitives, but in his own fierce way. This is impossible; for, try as he will, he cannot purge his consciousness of all the art of the world. He must be subject to a continual temptation to paint something as some one else has painted it; and the effort not to do this is so exhausting that he has little strength left to do anything positive. His largest picture in the Salon represents a woman putting fruit on a table. The walls and the table are plum-colour; a bright blue decorative pattern straggles inexplicably over both, but it is not more devoid of relief than the table, the single chair, or the woman herself. Through a window is seen a landscape with trees in blossom shaped like the trees in the earliest medieval illuminations. This landscape is a mere pattern except that it lacks symmetry. It would be rash to say that the picture is wrong in theory; but, whatever the theory of it may be, it fails in practice because it makes no appeal either to the eye or to the mind. If it is meant for pure decoration, it lacks both symmetry and material beauty. If it is meant to represent either action or some state of being, there are not enough facts in it to produce any kind of illusion. It is an incredible picture that could not be put to any kind of use, whereas the archaic art which the painter has tried, not to imitate, but to rival is not incredible and was nearly always designed for some particular purpose.

Lack of purpose, except the negative purpose of rebellion and protest, is the great defect of this revolutionary art. It is an attempt to use the methods of the impressionists for the purposes of decoration. But the revolutionaries do not know what they want to decorate, and they are at a loss for subject-matter important enough to endure the simplifications at which they aim. For in art and in literature only lofty themes of universal interest can be treated with extreme simplicity. But these revolutionaries paint ordinary studies of the nude, or *genre* scenes, or portraits in a style that might be rightly applied to images of the gods. Sometimes they attempt something mysterious or symbolic. One painter has a large picture representing nine nude persons all in monochrome, all gazing with the same primitive eyes at nothing in particular, while in front of them is a mustard-coloured hound and behind a landscape of hills in flat strata coloured green, red, dark yellow, and light yellow. The figures are all drawn to look as primitive as possible. But the real primitives did not try to make their pictures look primitive; rather they tried to make them as much like reality as they could. And this effort, combined with their purpose of decoration and expression, made their works interesting. The picture just described is only a more modern piece of sentiment-ality than the works of Ary Scheffer. It is vaguer in intention than those works, if a little less ugly in its general effect.

5. Bernard Berenson, letter to the *Nation* (New York)

12 November 1908, 461

Berenson (1865–1959), like Fry, was well known as a connoisseur of Italian art, yet he was prompted by a hostile review of
the French Salon d'Automne to publicly defend the drawings
of Matisse in the American journal the *Nation*. He bought a
landscape (illustrated in Benedict Nicolson, 'Roger Fry and
Post-Impressionism', *Burlington Magazine*, March 1951, xciii,
opp. p. 11) from Matisse's studio which he lent to the first Post-
Impressionist exhibition in 1910 (see headnote for no. 9).

SIR: In a note which appeared in your issue of October 29,
regarding the autumn Salon at Paris, there occur the two
following sentences:

> Some of the younger artists have surprisingly good and new work,
> along with direct insults to eyes and understanding. Such is Henri
> Matisse, who forgets that beholders are not all fools, and that it is not
> necessary to do differently from all other artists.

Will you allow one of the fools whom Matisse has thoroughly
taken in to protest against these phrases? They are more
hackneyed than the oldest mumblings in the most archaic extant
rituals. There is nothing so hoary in the sacrificial Vedas. They
have been uttered with head-shakings in Akkadian, in Egyptian, in
Babylonian, in Mycenæan, in the language of the Double-Ax, in
all the Pelasgic dialects, in proto-Doric, in Hebrew, and in every
living and dead tongue of western Europe, wherever an artist has
appeared whose work was not as obvious as the 'best seller' and
'fastest reader.' Of what great painter or sculptor or musician of the
last century has it not been said in the cant phrase of the
Boulevards – 'C'est un fumiste. Il cherche à épâter le monde'?

Henri Matisse seems to me to think of everything in the world
rather than of the need of 'doing differently from all other artists.'
On the contrary, I have the conviction that he has, after twenty
years of very earnest searching, at last found the great highroad

travelled by all the best masters of the visual arts for the last sixty centuries at least. Indeed, he is singularly like them in every essential respect. He is a magnificent draughtsman and a great designer. Of his color I do not venture to speak. Not that it displeases me – far from it. But I can better understand its failing to charm at first; for color is something we Europeans are still singularly uncertain of – we are easily frightened by the slightest divergence from the habitual.

Fifty years ago, Mr. Quincy Shaw and other countrymen of ours were the first to appreciate and patronize Corot, Rousseau, and the stupendous Millet. *Quantum mutatus ab illo!*[1] It is now the Russians and, to a less extent, the Germans, who are buying the work of the worthiest successors of those mighty ones.

NOTE

1 Virgil, *Aeneid* II, 274: 'how changed [things are] from what they were'?

6. Julius Meier-Graefe, *Modern Art*

2 vols, 1908, trans. Florence Simmonds
and George W. Chrystal, i, 211–12, 271; ii, 63–4

Meier-Grafe was the art editor for the Paris magazine, *Pan* and the translation of his *Modern Art* provided the first extensive historical account in English of the modern movement in painting. His passionate and eccentric style was both stimulating and bewildering for the British public and, though the two volumes were extensively illustrated, few people in Britain had seen the original paintings.

In the first extract Meier-Graefe deals with the 'expressionist' work of Van Gogh; in the second he deals with the influence of

Cézanne and in the third he speaks of the 'primitivism' of Gauguin.

I have dealt elsewhere with Van Gogh's anarchism, showing what seems to me his strong positive instinct, as opposed to the rhetorical anarchism of Morris, Crane, and others. His work is the strongest possible contrast to an indolent, state-supported art, meet to adorn the house of mediocrity. He destroys it. Here he may appear as the ruthless barbarian, casting off all regard for the law of the dwelling. The same hostility shows itself in Munch, another anarchist of equal sincerity. But what seems to the Philistine barbarism in Van Gogh, is often actually so in Munch. It must be evident that it is impossible to conceive of an interior in which Munch's most typical works would be in keeping, and this at once restricts his importance to the field of the extremest abstract art. Van Gogh merely negatives the contemporary domicile. In this, his pictures have the effect of blows with a club. But a setting where he would be harmonious, which he could adorn, is not only conceivable, but already in process of evolution, and here, again, his sacrifice is glorified with the nimbus of the peasant, who fertilises the earth anew with his own blood. It is improbable that the time will ever come when his pictures will be appreciated by the layman; it is more conceivable that pictures should cease to be produced altogether, than that Van Gogh's should become popular. But his portion in the development of the modern interior is already assured; it is indirect, but all the more penetrating for this reason; his tints and colours are elements, which serve and will serve in the most varied form. This gives him perhaps a greater importance than can be appreciated by a generation so near to William Morris as our own. Here, indeed, there is something new. The mind intent on the consciously decorative effort of our times found in Van Gogh, and not solely in his latest pictures, unhoped-for and very novel sustenance. It is indeed possible that this treasure conceals the one perfectly novel element of our essays in the formation of a style. If the connection seems slight we must remember in all humility that our efforts in this direction are in their infancy, and that this is the reason why this aspect of Van Gogh has hitherto served merely to complete the many-sided relations, which all progressive art will link with his wealth. Even his treatment of the coloured surface is calculated to deepen the teaching of the Japanese, so fruitful at present; it completes what Degas and Lautrec added to the importation, keeping the golden principle of simplification always in view. At the same time he achieves a splendour of effect beyond anything ever yet achieved by easel pictures. His masterpiece, *The Ravine*, a rendering of a remarkable rocky chasm near Arles, an intoxicating harmony of rich blue tones, is a technical model of incalculable value. Nature seems merely to have been used to enhance the richness of the tapestry-like effect by an accidental abnormal concatenation of strong lines, which disappear into an infinity of new planes. If it should prove feasible to transfer such works to large surfaces, and make them durable, we might almost cherish the illusion of having gained a decorative method equal to that of the old mosaicists, and combining the splendour of Gobelin with its distinction.

Modern decorative artists have not been unmoved by Van Gogh. His surfaces have proved helpful to the young Parisian painters, Denis, Ranson, Sérusier, and Bonnard, and his brush-stroke to the most important of modern ornamentists, Van de Velde. Van Gogh has sifted out from the great epoch of the Impressionists not all, but some highly important results, destined to a far-reaching influence even outside the sphere of abstract painting to which this school confined itself.

If we keep this connection in view and trace the road back from Van Gogh to his greatest exemplar, the beloved master of Barbizon appears in a new light deeply intertwined with all that moves us to-day. Van Gogh drew Millet into the radiant circle of Manet, Monet, and Cézanne, who were in danger of forgetting him, and reminded them what Millet's great fructifier, Daumier, had possessed of pictorial power.

And at the same time, this last of the great Dutchmen who had drifted to a foreign haven maintained his national tradition. He brought back to it what it had lent to the great French generation of 1830, remaining faithful to its noblest law: that we must follow Nature, and more especially our own nature....

No member of the school of Cézanne has succeeded in surpassing the master. But, where there is no teacher, it is inaccurate to talk of a school. It was not by spoken words that the seed was sown in this case. Nor is it Cézanne alone who leads the youth of France.

Renoir, Fantin, and, once again, Delacroix, divide their homage. If I have, nevertheless, spoken of the school of Cézanne in this connection, it is because certain essential aims of the younger men at least reveal the influence of Cézanne, and because this inter-relation is the sole bond of union between a number of very dissimilar painters. The three friends of Maurice Denis, to whom the following all too brief chapter is devoted, should not be grouped with Denis, Vallotton, and Gauguin's circle, to whom their relation is but superficial; they should be considered quite apart from this society. It is true that like these, they started from synthesis, and claimed at first to be purely decorative artists; each of them worked as an ornamentist, and even as an industrial artist. But this reaction with them was but a recoil, enabling them to rush forward more impetuously on the path of purely pictorial art. They have, as a fact, far more in common with those great masters we have called the pillars of modern painting, save that they lack all trace of that element of Courbet which is perceptible in these their predecessors. The animal strain is altogether foreign to their manner. As opposed to it, they might be called 'spirituels.' This gives them the aspect of decadents as compared with the others. And they are in fact decadents, in the same sense as their forerunners, and all modern painters are decadent more or less; and in a greater degree than the others, their painting lacks the strong support of a clearly defined tendency, and of a teacher. But tradition works in their highly developed instinct, and their taste enables them to profit by it. In their technique, however, they are more remote from the old masters, less methodical even than Renoir, who is said to have once despairingly confided to an acquaintance that he had no notion how to paint, and was inclined to give up art altogether, as he could not get beyond dilettantism – or than Cézanne, whose spleen led him to take his place in a student's class at Aix to learn drawing. All this is less incomprehensible than it sounds. It seems absurd in relation to our admiration for their works; but it seems natural to them in relation to their admiration for the old masters. Their modesty blinds them to the necessary compensations of development.

The old masters utter well-turned phrases; as compared with these, the words of the nineteenth-century leaders sound like suppressed exclamations; the younger men speak in interjections. And yet they echo back to us; that is the marvellous part of it. We may ask ourselves which is the greater miracle – the pictures evolved from the bearish vigour of Courbet, or the harmonies that breathe from the trembling essays of these young men . . . .

In his narrower significance, Gauguin is a continuation of the exotic element in French art from Degas and Lautrec; in a wider sense, he is an immeasurable extension of artistic boundaries in general. A continuation into barbarism, if you will, because he creates faces we cannot reckon as ours, because he does not restrict himself to the strange but recognised tradition given us by Japan, because he deals in and with forms the genealogy of which is not noted in our museums. He may be charged with having always wanted something else. He tells us in *Noã Noã* how he first sojourned with the Europeans in Tahiti, then in that part of the country where they rarely appear, and finally how he went into the wilderness, to be alone in an Elysian nature. Here he found courage to take a wife, not Titi, beloved of Europeans, but the chaste Tehura, who had never seen a white man. With her he shares his hut. And here an idyll unfolds itself, while in the background the old story of the conquest of the island by what we call European culture goes on. Tehura knows nothing of him, he knows nothing of her. Sex brings them nearer together. He tells her as much as he can. The child listens to him quietly and he admires her silence. Not until he has unbosomed himself completely does she speak to him in her turn, filling the old, empty European slowly with the knowledge, the legends, the poetry, the genius of the Maoris. They begin to love each other. One day he goes fishing with his neighbours. He is lucky, and the neighbours jest; when the tunny comes to a man's hook, he has a faithless Vahina at home . . . He does not think much of this, laughs with the others, but as he goes home, doubt torments him. Tehura is the same as ever; the thought of his age and of her fifteen years fills him with fear, and finally he confesses what the fish have told him. She answers not a word, rises slowly and goes softly to the door to see that no one is listening, and then she stands in front of him and prays aloud to Taaroa to save her. Mute before this naked majesty, he gazes at her, and when she prays him to strike her, because she has given him such evil thoughts, he sinks on his knees and together they offer up the fervid prayers of the heathen.

The book is not merely a unique poem in contemporary

literature, a legend of the Homeric stamp; it is also the history of Gauguin's art. Here it is more welcome to the European than in the painting of the artist of Tahiti. The poem adapts itself to our language, and the vivid episodes, the names with their wealth of vowels minister to our pleasure in splendour, without forcing us into exotic forms. The spirit is European; nothing, indeed, speaks more decisively for the European than Gauguin's flight from Europe.

In painting, on the other hand, this flight seems to have carried him to the utmost limits of representation. Here it is not the story-teller sojourning among us. The charm would compel us to set sail ourselves in these strange structures and share our food with the savages. Mistrust of the uncertain stirs within us, and habit hugs the fetters of time-honoured ennui. No listener to the story, however deeply moved thereby, really believed in the strange tale; nay more, his very emotion was increased by his consciousness of sitting as he listened in the old rocking-chair of Europe. We defend ourselves against the spell. It may be true that Nirvana lies smiling at us from afar, that delights are beckoning to us, things we have not and would fain have, conditions which may have prevailed among us too, when we were barbarians, but...

Every one is of Strindberg's opinion now, even the boldest of those who owe their culture to literature. They love chiaroscuro, twilight facts, which are altered by a change of illumination, the meaning of which is inspiring but obscure. When one appears who would break through the gloom and who offers us elements shining in all the undimmed lustre of their nature, they screen their eyes angrily with a hand, and judge by what they believe they see through their fingers. Of course all that remains is the detail so dear to criticism. The beauty has been shut out.

For all that Gauguin has done is beautiful, though we may say it is fragmentary, though we may not always grasp its objective meaning, though we may regret that in certain large panels the harmony of colour and line is not always so pure and strong as in Van Gogh's very much simpler pictures. There is a Gauguin in M. Fayet's possession in Paris, half-lengths of three savages, so exquisitely grouped and so pure in line, so masterly in the arabesque and so fine in colour that it suggests the avatar of a lovelier, more Grecian Giotto. The grace he found upon his island, by some incomprehensible connection caused him to find not motives, but means for the representation of the naked body in Nature, means which seem to us novel, because we have so long been unaccustomed to such naïve solutions by richly endowed artists. This man, who had nothing but his eyes with him in wilds, looked himself into an ordination of forms,' which people only bring into the world with them in periods of very exceptional brilliance. Had he given more, we should to-day be standing before an absolutely classic artist. Very often his fear of Europe drove him to extremes, where his power failed him; he was all his life a self-taught genius, and in certain minutiæ we are spoilt creatures. Sometimes his planes appear tame to us, just in those passages where the brush should have been wielded like a club. Van Gogh was brilliantly inspired, when he wished to collaborate with his friend; he was thinking of these languid planes, enframed by passages of the utmost boldness. But at times such tender, half-effaced charms spring from the languor, that we rejoice to have what we have.

Gauguin could do everything. He was a great lithographer, a great sculptor, and a skilful potter. When his medium is plastic, the danger of driving his synthesis into the barbaric is doubled or quadrupled. At the same time, the perversity of the European sometimes seduces him into making the primitive as wild and terrifying as possible.

All his life long Gauguin remained a great child, anxious to appear phenomenal at all hazards, more from a profound, fantastic ambition to be remarkable in his own eyes, than to impress others. This drove him to follow up every idea which could minister to this auto-suggestion. The artist in him took care of the rest instinctively. The unsuccessful exceptions in his work are atoned for by many splendid things, such as Schuffenecker's large relief *Soyez amoureuses vous serez heureuses*, of 1888, and the later and more harmonious panels belonging to M. Fayet, which are full of enchantment for those who are content to rely upon the eyes alone.

All Gauguin's sculptures are in wood or porcelain. He did everything himself, and seldom do sculptures reveal, as do his wooden surfaces, the joy of the artist in animating the material with every pressure of his hand. The eye glides over them without sinking in, and does not work, but is gently caressed.

7. Maurice Denis, 'Cézanne', trans. Roger Fry, *Burlington Magazine*

Part I: January 1910, xvi, 207–19; Part II: ibid., 275–80

When Fry became editor of the *Burlington Magazine* he decided to increase the amount of space devoted to modern art. One of his first efforts in this direction was to translate an article first printed in *L'Occident* by Maurice Denis (1870–1943), painter, theorist and friend of Cézanne. This Fry published in two parts with illustrations from Cézanne's work. Denis's article stressed Cézanne's 'classic' status in modern art and this article, together with Meier-Graefe's *Modern Art* (see no. 6), became the principal sources of information for the British when faced with the new art of Post-Impressionism.

INTRODUCTORY NOTE

Anyone who has had the opportunity of observing modern French art cannot fail to be struck by the new tendencies that have become manifest in the last few years. A new ambition, a new conception of the purpose and methods of painting, are gradually emerging; a new hope too, and a new courage to attempt in painting that direct expression of imagined states of consciousness which has for long been relegated to music and poetry. This new conception of art, in which the decorative elements preponderate at the expense of the representative, is not the outcome of any conscious archaistic endeavour, such as made, and perhaps inevitably marred, our own pre-Raphaelite movement. It has in it therefore the promise of a larger and a fuller life. It is, I believe, the direct outcome of the Impressionist movement. It was among Impressionists that it took its rise, and yet it implies the direct contrary of the Impressionist conception of art.

It is generally admitted that the great and original genius, – for recent criticism has the courage to acclaim him as such – who really started this movement, the most promising and fruitful of modern times, was Cézanne. Readers of the *Burlington Magazine* may therefore be interested to hear what one of the ablest exponents in design of the new idea has to say upon the subject. M. Maurice Denis has kindly consented to allow his masterly and judicious appreciation of Cézanne which appeared in *L'Occident*, Sept., 1907, to be translated for the benefit of a wider circle of English readers than has been reached by that paper. Feeling, as he did, that he had expressed himself therein once and for all, he preferred this to treating the subject afresh for the *Burlington Magazine*.

The original article was unillustrated, but seeing how few opportunities English readers have for the study of Cézanne's works, especially of his figure pieces, it has been thought well to include here some typical examples, excluding the better known landscapes and fruit pieces. It is possible that some who have seen only examples of Cézanne's landscapes may have been misled by the extreme brevity of his synthesis into mistrusting his powers of realizing a complete impression; they will be convinced, I believe, even in the reproduction by Cézanne's amazing portrait of himself, ... Before this supremely synthetic statement of the essentials of character one inevitably turns for comparison to Rembrandt. In . . . the *Portrait of a Woman*, we get an interesting light upon the sources of Cézanne's inspiration. One version of the El Greco which inspired this will be familiar to our readers from its appearance at the National Loan Exhibition. M. Maurice Denis discusses at length the position of El Greco in the composition of Cézanne's art. One point of interest, however, seems to have escaped him. Was it not rather El Greco's earliest training in the lingering Byzantine tradition that suggested to him his mode of escape into an art of direct decorative expression? and is not Cézanne after all these centuries the first to take up the hint El Greco threw out? The 'robust art of a Zurbaran and a Velazquez' really passed over this hint. The time had not come to re-establish a system of purely decorative expression; the alternative represent-ational idea of art was not yet worked out, though Velazquez perhaps was destined more than any other to show its ultimate range.

. . . *L'enfant au foulard blanc*, is another example of Cézanne's astonishing power of synthetic statement. The remaining illustrations, *The Bathers* and *The Satyrs*, ... show Cézanne in his more lyrical and romantic mood. He here takes the old traditional material of the nude related to landscape, the material which it might seem that Titian had exhausted, if Rubens had not found a fresh possibility therein. Rubens at all events seemed to have done all that was conceivable, so that Manet only saw his way to using the theme by a complete change of the emotional pitch. But here Cézanne, keeping quite closely within the limits established by the older masters, gives it an altogether new and effective value. He builds up a more compact unity by his calculated emphasis on rhythmic balance of directions.

ROGER E. FRY.

There is something paradoxical in Cézanne's celebrity; and it is scarcely easier to explain than to explain Cézanne himself. The Cézanne question divides inseparably into two camps those who love painting and those who prefer to painting itself the literary and other interests accessory to it. I know indeed that it is the fashion to like painting. The discussions on this question are no longer serious and impassioned. Too many admirations lend themselves to suspicion. 'Snobbism' and speculation have dragged the public into painters' quarrels, and it takes sides according to fashion or interest. Thus it has come about that a public naturally hostile, but well primed by critics and dealers, has conspired to the apotheosis of a great artist, who remains nevertheless a difficult master even for those who love him best.

I have never heard an admirer of Cézanne give me a clear and precise reason for his admiration; and this is true even among those artists who feel most directly the appeal of Cézanne's art. I have heard the words – quality, flavour, importance, interest, classicism, beauty, style.... Now of Delacroix or Monet one could briefly formulate a reasoned appreciation which would be clearly intelligible. But how hard it is to be precise about Cézanne!

The mystery with which the Master of Aix-en-Provence surrounded his life has contributed not a little to the obscurity of the explanations, though his reputation has benefited thereby. He was shy, independent, solitary. Exclusively occupied with his art, he was always restless and usually ill-satisfied with himself. He evaded up to his last years the curiosity of the public. Even those who professed his methods remained for the most part ignorant of him. The present writer admits that about 1890, at the period of his first visit to Tanguy's shop, he thought that Cézanne was a myth, perhaps the pseudonym of some artist well known for other efforts, and that he disbelieved in his existence. Since then he has had the honour of seeing him at Aix; and the remarks which he there gathered, collated with those of M.E. Bernard,[1] may help to throw some light upon Cézanne's aesthetics.

At the moment of his death, the articles in the press were unanimous upon two points; and, wherever their inspiration was derived from, they may fairly be considered to reflect the average opinion. The obituaries, then, admitted first of all that Cézanne influenced a large section of the younger artists; and secondly that he made an effort towards style. We may gather, then, that Cézanne was a sort of classic, and that the younger generation regards him as a representative of classicism.

Unfortunately it is hard to say without too much obscurity what classicism is.

Suppose that after a long sojourn in the country one enters one of those dreary provincial museums, one of those cemeteries abandoned to decay, where the silence and the musty smell denote the lapse of time; one immediately classifies the works exhibited into two groups: in one group the remains of the old collections of amateurs, and in the other the modern galleries, where the commissions given by the State have piled together the pitiful novelties bought in the annual salons according as studio intrigues or ministerial favour decides. It is in such circumstances that one becomes really and ingenuously sensitive to the contrast between ancient and modern art; and that an old canvas by some Bolognese or from Lebrun's atelier, at once vigorous and synthetic in design, asserts its superiority to the dry analyses and thin coloured photographs of our gold-medallists!

Imagine, quite hypothetically, that a Cézanne is there. So we shall understand him better. First of all, we know we cannot place him in the modern galleries, so completely would he be out of key among the anecdotes and the fatuities. One must of sheer necessity place him among the old masters, to whom he is seen at a glance to be akin by his nobility of style. Gauguin used to say, thinking of Cézanne: 'Nothing is so much like a *croûte* as a real masterpiece.' *Croûte* or masterpiece, one can only understand it in opposition to the mediocrity of modern painting. And already we grasp one of the certain characteristics of the classic, namely, *style*, that is to say synthetic order. In opposition to modern pictures, a Cézanne inspires by himself, by its qualities of unity in composition and colour, in short by its painting. The actualities, the illustrations to popular novels or historical events, with which the walls of our supposed museum are lined, seek to interest us only by means of the subject represented. Others perhaps establish the virtuosity of their authors. Good or bad, Cézanne's canvas is truly a *picture*.

Suppose now that for another experiment, and this time a less chimerical one, we put together three works of the same family, three *natures-mortes*, one by Manet, one by Gauguin, one by Cézanne. We shall distinguish at once the objectivity of Manet; that he imitates nature 'as seen through his temperament,' that he translates an artistic sensation. Gauguin is more subjective. His is a decorative, even a hieratic interpretation of nature. Before the Cézanne we think only of the picture; neither the object represented nor the artist's personality holds our attention. We cannot decide so quickly whether it is an imitation or an interpretation of nature. We feel that such an art is nearer to Chardin than to Manet and Gauguin. And if at once we say: this is a picture and a classic picture, the word begins to take on a precise meaning, that, namely, of an equilibrium, a reconciliation of the objective and subjective.

In the Berlin Museum, for instance, the effect produced by Cézanne is significant. However much one admires Manet's *La Serre* or Renoir's *Enfants Bérard* or the admirable landscapes of Monet and Sisley, the presence of Cézanne makes one assimilate them (unjustly, it is true, but by the force of contrast) to the generality of modern productions: on the contrary the pictures of Cézanne seem like works of another period, no less refined but more robust than the most vigorous efforts of the Impressionists.

Thus we arrive at out first estimate of Cézanne as reacting against modern painting and against Impressionism.

When he was first feeling his way out of the tradition of Delacroix, Daumier and Courbet, it was already the old masters of the museums that guided his steps. The revolutionaries of his day never came under the attraction of the old masters. He copied them, and one sees with surprise in his father's house at the Jas de Bouffan a large interpretation of a Lancret and a *Christ in Hades* after Navarete. We must, however, distinguish between this first manner, inspired by the Spanish and Bolognese, and his second fresh and delicately accented manner.

In the first period one sees what Courbet, Delacroix, Daumier and Manet became for him, and by what spontaneous power of assimilation he transmuted in the direction of style certain of their classic tendencies. No doubt he does not arrive at such realizations of placid beauty and plenitude as Titian's; but it is through El Greco that he touches Venice. 'You are the first in the decadence of your art,' wrote Baudelaire to Manet; and such is the debility of modern art that Cézanne seems to bring us health and promise us a renaissance by bringing before us an ideal akin to that of the Venetian decadence.

It is an instructive comparison, and one to which I would call attention, between Cézanne and the neurotic, somewhat deranged Greco, who, by an opposite effort, introduced into the triumphant maturity of Venetian art the system of discords and expressive deformations which gave its origins to Spanish painting. Out of this feverish decrepitude of a great epoch was born in turn the sane robust method of a Zurbaran and a Velazquez. But whilst El Greco indulged in refinements of naturalism and imagination out of lassitude with the perfection of a Titian, Cézanne transcribed his sensibility in bold and reasoned syntheses out of reaction against expiring naturalism and romanticism.

The same interesting conflict, this combination of style and sensibility, meets us again in Cézanne's second period, only it is the Impressionism of Monet and Pissarro that provides the elements, provokes the reaction to them and causes the transmutation into classicism. With the same vigour with which in his previous period he organised the oppositions of black and white, he now disciplines the contrasts of colour introduced by the study of open air light, and the rainbow iridescences of the new palette. At the same time he substitutes for the summary modelling of his earlier figures the reasoned colour-system found in the figure-pieces and *natures-mortes* of this second period, which one may call his 'brilliant' manner.

Impressionism – and by that I mean much more the general movement, which has changed during the last twenty years the aspect of modern painting, than the special art of a Monet or a Renoir – Impressionism was synthetic in its tendencies, since its aim was to translate a sensation, to realize a mood; but its methods were analytic, since colour for it resulted from an infinity of contrasts. For it was by means of the decomposition of the prism that the Impressionists reconstituted light, divided colour and multiplied reflected lights and gradations; in fact, they substituted for varying greys as many different positive colours. Therein lies the fundamental error of Impressionism. The *Fifre* of Manet in four tones is necessarily more synthetic than the most delicious Renoir, where the play of sunlight and shadow creates the widest range of varied half-tones. Now there is in a fine Cézanne as much simplicity, austerity and grandeur as in Manet, and the gradations retain the freshness and lustre which give their flower-like brilliance to the canvases of Renoir. Some months before his death Cézanne said: 'What I wanted was to make of Impressionism something solid and durable, like the art of the museums.' It was for this reason also that he so much admired the early Pissarros, and still more the early Monets. Monet was, indeed, the only one of his contemporaries for whom he expressed great admiration.

Thus at first guided by his Latin instinct and his natural inclination, and later with full consciousness of his purpose and his own nature, he set to work to create out of Impressionism a certain classic conception.

In constant reaction against the art of his time, his powerful individuality drew from it none the less the material and pretext for his researches in style; he drew from it the sustaining elements of his work. At a period when the artist's sensibility was considered almost universally to be the sole motive of a work of art, and when improvisation – 'the spiritual excitement provoked by exaltation of the senses' – tended to destroy at one blow both the superannuated conventions of the academies and the necessity for method, it happened that the art of Cézanne showed the way to substitute reflexion for empiricism without sacrificing the essential *rôle* of sensibility. Thus, for instance, instead of the chronometric notation of appearances, he was able to hold the emotion of the moment even while he elaborated almost to excess, in a calculated and intentional effort, his studies after nature. He *composed* his *natures-mortes*, varying intentionally the lines and the masses, disposing his draperies according to premeditated rhythms, avoiding the accidents of chance, seeking for plastic beauty; and all this without losing anything of the essential *motive* – that initial motive which is realised in its essentials in his sketches and water colours. I allude to the delicate symphony of juxtaposed gradations, which his eye discovered at once, but for which at the same moment his reason spontaneously demanded the logical support of composition, of plan and of architecture.

There was nothing less artificial, let us note, than this effort towards a just combination of style and sensibility. That which others have sought, and sometimes found, in the imitation of the old masters, the discipline that he himself in his earlier works sought from the great artists of his time or of the past, he discovered finally in himself. And this is the essential characteristic of Cézanne. His spiritual conformation, his *genius*, did not allow him to profit directly from the old masters: he finds himself in a situation towards them similar to that which he occupied towards his contemporaries. His originality grows in his contact with those whom he imitates or is impressed by; thence comes his persistent *gaucherie*, his happy *naïveté*, and thence also the incredible clumsiness into which his sincerity forced him. For him it is not a question of imposing style upon a study as, after all, Puvis de Chavannes did. He is so naturally a painter, so spontaneously classic. If I were to venture a comparison with another art, I should say that there is the same relation between Cézanne and Veronese as between Mallarmé of the *Herodiade* and Racine of the *Berenice*. With the same elements – new or at all events refreshed, without anything borrowed from the past, except the necessary forms (on the one hand the mould of the Alexandrine and of tragedy, on the other the traditional conception of the composed picture) – they find, both poet and painter, the language of the Masters. Both observed the same scrupulous conformity to the necessities of their art; both refused to overstep its limits. Just as the writer determined to owe the whole expression of his poem to what is, except for idea and subject, the pure domain of literature – sonority of words, rhythm of phrase, elasticity of syntax – the painter has been a painter before everything. Painting oscillates perpetually between invention and imitation: sometimes it copies and sometimes it imagines. These are its variations. But whether it reproduces objective nature or translates more specifically the artist's emotion, it is bound to be an art of concrete beauty, and our senses must discover in the work of art itself – abstraction made of the subject represented – an immediate satisfaction, a pure aesthetic pleasure. The painting of Cézanne is literally the essential art, the definition of which is so refractory to criticism, the realization of which seems impossible. It imitates objects without any exactitude and without any accessory interest of sentiment or thought. When he imagines a sketch, he assembles colours and forms without any literary preoccupation; his aim is nearer to that of a Persian carpet weaver than of a Delacroix, transforming into coloured harmony, but with dramatic or lyric intention, a scene of the Bible or of Shakespeare. A negative effort, if you will, but one which declares an unheard of instinct for painting.

He is the man who paints. Renoir said to me one day: 'How on earth does he do it? He cannot put two touches of colour on to a canvas without its being already an achievement.'

It is of little moment what the pretext is for this sampling of colour: nudes improbably grouped in a non-existent landscape, apples in a plate placed awry upon some commonplace material – there is always a beautiful line, a beautiful balance, a sumptuous sequence of resounding harmonies. The gift of freshness, the spontaneity and novelty of his discoveries, add still more to the interest of his slightest sketches.

'He is' said Sérusier, 'the pure painter. His style is a pure style; his poetry is a painter's poetry. The purpose, even the concept of the object represented, disappears before the charm of his coloured forms. Of an apple by some commonplace painter one says: I should like to eat it. Of an apple by Cézanne one says: How beautiful! One would not peel it; one would like to copy it. It is in that that the spiritual power of Cézanne consists. I purposely do not say idealism, because the ideal apple would be the one that stimulated most the mucous membrane, and Cézanne's apple speaks to the spirit by means of the eyes.'

'One thing must be noted,' Sérusier continues: 'that is the absence of subject. In his first manner the subject was sometimes childish: after his evolution the subject disappears, there is only the *motive*.' (It is the word that Cézanne was in the habit of using.)

That is surely an important lesson. Have we not confused all the methods of art – mixed together music, literature, painting? In this, too, Cézanne is in reaction. He is a simple artisan, a primitive who returns to the sources of his art, respects its first postulates and necessities, limits himself by its essential elements, by what constitutes exclusively the art of painting. He determines to ignore everything else, both equivocal refinements and deceptive methods. In front of the *motive* he rejects everything that might distract him from painting, might compromise his *petite sensation* as he used to say, making use of the phraseology of the aesthetic philosophy of his youth: he avoids at once deceptive representation and literature.

The preceding reflections allow us to explain in what way Cézanne is related to Symbolism. Synthetism, which becomes, in contact with poetry, Symbolism, was not in its origin a mystic or idealist movement. It was inaugurated by landscape-painters, by painters of still-life, not at all by painters of the soul. Nevertheless it implied the belief in a correspondence between external forms and subjective states. Instead of evoking our moods by means of the subject represented, it was the work of art itself which was to transmit the initial sensation and perpetuate its emotions. Every work of art is a transposition, an emotional equivalent, a caricature of a sensation received, or, more generally, of a psychological fact.

'I wished to copy nature,' said Cézanne, 'I could not. But I was satisfied when I had discovered that the sun, for instance, could not be *reproduced*, but that it must be *represented* by something else ... by colour.' There is the definition of Symbolism such as we understood it about 1890. The older artists of that day, Gauguin above all, had a boundless admiration for Cézanne. I must add that they had at the same time the greatest esteem for Odilon Rédon. Odilon Rédon also had searched outside of the reproduction of nature and of sensation for the plastic equivalents of his emotions and his dreams. He, too, tried to remain a *painter*, exclusively a painter, while he was translating the radiance and gloom of his imagination.

If I have insisted on the name of Rédon in this connexion it is not merely to render the homage due to this artist and to acquit the gratitude of a generation, but that we may draw from the comparison of these two masters a still further precision in our definition of Cézanne. Yes, Rédon stands as the origin of Symbolism so far as the *plastic* expression of the ideal is concerned; and on the other hand the example of Cézanne taught us to transpose the data of sensation into the elements of a work of art. Rédon's subject is rather subjective, Cézanne's rather objective, but both of them express themselves by a method which aims at the creation of a concrete object, at once artistic and representative of a response to sensation. Complex as his epoch, the artist we are endeavouring to explain found then in this method his equilibrium, the profound unity of his efforts, the solution of his antinomies.

It is a touching spectacle that a canvas of Cézanne presents; generally unfinished, scraped with a palette-knife, scored over with *pentimenti* in turpentine, many times repainted, with an *impasto* that approaches actual relief. In all this evidence of labour, one catches sight of the artist in his struggle for style and his passion for nature; of his acquiescence in certain classic formulæ and the revolt of an original sensibility; one sees reason at odds with inexperience, the need for harmony conflicting with the fever of original expression. Never does he subordinate his efforts to his technical means; 'for the desires of the flesh,' says St. Paul, 'are contrary to those of the spirit, and those of the spirit are contrary to those of the flesh, they are opposed one to another in such wise that ye do not that which ye would.' It is the eternal struggle of reason with sensibility which makes the saint and the genius.

Let us admit that it gives rise sometimes, with Cézanne, to chaotic results. We have unearthed a classic spontaneity in his very sensations, but the realization is not reached without lapses. Constrained already by his need for synthesis to adopt disconcerting simplifications, he deforms his design still further by the necessity for expression and by his scrupulous sincerity.[2] It is herein that we find the motives for the *gaucherie* with which Cézanne is so often reproached, and herein lies the explanation of that practice of naïveté and ungainliness common to his disciples and imitators.

True, tradition is not an affair of the correctness and rhetoric of the art school, as is believed by certain artists who, under pretext of following Leonardo and Titian, make us regret Cabanel and Benjamin Constant. But it would be just as puerile to glorify Cézanne for his negligences and imperfections. We must not become the dupes of the spirit of paradox and anarchic subtlety. People now judge contemptuously a work which shows patient execution; they admire only sketches and those especially in which the summary invention and rapid handling imply a sort of nihilism in art; this is the very superstition of the unfinished. Doubtless Suarès is right in saying: 'In times of decadence everyone is an anarchist, both those who are and those who boast that they are not. For each finds his law within himself. . . . We love order passionately, but it is the order we desire to make, not the order we receive.' The works of artists of other ages remain for us a fixed standard: let us seek no other. It is because some enthusiastic critics have preferred Cézanne to Chardin and Veronese that it is right to recognize his lapses and avow in all simplicity that he has suffered the reaction of our age of disorder. Nevertheless such is the power of his invention and the sincerity of his gesture that his ungainliness scarcely disturbs us and usually disappears in the general harmony. With qualities as beautiful, Chardin and Veronese had the accomplishment and science to go further in the execution of the work of art. They played with difficulties insurmountable for us: their supple fancy accommodated itself to the laws of perspective and anatomy that we reject as the worst restrictions. They knew how to trace a straight line or a regular curve with the point of the brush. We cannot but regret the old order. The same shocks which formerly overthrew the French constitution favoured romantic influences, the first origins of the decadence of the crafts and of the intellectual anarchy in which we are struggling. Let us then admire Cézanne, who has shown us the possibility of a classic Renaissance and given us works of such nobility of style at a time when, in the words of Gustave Moreau, 'all that is good is a failure.'

What astonishes us most in Cézanne's work is certainly his research for form, or, to be exact, for deformation. It is there that one discovers the most hesitation, the most *pentimenti* on the artist's part. The large picture of the *Baigneuses*, left unfinished in the studio at Aix, is from this point of view typical. Taken up again, numberless times during many years, it has varied but little in general appearance and colour, and even the disposition of the brush-strokes remains almost permanent. On the other hand the dimensions of the figures were often readjusted; sometimes they were life-size, sometimes they were contracted to half; the arms, the torsos, the legs were enlarged and diminished in unimaginable proportions. It is just there that lies the variable element in his work; his sentiment for form allowed neither of silhouette nor of fixed proportions.

For, to begin with, he did not comprehend drawing by line and contour. In spite of the exclamation reported by M.V. during the sittings for his portrait, 'Jean Dominique is strong!' it is certain that he did not love M. Ingres. He used to say, 'Degas is not enough of a painter; he has not enough of *that!*' – and, with a nervous gesture, he imitated the stroke of an Italian decorator. He often talked of the caricaturists, of Gavarni, of Forain, and above all of Daumier. He liked exuberance of movement, relief of muscular forms, impetuosity of hand, bravura of handling. He used to draw from Puget. He demanded always ease and vehemence in execution. He preferred, one cannot doubt, the *chic* drawing of the Bolognese to the conciseness of Ingres.

On the walls of Jas de Bouffan, covered up now with hangings, he has left improvizations, studies painted as the inspiration came, and which seem carried through at a sitting. They make one think, in spite of their fine pictorial quality, of the fanfaronnades of Claude in Zola's *L'Œuvre*, and of his declamations upon 'temperament.' The models of his choice at this period are engravings after the Spanish and Italian artists of the seventeenth century. When I asked him what had led him from this vehemence of execution to the patient technique of the separate brush-stroke, he replied, 'It is because I cannot render my sensation at once; hence I put on colour again, *I put it on as best I can*. But when I begin I endeavour always to paint with a full impasto like Manet, *giving the form with the brush*.'

'There is no such thing as line,' he said, 'no such thing as modelling, there are only contrasts. When colour attains its richness form attains its plenitude.'[3]

Thus, in his essentially concrete perception of objects, form is not separated from colour; they condition one another, they are indissolubly united. And in consequence in his execution he wishes to realize them as he sees them, by a single brush-stroke. If he fails it is certainly in part from the imperfection of his craft, of which he used to complain, but also and above all from his scruples as a colourist, as we shall see presently.

All his faculty for abstraction – and we see how far the painter dominates the theorist – all his faculty for abstraction permits him to distinguish only among notable forms 'the sphere, the cone and the cylinder.' All forms are referred to those which he is alone capable of thinking. The multiplicity of his colour schemes varies them infinitely. But still he never reaches the conception of the circle, the triangle, the parallelogram; those are abstractions which his eye and brain refuse to admit. *Forms* are for him *volumes*.

Hence all objects were bound to tell for him according to their relief, and to be situated according to planes at different distances from the spectator within the supposed depth of the picture. A new antinomy, this, which threatens to render highly accidental 'that plane surface covered with colours arranged in a determined order.' Colourist before everything, as he was, Cézanne resolves this antinomy by chromatism – the transposition, that is, of values of black and white into values of colour.

'I want,' he told me, following the passage from light to shade on his closed fist – 'I want to do with colour what they do in black and white with the stump.' He replaces light by colour. This shadow is a colour, this light, this half-tone are colours. The white of this table-cloth is a blue, a green, a rose; they mingle in the shadows with the surrounding local tints; but the crudity in the light may be harmoniously translated by dissonant blue, green and rose. He substitutes, that is, contrasts of tint for contrasts of tone, He disentangles thus what he used to call 'the confusion of sensations.' In all this conversation, of which I here report scraps, he never once mentioned the word values. His system assuredly excludes relations of values in the sense accepted in the schools.

Volume finds, then, its expression in Cézanne in a gamut of tints, a series of touches; these touches follow one another by contrast or analogy according as the form is interrupted or continuous. This was what he was fond of calling *modulating* instead of modelling. We know the result of this system, at once shimmering and forcible; I will not attempt to describe the richness of harmony and the gaiety of illumination of his pictures. It is like silk, like mother-of-pearl and like velvet. Each *modulated* object manifests its contour by the greater or less exaltation of its colour. If it is in shadow its colour shares the tints of the background. This background is a tissue of tints sacrificed to the principal motive which they accompany. But on any and every pretext the same process recurs of chromatic scales where the colours contrast and interweave in tones and half-tones. The whole canvas is a tapestry where each colour *plays* separately and yet at the same time fuses its sonority in the total effect. The characteristic aspect of Cézanne's pictures comes from this juxtaposition, from this mosaic of separate and slightly fused tones. 'Painting,' he used to say, 'is the registration of one's coloured sensations' (E. Bernard). Such was the exigence of his eye that he was compelled to have recourse to this refinement of technique in order to preserve the quality, the flavour of his sensations, and satisfy his need of harmony. Bachaumont in 1767 wrote of Chardin; 'His method of painting is singular. He poses his colours one after another almost without mixing them, in such a way that his work somewhat resembles a mosaic or patchwork like the needlework tapestry called cross-stitch.'

The fruit-pieces of Cézanne and his unfinished figures afford the best examples of this method, the idea of which was perhaps taken from Chardin: a few decisive touches declare the roundness of the form by their juxtaposition with softened tints, the contour does not come till the last, as a vehement accent, put in with turpentine to underline and isolate the form already realized by the gradation of colour.

In this assemblage of tints with an aim at grandeur of style, perspective disappears; values too (in the school of art sense) and values of atmosphere are attenuated and equalized. The decorative effect and the balance of the composition appear all the more complete owing to this sacrifice of aerial perspective. Venetian painting with a more enveloping chiaroscuro offers frequently this fine aspect of unity of plane.[4] It is curious that it is this which most struck the first symbolists, Gauguin, Bernard, Anquetin – those, in fact, who were the first to love and imitate Cézanne. Their synthetic system admitted only flat tints and a hard contour; thence arose a whole series of decorative works which I, certainly, do not wish to decry; but how much more synoptic, how much more concrete and vital were the syntheses of Cézanne!

Synthesis does not necessarily mean simplification in the sense of suppression of certain parts of the object; it is simplifying in the sense of *rendering intelligible*. It is, in short, creating a hierarchy: submitting each picture to a single rhythm, to a dominant; sacrificing, subordinating – generalizing. It is not enough to *stylize* an object (as they say in the school of Grasset), to make some sort of copy of it, and then to underline the external contour with a thick stroke. Simplification so obtained is not synthesis.

'Synthesis,' says Sérusier, 'consists in compressing all forms into the small number of forms which we are capable of thinking – straight lines, certain angles, arcs of circles and ellipses; outside these we are lost in the ocean of variety.' That, no doubt, is a mathematical conception of art; it does not, however, lack grandeur. But that is not, whatever Sérusier says, the conception of Cézanne. He certainly does not lose himself in the ocean of variety;

he knows how to elucidate and condense his impressions; his formulae are luminous and concise; they are never abstract.

And this, again, is one of the points wherein he touches the classics; he never compromises by abstraction the just equilibrium between nature and style. All his labour is devoted to preserving his sensation; but this sensation implies the identity of colour and form; his sensibility implies his style. Naturally and instinctively he unites, in his spirit if not on his canvas, the grace and brilliance of modern colourists with the robustness of the old masters. Doubtless the realization is not reached without labour nor without lapses. But the order which he discovers is for him a necessity of expression.

He is at once the climax of the classic tradition and the result of the great crisis of liberty and illumination which has rejuvenated modern art. He is the Poussin of Impressionism. He has the fine perception of a Parisian, and he is splendid and exuberant like an Italian decorator. He is orderly as a Frenchman and feverish as a Spaniard. He is a Chardin of the decadence and at times he surpasses Chardin. There is something of El Greco in him and often the healthfulness of Veronese. But such as he is he is so naturally, and all the scruples of his will, all the assiduity of his effort have only aided and exalted his natural gifts.

The attempt here made is to *define* the work of the painter: not to express its poetry. All the magic of words would not suffice to translate, for one who has never had it, the unforgettable impression which the sight of a fine Cézanne arouses. The charm of Cézanne cannot be described; nor could one tell of the nobility of his landscapes, the freshness of his chords of green, the purity and profoundity of his blues, the delicacy of his carnations, the velvety brilliance of his fruit. Few artists have had so original a sensibility – but that has been said of so many others in our time that it is better not to insist on it; it is the most ordinary praise one can give an artist. He liked to speak, with an appearance of modesty, of his 'little sensation,' of his 'little sensibility.' He complained that Gauguin had taken from his and 'l'eut promenée dans tous les paquebots.' In truth his art is so concise and so natural, so living and so spontaneous, that it is difficult to get inspiration from his technical methods without carrying off with them something of himself as well. For Félibien, speaking of painters, the sensation is: 'The application of things to the spirit *or* the judgment which the spirit passes on them.' The two operations, the *Aspect* and *Prospect*, as Poussin says, are no longer separate with Cézanne. To organize one's sensations[5] was a discipline of the seventeenth century; it is the preconceived limitation of the artist's receptivity. But the true artist is like the true *savant*, 'a child-like and serious nature.'[6] He accomplishes this miracle – to preserve amidst his efforts and his scruples all his freshness and naïveté.

NOTES

1 Pub. in *L'Occident*, July 1904. [Denis's note]
2 I have tried to show in an article in *les Arts de la Vie* (July 1904) that the *gaucherie* of the Primitives consists in painting objects according to our usual knowledge of them, instead of painting them as the Moderns do, according to a preconceived idea of the picturesque or artistic. The picturesque being the element of nature recognized as proper to painting: it follows that the moment an artist departs from the admitted formulæ and paints with naïve sincerity he incurs the reproach of ignorance and gaucherie. [Denis's note]
3 Quoted by E. Bernard.
4 There was discovered in 1905 in the Scuola di S. Rocco a fragment of a frieze by Tintoretto folded back against the wall when it was first put up, because it was too large for the space; this had preserved all the freshness of its colour. In it were apples painted in pale green and bright red on a ground of leaves of Veronese green. *It is all colour.* One would call it a Cézanne. Perhaps it lacks the finishing touch of umber which would have sobered it, but, such as it is, this precious fragment indicates in Tintoretto an effort at chromatism altogether similar to that which I have explained in Cézanne. [Denis's note]
5 'There are two things in the painter: the eye and the brain; each should help the other; one should work at their mutual development; for the eye by the vision of nature, for the brain by the logic of organized sensations which affords the means of expression.' (Cited by E. Bernard, l.c.) Poussin wrote to M. de Chantelon: 'My nature compels me to seek out and love things that are well ordered.' [Denis's note]
6 E. Renan. [Denis's note]

8. Robert Dell, 'Introduction' to *Modern French Artists*

Brighton, June 1910, 3–8

Dell, who lived in France, was art correspondent for the *Burlington Magazine*. He was asked to supply pictures for the recently opened Brighton Galleries which included the work not only of plein-airistes and Impressionists but of those painters who were variously named neo-Impressionists, Symbolists, Pointelistes (sic) and Intimistes.

The pictures which created most interest were Gauguin's *Les Boeufs* (now called *Christmas Night*) (1984: private collection, New York) and Derain's *Westminster Bridge, London,* (1905: Max Kaganovitch collection, Paris). The exhibition also included a portrait by Cézanne of *Albin Valabrègue,* two small still-life studies by Matisse and works by Cross, Denis, Friesz, Rouault, Sérusier, Signac, Vallotton and Vlaminck.

The aim of the present Exhibition is to give, by chosen examples of the various existing schools, some general idea of contemporary French painting in its different forms of expression. This has not, so far as I know, yet been attempted in England, and Brighton is to be congratulated on the possession of a Director of its Public Art Gallery sufficiently enterprising to conceive so ambitious a scheme. There have been many exhibitions of French art in England, some restricted to a particular school, others of a more general character, but I do not remember any which was really comprehensive of all the contemporary schools of French painting. Even that held in 1908 at the Franco-British Exhibition was very far from being really representative of French art either in the 19th or the 20th century. The retrospective exhibition of pictures on that occasion was sadly inadequate and contemporary French painting was represented almost entirely by exhibitors at the two official Salons, to the exclusion of the younger schools.

Here a retrospective exhibition has not been attempted; there was not the space for an adequate representation of the past – even the immediate past – as well as the present. Only a few examples of deceased artists have been included, and these are mostly artists who have died quite recently. Even in the representation of the present there are gaps that one would have liked to see filled, and certain artists are less adequately represented than they ought to be; nobody is more conscious of that fact than myself, and since, being in Paris, I was necessarily directly responsible for the final choice of the majority of the pictures, it is to my charge that any shortcomings must be laid, Certain extenuating circumstances may be pleaded. We have had to face the competition of other important exhibitions outside France, notably of the International Exhibition at Brussels and of the exhibition of French art now being held at Buenos Ayres. Moreover at this season both the Salons are open and numerous other exhibitions are being held in Paris, so that many of the painters who were invited to exhibit had empty studios. We had hoped to fill many of the gaps by the aid of dealers and private collectors and, as regards the former, our hope was to a great extent fulfilled. The catalogue shows how much we owe to their kindness; it is a considerable sacrifice on their part to lend in the height of the Paris season, when they want all the pictures they can get.

I regret to be obliged to say that the response of the French private collectors, on the other hand, has been disappointing. Most of the requests were met with a categorical refusal, even in cases where a collector was asked to lend one of numerous examples of a particular artist in his possession. We have, therefore, all the more reason to be grateful to the few who have shown a more generous spirit, to M. Aubrey, M. Christain Cherfils. Madamoiselle Diéterle, M. Théodore Duret, Herr Walther Epstein, Madame de la Faille de Leverghem, M. Paul Gallimard, M. Alphonse Kann, Dr. Keller, Madame Menard-Dorian, Comte Robert de Montesquiou, M. Swann, etc.

One can, however, say with some confidence that, on the whole, the various contemporary schools which have issued from the Impressionist movement are fairly adequately represented; this is all the more satisfactory since these schools are the least known in England. The original Impressionists – Manet, Monet, Pissarro, Sisley, Renoir, Guillaumin, etc. – are now well-known to a large number of Englishmen, but their successors and artistic descendants hardly at all. Yet nobody can say that he knows contemporary French painting unless he knows the works of the Neo-Impressionists, and an exhibition of French art which did not give a large place to those works would not be at all representative. They have a place of honour at the International Exhibition at Brussels and will have one next year at the International Exhibition at Rome; it is time that the English public ceased to ignore the existence of a school of painting which is already represented in the public galleries of Germany.

Many visitors to the present exhibition, brought face to face for the first time with pictures so different from those that they have been accustomed to see on the walls of Burlington House, will be tempted to laugh at them as mere eccentricities or denounce them as outrages. They would do well to suspend judgment. They are not asked to pretend to like what they do not; but they are asked not to jump to the conclusion that the unaccustomed is necessarily ridiculous. Every successive new movement in art has been found ridiculous at first; every pioneer has been told that he outraged all the accepted canons of art, as he probably did. In the end the accepted canons of art have to go to the wall and the heresy of the past, as always, has become the orthodoxy of the present. Nowadays the Neo-impressionists are told to look at Constable or Turner, at Corot or Millet or Rousseau or Daubigny; even Manet and the early Impressionists are admitted into the fold of respectability and their works competed for by American millionaires, but those who have carried their theories further are still anathema to many. Yet our fathers and grandfathers laughed at Constable and Turner and the school of 1830 and the early Impressionists as much as some of us laugh at the Neo-impressionists. Corot and Rousseau were regarded by the orthodox in art of their time as dangerous innovators; even Ruskin, who was the first fully to appreciate Turner, in his turn denounced Whistler. Daumier was looked upon as a trifler by the artistic pontiffs of his day and, when Couture, Manet's master, wished to warn his pupil of the consequences of his aberrations, he told him that he would never be anything more than the Daumier of his time. There can hardly be a painter nowadays who would not be flattered by the suggestion that he might become the Daumier of his time; few would take it as a compliment if they were told that they might hope to emulate Couture.

As for the early Impressionists, the bitterness with which they were attacked has not been exceeded even in theological controversy. Manet and his friends could not sell for a hundred francs apiece pictures which now fetch hundreds and even thousands of pounds. Those who had no private fortune suffered want and privation rather than surrender their ideals; Sisley died in poverty so lately as 1899, and would have starved at one time, but for the help of friends, notably of M. Durand-Ruel, one of the first to appreciate the possibilities of Impressionism. Cézanne has only begun, since his death in 1906, to be appreciated by more than a tiny minority. Not only the Impressionists, but all artists who did not accept the academic dogmas were excommunicated by officialism. The famous 'Salon des refusés,' or exhibition of painters rejected by the Salon, in 1863 included works by Bracquemont, Cals, Cazin, Chintreuil, Fantin-Latour, Harpignies, Jongkind, Jean-Paul Laurens, Alphonse Legros, Manet, Pissarro, Antoine Vollon and Whistler; one of these works was the celebrated *Dejeuner sur l'herbe* of Manet, now in the Moreau-Nélaton collection in the Louvre. Three years later Emile Zola, one of the very first to recognise the genius of Manet, had to leave the staff of the daily paper, *L'Evenement*, on account of his articles eulogising the work of the painter; for some time afterwards no paper would print his art criticisms. Zola lived to see Manet's *Olympia* placed in the Luxembourg by public subscription, and his prophecy of 7th May, 1866, that Manet would one day be represented in the Louvre is now fulfilled, like another celebrated prophecy of his; for the gift of prophecy is the insight of genius. Manet is in the Louvre, and Zola lies in the Panthéon; so true it is that we build the sepulchres of the prophets whom our fathers stoned.

In the modest conviction that we are not wiser than our fathers, we may profitably learn from the history of the past the unwisdom of hasty condemnation of what is new or unusual. It does not, of course, follow that every new movement or departure from accepted canons, in art or anything else, is necessarily valuable or durable. But it is well to suspend judgment until study and reflection have provided us with materials for forming a sound judgment; and it is not well to ridicule what is not immediately comprehensible. Nor should we assume that the artist is not serious, that he is trying to be original at all costs and only succeeding in being eccentric. That is what Ruskin said, in effect, about Whistler, but he was quite mistaken. Whistler was not, as some of his injudicious admirers have tried to convince us, among the greatest masters, but he was much more than a man who threw a paint-pot in the face of the public. There are, of course, artists whose only aim is to be considered original, who delight to 'épater le bourgeois,' as the French say; naturally so, since notoriety is easier to win than fame. But such are certainly rarer exceptions among Neo-impressionists than are the bidders for a cheap popularity among the more academic.

A great mistake is often made about the artistic movements of the 19th century and in particular the Impressionist movement. It is supposed that, because they were a reaction against the artistic canons of their time, they were a-breaking away from all tradition, that their leaders were mere anarchists who despised the past and wished to make a clean cut with it. Nothing could be more untrue or, indeed, more impossible. Even had they wished to do so, the leaders of these movements could not have ignored the past; no entirely new start has ever yet been made in the world or ever will be, every new movement owes something to what has gone before.

It must be remembered that what was counted as the true tradition when Corot and Courbet first began to paint directly from nature was not only stereotyped and degenerate; it was also by no means very ancient and was not a natural expression of the French spirit. The natural evolution of French artistic tradition had been checked by the Italianising influence of the official artistic organisation and, in particular, of the French school at Rome. With the great painters of the 18th century the French spirit burst out into complete freedom. Watteau, the greatest of them all and one of the greatest artists of all time, was never a pupil of the school of Rome; his masters were Rubens and Titian, but his art was the expression of his own individual genius. Fragonard was a pupil of the school of Rome but soon deserted classicism with admirable results; who would choose his *Coresus and Callirhoë* in the Louvre in preference to the *Foire de Saint-Cloud*? With the Directorate came a reaction to a pseudo-classicism, which became the stereotyped tradition of official art. Watteau was despised as a painter of *genre*. Fragonard died poor and forgotten; pomposity reigned supreme. Painters were called upon to confine themselves to the grand manner and to noble subjects, and a noble subject

81

usually meant something in Roman togas. The painting of landscape or of everyday life was looked upon as an inferior art. The classical school possessed at least two great artists. David and his greater pupil, Ingres, but it is by their portraits rather than their tiresome historical compositions that they will live. Ingres, moreover, traces his artistic descent to Poussin and in his turn was the artistic ancestor of Degas, who has never ranked himself among the Impressionists, although he has much in common with them. One has only to compare such a picture as *La Source* in the Louvre, which Ingres painted in 1856, with a nude by Degas to see the affinity.

The first reaction against the classical school was that of the Romanticists, of whom Delacroix and Géricault were the chief, but the romantic movement soon degenerated. Out of it came the 'Troubadour' and Mediaeval' schools, who, leaving Greek and Roman antiquity, devoted themselves to subjects supposed to illustrate the Middle Ages, a period which they extended considerably; their pictures reach the lowest depths of banality and unfortunately their following is not yet extinct. Delacroix has had great influence on French art in the nineteenth century; his technique in some respects anticipated that of the Impressionists. Corot and Courbet have had even greater influence, especially Courbet. They were the first French painters to paint directly from nature, and it is certain that Corot, at any rate, was to a great extent influenced by Constable and Bonington, who were at first better appreciated in France than in England. From Corot and Courbet sprang the now famous school of 1830 or 'Barbizon School,' to which the early Impressionist landscape painters owed much. But the peculiar technique of the Impressionists differs entirely from that of the school of 1830 and its originator was Edouard Manet. He was the real founder of Impressionism. Turner, who is often called an Impressionist, and who was one in a sense, seems to have had little if any influence on the early French Impressionists and is not an Impressionist in the technical sense of the term. The French painters who have been influenced by him, of whom Félix Ziem is one of the best-known, do not belong to the Impressionist school.

The name 'Impressionist,' by the way, was first given to Manet and his friends by an opponent in 1874. In that year an exhibition of thirty painters was held in the Boulevard des Capucines; it included a picture by Claude Monet with the title: *Impression; soleil levant*, and the writer of a bitter attack on the exhibition in *La Charivari* headed his article: *Exposition des Impressionistes*. The name stuck and was eventually adopted by the group, who used it officially for the first time on the occasion of the exhibition held in M. Durand-Ruel's gallery in the rue Le Peletier in 1877. ("Manet and the French Impressionists," by Théodore Duret. London: Grant Richards. Pages 113–116.)[1]

The origin of Impressionism, like that of the Barbizon school, was merely a revolt against pseudo-classical convention and a desire to paint things as they are. Manet, who was essentially a figure-painter, revolted from the painting of pompous historical subjects and of the nude from professional models, supposed to come up to a conventional standard of beauty. He determined to begin painting ordinary men and women. M. Duret, in the admirable book which I have already cited and from which I take these facts, tells us that Manet was specially influenced by the Venetians of the Renaissance and by the great Spanish masters, Velasquez, Goya and Greco. The Spanish influence is, indeed, very marked in some of his earlier pictures, but he gradually evolved a method of his own, profoundly personal, in which any influences that remain are only secondary.

The impressionist landscape painters were first attracted to the direct study of nature by Corot or Courbet, but the principles of their technique, as has already been said, owe their origin to Manet. Not that Manet or any other Impressionist first elaborated a theory and then sat down to apply it. Action always precedes theory, and so it was with the Impressionists; they arrived at a peculiar method by an almost unconscious evolution, after endless experiments. A complete technical account of their methods cannot be attempted here; it is enough to say that its chief points are the discarding of the convention of a fixed opposition of light and shade ('Chiaroscuro') in favour of an opposition of different tones; the practice of painting in bright sunlight with brilliant tones and of representing shadows not as the absence of light (which they are not) but as less intense light; the use only of the seven colours of the spectrum juxtaposed in touches (division of tones) leaving them to harmonise at a certain distance in the eye of the beholder. The Impressionist was concerned with effects of light and atmosphere. Recognising that light creates colour, he refused to give natural objects the fixed colours which had hitherto been supposed to belong to them and painted them as he saw them, now of one colour, now of another.

The Impressionist pictures were startling, because they seemed unnatural to those who had not closely observed nature and were accustomed to the conventions of the landscapes which they knew. The revolution in technique was accompanied by a revolution in the choice of subjects; the Impressionists discarded the conventional idea of beauty as something fixed and absolute, they knew no distinction between noble and ignoble subjects, nothing for them was common or unclean. Here, surely, the great masters of the past are with them. 'Art,' as Whistler said, 'is occupied in seeking and finding the beautiful in all conditions and in all times.' The Impressionist landscape painters so dealt with nature; they did not concern themselves with formal composition nor did they seek for specially picturesque subjects. 'The open-air Impressionist,' to quote M. Duret again, 'was concerned only with some momentary effect of atmosphere, of light or of foliage, which has struck him; that effect gave him subject; he was no longer preoccupied with the scene in which he had surprised it.'

Innovators as they seemed to be and in fact were, the Impressionists had more in common with genuine French artistic tradition than the pseudo-classicists. They could trace their artists lineage even to Claude and Poussin and to earlier French painters. The great artists of the 18th century – Watteau, Chardin, Fragonard – were truly their precursors. The account given by Bachaumont in 1776 of Chardin's method of painting[2] shows that it closely resembled that of Cézanne. Watteau's famous *Embarquement pour Cythere* is, as M. Camille Mauclair has pointed out, an Impressionist canvas, both in its lighting and in its adoption of the division of tones. There is at least one picture of Fragonard, the beautiful *Liseuse* in Dr. Tuffier's collection, which is fundamentally Impressionist; it might be a Renoir.

Need it be said that the last word in art was not spoken by the Impressionists? Impressionism is not infallible; to elevate it into a dogma, to hold it up as the one and only correct method of painting would to be to fall into the error of those who condemned it. One has only to look in the present exhibition at the work of the various schools which have issued from Impressionism to see that already new methods are being evolved. Even the Neo-impressionists proper – those who follow most closely the first Impressionists – have departed from their method to some extent;

for instance they show a tendency to ignore composition less than their predecessors, and they do not all adhere strictly to the division of tones. With some of them there is a tendency towards a return to classicism, not the classicism of David and his school, but in the sense of the research of style or synthetic order. As M. Maurice Denis has remarked, 'Impressionism was synthetic in its tendencies, since its aim was to translate a sensation, to realize a mood; but its methods were analytic, since colour for it resulted from an infinity of contrasts.'[3] This, in M. Denis' opinion, was the error of Impressionism. The reaction to classicism in the sense mentioned was initiated by Cézanne, 'the Poussin of Impressionism,' as M. Denis calls him. The two articles by M. Denis published in the *Burlington Magazine* of January and February last should be read in order to appreciate the influence of Cézanne on contemporary French painting. It is from Cézanne too that the Symbolist school derives, the school to which M. Denis himself belongs, although his work shows also the influence of Puvis de Chavannes, while remaining intensely personal. M. Odilon Redon was another of the pioneers of the school, of which Gauguin and M. Henri Matisse are prominent examples.

Although Symbolism is a development of Impressionism, it is also a reaction against it. It recognises that nature cannot be reproduced but can only be represented; it is an attempt, in the words of Mr. Roger Fry, at 'that direct expression in painting of imagined states of consciousness which has for long been relegated to music and poetry.'[4] It is impossible here adequately to describe its aims and methods; the reader is referred to the two articles by M. Maurice Denis already cited, in which the Symbolist idea is set forth by one of its ablest exponents in painting. M. Denis quotes a suggestive remark of Cézanne: 'I wished to copy nature; I could not. But I was satisfied when I had discovered that the sun, for instance, could not be *reproduced*, but that it must be *represented* by something else – by colour.' Symbolism is at one with Impressionism in its neglect of subject, its total rejection of literary interest in painting. 'Instead of evoking our moods,' says M. Denis, 'by means of the subject represented, it was the work of art itself which was to transmit the initial sensation and perpetuate its emotions.' Symbolist painting is primarily decorative, and we must not look in it for exact reproduction of nature or an attempt at such reproduction. Some of the Symbolists simplify to the furthest possible extent and even perspective disappears in their pictures; they, thus, approach the Primitives, who were perhaps after all the true realists. M. Maurice Denis, at least, has maintained that the apparant *gaucherie* of the Primitives 'consists in painting objects according to our usual knowledge of them, instead of painting them as the Moderns do, according to a preconceived idea of the picturesque or artistic. The *picturesque* being the element of nature *recognised* as proper to painting, it follows that the moment an artist departs from the admitted formulae and paints with naïve sincerity, he incurs the reproach of ignorance of *gaucherie*.' If this be so, it is easy to understand how Symbolism has been the outcome of Impressionism, for the aim of the Impressionists was to do just what M. Denis says that the Primitives did.

Another development of Impressionism in an opposite direction is the *Pointelliste* method, which consists in carrying the system of the division of colour to its extreme limit, applying the primary colours, pure without any admixture, in points and minute touches, Sourat, who died in 1891 at the age of thirty-two, was the first to adopt this method; it is represented in the present exhibition by some living painters and by Henri-Edmond Cross, who died, after a long and painful illness, a few weeks before the opening of the exhibition. Another group of artists, although they derive from the Impressionists, derive also from Gustave Moreau; they have sometimes been called Idealists. They have not adhered strictly to the method of Moreau, who had considerable affinity with the English Pre-raphaelites, but his influence is visible enough in their work. To another group of Neo-impressionists the name of *Intimistes* has been given; they have discarded the bright colours of the Impressionists and paint usually in grey tones.

Of all these groups the Symbolists will probably be found the most startling by those who are not accustomed to them. Their apparent *gaucherie* and their disregard of perspective are perplexing at first to the modern eye; although we are accustomed to overlook these drawbacks, as they seem to us, in the Primitives, we find them intolerable in modern painting. Yet there can be no doubt that, at any rate where wall decoration is concerned, perspective is itself a drawback. it is really to a large extent the existence of perspective in modern wall decorations that makes them so much less satisfactory than Primitive decorations; perspective is an attempt to get rid of the wall, whereas the best decoration recognises the existence of the wall and starts from that fact. It will be found that the works of the Symbolists are wonderfully decorative, when one accepts and gets used to their point of view.

The schools that have been the outcome of Impressionists, form only one section of the exhibition and, if this introduction is mainly devoted to them, it is because it would seem that, being little known in England, they need to be introduced and explained, rather than the other schools represented. It is not because these other schools are considered of no account. Some of the leading painters of the two great Salons, the Société des Artistes Français and the Société Nationale des Beaux Arts, have contributed to the exhibition and one can appreciate the Neo-impressionists without depreciating them. Among them are some, MM. Besnard and Raffaëlli for instance, who are often counted as Impressionists and have, in fact, much in common with the Impressionist school. Others, such as M. Alexis Vollon, follow another tradition, but are equally free from the influence of that pseudo-classicism against which Impressionism was a protest. All true lovers of the art are catholic in their taste and, whatever their personal preferences, can admire and recognise merit wherever it is to be found. The present exhibition will have served its purpose if it gives even an imperfect notion of the marvellous vitality and variety of contemporary French painting.

NOTES

1 Those who wish for further information on the Impressionist movement should read this, the standard work on the subject, the English edition of which has just been published; it is excellently translated by Mr. Crawford Flitch. A copy is in the Reference Library. The original French edition is now out of print and is very difficult to obtain. [Dell's note]
2 Quoted by M. Maurice Denis, *Burlington Magazine*, January, 1910, p. 213. [Dell's note]
3 *Burlington Magazine*, February, 1910, p. 279. [Dell's note]
4 *Burlington Magazine*, January, 1910, p. 207. [Dell's note]

9. Unsigned review (probably by Roger Fry), 'Modern French Pictures at Brighton'

The Times, 11 July 1910, 12

Most of the reviews of this exhibition were hostile and dismissive, but *The Times* printed this sympathetic and conciliatory article. The writer suggests the term 'expressionists' as a label for the new movement – a name which Fry himself favoured before he adopted the term 'Post-Impressionists'.

In the Public Art Galleries at Brighton there is now to be seen an exhibition of modern French pictures got together by Mr. H.D. Roberts, the director, through the agency of Mr. R.E. Dell, which, as Mr. Dell justly boasts in his interesting introduction to the catalogue, is more fully representative of contemporary French art than any exhibition that has been held elsewhere in England. Never before have works in the latest movement of French painting been publicly shown in any numbers, and Mr. Dell is at some pains to explain the aims of this movement, which still arouses surprise and resentment even in France.

It has not yet got any single name for itself, and indeed it really consists of several movements, some reacting against impressionism, some developed out of it. There are symbolists and primitives, and intimists and neo-impressionists; and it is difficult to define any of them. But of all it may be said that their aim is rather to present a mental image, more or less controlled by theories, than to set down a direct impression of a real scene. They would first of all eliminate from their pictures all facts that are irrelevant to what they wish to express; and besides this they claim the liberty to deal as they choose with the facts that remain, still in the interests of expression. Perhaps, therefore, we should call them expressionists rather than impressionists, since they have given up the impressionist curiosity about new aspects of reality and new methods of representing them for a curiosity about new methods of expressing the emotions aroused by reality. This is an entirely logical development, but the results will surprise most people. At first sight they will take these pictures for the work, or rather the play, of charlatans or mountebanks, of men who paint thus because they cannot do any better, and who make a parade of their incapacity.

Yet there is no doubt that Cézanne and Gauguin were, and that M. Matisse and M. Maurice Denis are, accomplished artists; and it is worth while to inquire what their aim is, whether they have accomplished it, and whether it is worth accomplishing. Most people will say at once that their pictures have no resemblance to reality; but most people get their ideas of reality from pictures, and often from bad pictures. The representation of reality has varied so much at different times that it is dangerous to dogmatize about it. These painters deliberately use representation as a means of expression, and they do not care how little they represent, if they succeed in expression. Cézanne, who was perhaps the founder of the movement, has only one picture in the exhibition—that a portrait of M.A. Valabrègue. One could wish that some of his nudes were shown. At first sight this portrait looks botchy in execution and wooden in pose; but forget your ideas of what execution and pose ought to be, and you will become aware of a peculiar intensity in the picture as you look at it. The artist seems determined to keep free of rhetoric and commonplace. He will say nothing about his subject merely because other artists would have said it. He will paint only his own idea of the man as simply as possible. And that idea is certainly both clear and directly expressed. He does not give us a brilliant snap-shot, but a representative expression and attitude. You may think the picture ugly, but it stays in your mind – a sign that it has been clearly conceived in the mind of the artist. The simplification may be excessive – Cézanne cannot make so many facts expressive as Titian or even Giorgione – but the essentials remain. We know almost as much about the sitter as painting can tell us, and it tells us nothing that we do not want to know.

The Gauguins, again, are not fully representative. There is a pleasant little landscape with no novelty in it; a still life, "Fruits of Tahiti," which is very rich and strong in colour; and a curious picture of oxen passing through a town in the snow. This last work looks so primitive that most people will be reminded of the animals in a Noah's ark. Gauguin here is not attempting any general illusion of reality, but, after the primitive method, setting down those facts that interest him and working them all into a clear and harmonious design. He falls short of the great Italian primitives in beauty of material; and this failing is common to many of the painters of the new movement. The roughness and untidiness of their paint will probably shock most spectators more even than their forms and colours; and not without reason, for a beautiful quality of paint, like verbal beauty in literature, is a symptom of successful expression. If paint is clumsily applied it shows effort rather than accomplishment. Further, the more facts are eliminated from a picture the more it must depend upon beauty of material for its effect. Those painters of the new movement who reduce everything to rhythm and colour need more of the coach-painter's art then most of them possess. Their works inherit the rough handling of the impressionists, which is permissible when it is the result of an eager grappling with fact, but becomes meaningless in a picture as empty of fact as a lyrical poem. There is too much of this roughness in *The Willows* of M. Othon Friesz, a picture that looks like a mere background, and in the same painter's *Indolence*, an interesting landscape peopled by nude figures with a primitive intensity of pose. The painter here has made a brave effort to simplify everything to its essential elements; but he has not justified this simplicity by a lyrical charm of execution. If pictures are to remind us of music, they must have a material beauty equivalent to the beauty of sound.

It is also a defect of many of these painters that their themes are too trivial to endure extreme simplification. There must be grandeur of conception where there is so little representation. M. Maurice Denis is aware of this fact. He tries to achieve the grandeur of Piero della Francesca in his *Homage to the Child*; and, like Piero, to combine this grandeur with a true representation of light. The result is interesting, and even beautiful. But he has not got all Piero's power of giving character and structure with great economy of modelling; and his figures are a little vapid. In his *Nausicaa* a number of girls are playing on a sunlit shore, while in the foreground Odysseus sleeps in the shadow of a low shrub. The essence of the picture is the contrast between the movement of the sunlit girls and the weary and darkened slumber of the man. Everything is very skilfully simplified to emphasize this contrast, and there is considerable beauty of material, so that the archaic *gaucherie* of movement is not over-emphasized by *gaucherie* of execution.

All these painters tend to use colour rather as an equivalent of the force of light than as a direct representation of it. This method is carried to an extreme in M. André Derain's views of London. There are pink trees on the Embankment, the road is green, and the sidewalks pink; the traffic is blue and the sky yellow and green. One can see no clear reason, scientific or aesthetic, for the choice of these particular colours. But the picture has rhythm and movement; it is amusing, and it would look far less ugly on the walls of a room than most photographic landscapes.

The pointellistes are rather scientific than aesthetic in their principles, and they have ceased to be novel. Their aim is to give the force of light by a mosaic of bright colours all reacting on each other, so that a shimmering glitter and sparkle is the result. Certainly the *Afternoon in a Provencal Garden* of the late H.E. Cross does glitter and sparkle; but the composition of the picture, as of many pointelliste works, is merely photographic; and the illusion of light only emphasizes its photographic quality. There is the same defect in M. Signac's *Venice*; but his *Notre Dame de la Garde, Marseilles*, is far superior in design and it would be impossible to represent the pale brilliance of early morning so well by any other means without the mastery of a Rubens. No painting could be higher in key than this, yet it is not harsh or chalky, and the method is justified here by its results, since, when the picture is seen at a proper distance it is not obtrusive. It is a pity that M. Matisse, who has gone further than any one else in the expression of rhythm and the simplification of fact, is represented only by two slight pictures of still life. These are not enough to give any idea of his adventurous talent. The whole tendency of modern painting, so far as it has a tendency, is to become more archaic in the interests of expression. It drops facts, as the woman who was pursued by wolves dropped her children. In modern conditions no painter can well attain to the mastery of a Titian or a Velasquez, that mastery which can make complete representation expressive. Therefore, the painters who have most to express sacrifice more and more of representation to expression. And that is the reason why they are so unintelligible to the public, who miss in their works those facts which they expect to find in a picture. But the simplifying tendency cannot be carried much further than M. Matisse carries it. In many ways he is more primitive than the early Sienese in his reaction against photographic realism. One can only hope that a new Giotto will arise and once more unite representation with expression, so that European painting may start again upon a course of steady progress.

The private view for 'Manet and the Post-Impressionists' was held on Guy Fawkes day, Saturday 5 November 1910, and opened to the public on Monday 8 November. The significance of the date was not lost on those who saw in Post-Impressionism a threat to the English art-establishment. The pictures had been gathered hurriedly from French dealers to fill a gap in the programme of the Grafton Galleries, and the term 'Post-Impressionism' was the last-minute suggestion of a journalist which Fry took up and used in the title of the show. The phrase 'Post-Impressionism' appeared for the first time in print in an advertisement carried by *Art News* (15 October 1910, 5).

The critical response was enormous and almost without exception every London daily paper, evening paper and journal carried articles, photographs and comment on the exhibition, to be followed rapidly by the provincial press. Almost all criticism in the first month (with the exception of articles by Roger Fry and Frank Rutter) was hostile, but by the beginning of December the tide began to turn in favour of the new 'movement'. Ignorance about Cézanne, Van Gogh, Gauguin, Matisse and the other artists was almost total and journalists depended heavily upon the few details supplied by Desmond MacCarthy in the introduction to the catalogue (no. 10), and in a few instances upon Maurice Denis's articles on Cézanne in the *Burlington Magazine* (no. 7) and Meier-Graefe's *Modern Art* (no. 6).

The catalogue for the exhibition was incomplete and inaccurate and though it listed 154 items there were in fact many more. Benedict Nicolson has identified many of them (see 'Post-Impressionism and Roger Fry', *Burlington Magazine*, March 1951, xciii, 13). There were 21 pictures by Cézanne, 37 works by Gauguin, of which his *l'Esprit Veille* (or *Manaó Tupapau*) (1892: Albright-Knox Art Gallery, Buffalo) created most hostility. Van Gogh was represented by 20 paintings of which *Wheat Field under Threatening Skies with Crows* (Rijksmuseum Vincent van Gogh, Amsterdam) and *Girl with a Cornflower* (now known as *Head of a*

Boy with a Carnation Between his Teeth: location unknown) drew most interest; there were just two seascapes by Seurat, nine pictures by Vlaminck, six by Rouault, five by Maurice Denis, three by Derain and three by Friesz. Neither of the two pictures by Picasso – his *Nude Girl with a Basket of Flowers* and a portrait of *Clovis Sagot* – represented his most recent, cubist manner. The greatest stir was created by Matisse's *La Femme aux Yeux Vertes* (1909: San Francisco Museum of Art), though he also sent two landscapes, one entitled *Trees near Melun* (1901: Belgrade Museum of Art and Archaeology), lent by Bernard Berenson, and some recent sculpture including a *Reclining Women* (probably *Reclining Nude-1*: Musée d'Art Moderne, Paris).

---

10. Desmond MacCarthy, 'The Post-Impressionists'

Introduction to the catalogue of the exhibition 'Manet and the Post-Impressionists', Grafton Galleries, 8 November 1910–14 January 1911

Desmond MacCarthy (1877–1952) was the secretary of the first Post-Impressionist exhibition. Together with Fry he went to Paris in the summer of 1910 to choose pictures for the exhibition and went on alone to find more in Munich and Holland. His introduction, composed from notes supplied by Roger Fry, was much quoted in contemporary reviews and in particular his remark about a rocking horse having more of the 'true horse' about it than a 'photograph of the Derby winner'.

The pictures collected together in the present Exhibition are the work of a group of artists who cannot be defined by any single term. The term 'Synthesists,' which has been applied to them by learned criticism, does indeed express a quality underlying their diversity; and it is the principal business of this introduction to expand the meaning of that word, which sounds too like the hiss of an angry gander to be a happy appellation. As a definition it has the drawback that this quality, common to all, is not always the one most impressive in each artist. In no school does individual temperament count for more. In fact, it is the boast of those who believe in this school, that its methods enable the individuality of the artist to find completer self-expression in his work than is possible to those who have committed themselves to representing objects more literally. This, indeed, is the first source of their quarrel with the Impressionists: the Post-Impressionists consider the Impressionists too naturalistic.

Yet their own connection with Impressionism is extremely close; Cézanne, Gauguin and Van Gogh all learnt in the Impressionist school. There are pictures on the walls by these three artists, painted in their earlier years, which at first strike the eye as being more impressionist than anything else; but nevertheless, the connection of these artists with the Impressionists is accidental rather than intrinsic.

By the year 1880 the Impressionists had practically won their battle; nor is it likely any group of artists will ever have to fight so hard a one again. They have conquered for future originality, if not the right of a respectful hearing, at least of a dubious attention. By 1880 they had convinced practically everybody whose opinion counted, that their methods and ideas were at any rate those of artists, not those of cranks and charlatans. About this date the reaction against Impressionism, which this Exhibition represents, began to be distinctly felt. The two groups had one characteristic in common: the resolve of each artist to express his own temperament, and never to permit contemporary ideals to dictate to him what was beautiful, significant, and worthy to be painted. But the main current of Impressionism lay along the line of recording hitherto unrecognised aspects of objects; they were interested in analysing the play of light and shadow into a multiplicity of distinct colours; they refined upon what was already illusive in nature. In the pictures of Seurat, Cross, and Signac exhibited, this scientific interest in the representation of colour is still uppermost; what is new in these pictures is simply the method of representing the vibration of light by painting objects in dots and squares. The Post-Impressionists on the other hand were not concerned with recording impressions of colour or light. They were interested in the discoveries of the Impressionists only so far as these discoveries helped them to express emotions which the objects themselves evoked; their attitude towards nature was far more independent, not to say rebellious. It is true that from the earliest times artists have regarded nature as 'the mistress of the masters'; but it is only in the nineteenth century that the close imitation of nature, without any conscious modification by the artist, has been proclaimed as a dogma. The Impressionists were artists, and their imitations of appearances were modified, consciously and unconsciously, in the direction of unity and harmony; of course, as artists, they selected and arranged. But the receptive, passive attitude towards the appearances of things often hindered them from rendering their real significance. Impressionism encouraged an artist to paint a tree as it appeared to him at the moment under particular circumstances. It insisted so much upon the importance of his rendering his exact impression that his work often completely failed to express a tree at all; as transferred to canvas it was just so much shimmer and colour. The 'treeness' of the tree was not rendered at all; all the emotion and associations such as trees may be made to convey in poetry were omitted.

This is the fundamental cause of quarrel between the Impressionists and the group of painters whose pictures hang on these walls. They said in effect to the Impressionists: 'You have explored nature in every direction, and all honour to you; but your methods and principles have hindered artists from exploring and expressing that emotional significance which lies in things, and is the most important subject matter of art. There is much more of that significance in the work of earlier artists who had not a tenth part of your skill in representing appearance. We will aim at that; though by our simplification of nature we shock and disconcert our contemporaries, whose eyes are now accustomed to your revelation, as much as you orginally disconcerted your contemporaries by your subtleties and complications.' And there is no denying that the work of the Post-Impressionists is sufficiently disconcerting. It may even appear ridiculous to those who do not recall the fact that a good rocking-horse often had more of the true horse about it than an instantaneous photograph of a Derby winner.

The artists who felt most the restraints which the Impressionist attitude towards nature imposed upon them, naturally looked to the mysterious and isolated figure of Cézanne as their deliverer. Cézanne himself had come in contact with Manet and his art is derived directly from him. Manet, it is true, is also regarded as the father of Impressionism. To him Impressionism owes nearly all its power, interest and importance. He was a revolutionary in the sense that he refused to accept the pictorial convention of his time. He went back to seventeenth-century Spain for his inspiration. Instead of accepting the convention of light and shade falling upon objects from the side, he chose what seemed an impossibly difficult method of painting, that of representing them with light falling full upon them. This led to a very great change in the method of modelling, and to simplification of planes in his pictures which resulted in something closely akin to simple linear designs. He adopted, too, hitherto unknown oppositions of colour. In fact he endeavoured to get rid of chiaroscuro.

Regarded as a hopeless revolutionary, he was naturally drawn to other young artists, who found themselves in the same predicament; and through his connection with them and with Monet he gradually changed his severe, closely constructed style for one in which the shifting, elusive aspects of nature were accentuated. In this way he became one of the Impressionists and in his turn influenced them. Cézanne, however, seized upon precisely that side of Manet and his followers, which Monet and the other Impressionists ignored. Cézanne, when rendering the novel aspects of nature to which Impressionism was drawing attention, aimed first at a design which should produce the coherent, architectural effect of the masterpieces of primitive art. Because Cézanne thus showed how it was possible to pass from the complexity of the appearance of things to the geometrical simplicity which design demands, his art has appealed enormously to later designers. They recognise in him a guide capable of leading them out of the *cul de sac* into which naturalism had led them. Cézanne himself did not use consciously his new-found method of expression to convey ideas and emotions. He appealed first and foremost to the eye, and to the eye alone. But the path he indicated was followed by two younger artists, Van Gogh and Gauguin with surprising results. Van Gogh's morbid temperament forced him to express in paint his strongest emotions, and in the methods of Cézanne he found a means of conveying the wildest and strangest visions conceived by any artist of our time. Yet he, too, accepts in the main the general appearance of nature; only before every scene and every object he searches first for the quality which originally made it appeal so strangely to him: *that* he is determined to record at any sacrifice.

Gauguin is more of a decorator. He felt that while modern art had opened up undiscovered aspects of nature, it had to a great extent neglected the fundamental laws of abstract form, and above all had failed to realise the power which abstract form and colour can exercise over the imagination of the spectator. He deliberately chose, therefore, to become a decorative painter, believing that this was the most direct way of impressing upon the imagination the emotion he wished to perpetuate. In his Tahitian pictures by extreme simplification he endeavoured to bring back into modern painting the significance of gesture and movement characteristic of primitive art.

The followers of these men are pushing their ideas further and further. In the work of Matisse, especially, this search for an abstract harmony of line, for rhythm, has been carried to lengths which often deprive the figure of all appearance of nature. The general effect of his pictures is that of a return to primitive, even perhaps of a return to barbaric, art. This is inevitably disconcerting; but before dismissing such pictures as violently absurd, it is fair to consider the nature of the problem which the artist who would use abstract design as his principle of expression, has to face. His relation to a modern public is peculiar. In the earliest ages of art the artist's public were able to share in each successive triumph of his skill, for every advance he made was also an advance towards a more obvious representation of things as they appeared to everybody. Primitive art, like the art of children, consists not so much in an attempt to represent what the eye perceives, as to put a line round a mental conception of the object. Like the work of the primitive artist, the pictures children draw are often extraordinarily expressive. But what delights them is to find they are acquiring more and more skill in producing a deceptive likeness of the object itself. Give them a year of drawing lessons and they will probably produce results which will give the greatest satisfaction to them and their relations; but to the critical eye the original expressiveness will have vanished completely from their work.

The development of primitive art (for here we are dealing with men and not children) is the gradual absorption of each newly observed detail into an already established system of design. Each new detail is hailed with delight by their public. But there comes a point when the accumulations of an increasing skill in mere representation begin to destroy the expressiveness of the design, and then, though a large section of the public continue to applaud, the artist grows uneasy. He begins to try to unload, to simplify the drawing and painting, by which natural objects are evoked, in order to recover the lost expressiveness and life. He aims at *synthesis* in design; that is to say, he is prepared to subordinate consciously his power of representing the parts of his picture as plausibly as possible, to the expressiveness of his whole design. But in this retrogressive movement he has the public, who have become accustomed to extremely plausible imitations of nature, against him at every step; and what is more, his own self-consciousness hampers him as well.

The movement in art represented in this exhibition is widely spread. Although, with the exception of the Dutchman, Van Gogh, all the artists exhibited are Frenchmen, the school has ceased to be specifically a French one. It has found disciples in Germany, Belgium, Russia, Holland, Sweden. There are Americans, Englishmen and Scotchmen in Paris who are working and experimenting along the same lines. But the works of the Post-Impressionists are hardly known in England, although so much discussed upon the Continent. The exhibition organised by Mr. Robert Dell at Brighton last year has been our only chance of seeing them. The promoters of this exhibition have therefore thought it would be interesting to provide an opportunity for a greater number to judge these artists. The ladies and gentlemen on the Honorary Committee, though they are not responsible for the choice of the pictures, by lending their names have been kind enough to give this project their general support.

11. Robert Ross, 'The Post-Impressionists at the Grafton: The Twilight of the Idols'

Morning Post, 7 November 1910, 3

Robert Ross (1869–1918) was art critic for a number of journals including the *Morning Post* (1908–12). He is probably best remembered as the editor of Oscar Wilde's works.

A date more favourable than the Fifth of November for revealing the existence of a wide-spread plot to destroy the whole fabric of European painting could hardly have been chosen. On Saturday accordingly the Press was invited to the Grafton Gallery – an admirable substitute for the vaults of Westminster – where the new Guido Fawkes, his colleagues, and alleged predecessors are exhibiting their gunpowder. Mr. Roger Fry, I regret to say, has acted the part of Catesby, while a glance at the names of the honorary committee reveal that more than one member of the Upper House is implicated. It is the way of modern conspiracies; we all join them sooner or later. To-day, which is the private view, it will be decided whether the anticipated explosion is going to take place. Mr. Edmund Gosse in one of his delightful prose portraits has recalled how the poet Mallarmé first came to England in order to try and interest people in Manet's designs for Edgar Poe's *Raven*. 'They were received by us,' he confesses, 'with undying laughter.' Now, Manet has been long accepted for one of the milestones, one of the parting points, in the history of Nineteenth Century painting, was so accepted long before his pictures were removed from the purgatorial Luxembourg to the Louvre. Here at the Grafton the presence of some of his pictures (none, it will be admitted, of first-rate importance) indicate that his name is invoked as the artistic parent – the not very respected parent, we gather – of the Post-Impressionists, or 'Synthesists,' as we must learn to call them. Whether Mr. Gosse and his friends were wrong about the designs for the *Raven* or not (and I am inclined to think they were justified) there is no doubt whatever that the vast majority of the pictures at the Grafton, notably those of MM. Matisse, Maurice Denis, and Maurice de Flamineck, will be greeted by the public with a more damning and more permanent ridicule. When the first shock of merriment has been experienced there must follow, too, a certain feeling of sadness that distinguished critics whose profound knowledge and connoisseurship are beyond question should be found to welcome pretension and imposture. It is only comparable to the no less deplorable credulity evinced by serious men of science in the chicanery of spiritualism, automatic writing, and the narratives of the neuropath. Public taste has been so often wrong. Its present idols become the paving stones or macadam of the morrow. The pariahs of one generation are prophets for its successor. With these lamentable precedents perhaps it is more regrettable than extraordinary that any charlatanism in art or literature may now enjoy the privilege of examination such as should be only accorded to serious new developments. And it is an error to suppose that because posterity momentarily reverses a hostile contemporary opinion the original opinion was wrong. Fortified with such minatory principles anyone can derive from the exhibition of the Post-Impressionists at the Grafton instruction pleasure, and amusement.

The preface to the catalogue sets forth in straightforward language the historical relation of the 'Synthesists' to Manet. 'The Post-Impressionists were not concerned with recording impressions of colour, light, or anything else exactly. They were interested in the discoveries of the Impressionists only so far as these discoveries helped them to express emotions which the objects themselves evoked; their attitude towards Nature was far more independent, not to say rebellious.' 'Anything else exactly' is particularly naïve. I can only reply to the frank admission of the apologist that the emotions of these painters (one of whom, Van Gogh, was a lunatic) are of no interest except to the student of pathology and the specialist in abnormality. At Broadmoor there are a large number of post-impressionists detained during his Majesty's pleasure. Their works are, however, already the property of the State. The National Arts Collections Fund may sleep in peace.

In the Octagonal Gallery, side by side with the Manets, may be seen a number of works by Cezanne, who 'seized upon precisely that side of Manet and his followers which Manet and the

Impressionists ignored.' To my uninitiated eyes they appear sketches or underpainting of pictures by somone who, if he cannot draw very well, sees though he does not seize the true aspects of Nature at rather commonplace moments. We are told "that he aimed at design which should produce the coherent architectural effect of the masterpieces of primitive art." All I can say is that he failed; whether from insufficient knowledge of the manipulation of paint or an entire misunderstanding of the aims or methods of the primitives I do not profess to judge. The French are too progressive, too curious for new experiments, ever to succeed in the archaistic rehandling of archaic formulas. M. Denis in one room illustrates the impossibility of a Frenchman assimilating the sentiment of the Primitives; Cezanne, in the other, the hopeless attenspt to parody their objectivity. Cezanne is neither coherent nor architectural. Even the great Puvis, who is both, was never more than haunted by the Primitives. In the centre of this room is, however, one of the really beautiful things in the exhibition, a terra-cotta nude by Maillol, who has much in common with Mr. Epstein. At all events he produces an effect similar to that of naturalistic Egyptian sculpture, though the insistence on the feminism of the model is alien to the spirit of all ancient art not deliberately pornographic. Of the Manets, *Un bar aux Folies-Bergère*, for all its fine painting (note the flowers in the tumbler), is eloquent of the danger, foreseen by Reynolds, of resting pedantically satisfied with your own period and ignoring what he called the universal. What a distressing possession, how unpleasing, how 'fashionable.' But Manet is a great painter and a great draughtsman.

In the Large Gallery may be found some justification of the exhibition, let me say of the enthusiasm, about the new school of paint, if new school there be. Gauguin is an artist with a fresh idea, a curious technique, and a fantastic vision. He, too, can draw and paint. His fantastic vision includes, conforms even to, obvious types of beauty, a rare quantity in modern art. *L'arlesienne* (No. 37) may not be a great picture, but of its charm and accomplishment there can be no two opinions. His colour is a little odd, barbaric rather than primitive; but when you learn that he emigrated to Tahiti, which forms the subjects and backgrounds of his pictures, you realise that he does not hypnotise his admirers by mere strangeness and unfamiliarity, on which the others chiefly rely. He has some of the pattern and invention of Beardsley. I am indebted to Mr. Walter Sickert for pointing out to me the influence of Cézanne in Gauguin's European pictures; in *Femme sous les Palmiers* (No. 32) and of Pissaro in *Bretagne* (No. 26). However delightful these earlier works, how fortunate that he did get away to Tahiti. *Paysage Bretonne* (No. 87), and *Le gardeur d'oies* (No. 88, wrongly catalogued Van Gogh) are by one who, if not a master, counts in the hierarchy of artists. Whether he went to a South Sea island in order to become a real primitive, he certainly succeeded. In No. 28, *Vue sur la Martinique*, you get the identical point of view (or superficially the identical point of view) of the naturalistic fresco painters in the Egyptian Tombs. The jolly little black pigs only rendered in silhouette, yet so real and living, and the brown-skinned girl crouching in the foreground are as much copied from despised Nature as the famous 'frieze of geese,' familiar to all students of Egyptology. On the north wall of the Large Gallery are a group of later Gauguins, in which you observe that his peculiar style, which may please or displease, undoubtedly has the 'decorative' qualities claimed for the school. Like all good decorations, it is based on something seen, like the famous Britons on the Roman drop curtain; not mere emotion. Here is Byzantinism vindicated. He would not have got that effect by gazing at Russian eikons or early mosaics. The Gauguins should have been hung together instead of being used for salting the exhibition, Young artists need scarcely be told that until there is a Tahiti village at Shepherd's Bush it is premature to 'gauguinise' the European landscape or the Aryan race. Let them look for their own grotesques, like the carvers of the miserere seats. I do not profess to see any connection between Van Gogh and Gauguin, except that their names, in a Shakespearean sense, both begin with G. Van Gogh is the typical matoid and degenerate of the modern sociologists. *Jeune Fille au Bleuet* (No. 67) and the *Cornfield with Blackbirds* (No. 71) are the visualised ravings of an adult maniac. If that is art it must be ostracised, as the poets were banished from Plato's republic. A later blossom of an unsavoury stock has not only dispersed with chiaroscuro (one of the achievements of Cézanne, I learn) but has dispensed with painting. The only primitives he resembles are the woolwork trophies of our great-grandmothers, though there is less form, less art. To the negative silliness of covering canvas with pigment *La Femme aux yeux verts*

(No. 111) he has added the primitive and more pristine offence of "modelling." If Van Gogh belongs to the School of Bedlam, M. Matisse follows the Broadmoor tradition in a predilection for mere discords of pigment; needless to say, quite a different thing from discordant colouring. To dispense with light and shade, with technique altogether, may be childish, as the catalogue claims; it has no relation to the noble failures and successes of the elaborate Primitives. If Corot was right when he said 'painting is not as difficult as you make out,' it cannot be as easy as M. Matisse would have us think. To discuss the 'pavement art' of MM. Denis Signac, Henri-Edmond Cross, and Georges Seurat and others would be waste of time, for anyone sufficiently idle and mindless could reproduce them. One is sorry to find M. Vallotton (No. 102 and others) in such company: M. Jean Puy (No. 125), M. Marquet (114), or the mild French Blake and water, Odilon Redon, who has joined every movement for the last twenty years. These are strayed sheep from the gracious uplands and valleys of art. Whitman has been called an artist in his very rejection of art. And there have constantly appeared in the world of painting geniuses such as El Greco and Monticelli who ignore the grammar, the architecture, and the limitations of their medium, as Whitman has done in the case of language. Both, it must be noted, were proficient craftsmen, and conformed in their early work to the discipline of study. Gauguin, too, is a genius of this kind. He submitted himself, however, to Nature, exotic Nature, it may have been – distorted it certainly was when filtered through Cézanne, who candidly admitted his own failure, an almost unnecessary avowal. Once in his life every artist must propitiate the implacable goddess or he can never be initiated to the Eleusinian *simplicities* of Art. The relation of M. Henri Matisse and his colleagues to painting is more remote than that of the Parisian Black Mass or the necromantic orgies of the Decadents to the religion of Catholics. Did Ingres when he quarrelled with Delacroix foresee the development of Romanticism into Post-Impressionism? It is not impossible. If so, he is indeed the great man he is claimed to be by artists and critics. I can now understand why the German Emperor dismissed a high official from a Berlin Gallery. And if the movement is spreading (another boast of the catalogue) it should be treated like the rat plague in Suffolk. The source of infection (*e.g.*, the pictures) ought to be destroyed.

12. Unsigned review, 'Paint Run Mad: Post-Impressionists at Grafton Galleries'

Daily Express, 9 November 1910, 8

The incredulity felt by the critic of the *Daily Express* was shared by many journalists for whom the work at the first Post-Impressionist exhibition was a source of mystery and irritation. Nevertheless the paintings provided sensational copy.

There are more shocks to the square yard at the exhibition of the Post-Impressionists of France, at the Grafton Galleries, than at any previous picture show in England. It is paint run mad.

The aim of the Post-Impressionists is not to paint objects as they strike the human eye, but simply to convey an emotion by means of paint. As regards the negative proposition, they are invariably successful; but one may cavil at the fact that the emotion expressed is too often that of a violent bilious headache.

In the first room the visitor is let down gently with several splendid Manets, including that painter's famous *Bar at the Folies-Bergère* – which, by the way, is insured for £10,000 – surely a record sum for a modern picture. In the large gallery the eye meets Gauguin's primitive, almost barbaric, studies of Tahitian women – bizarre, morbid, and horrible.

In a typical canvas hideous brown women, with purple hair and vitriolic faces, squat in the midst of a nightmare landscape of drunken palm trees, crude green grass, vermilion rocks, and numerous glaringly coloured excrescences impossible to identify.

His *Sacred Mountain* equally well represents a cornfield or an aeroplane disaster, and one leaves it with a strong suspicion that it might look better the other way up!

It is astonishing to find that van Gogh's raging golden sea against a royal blue sky, a vermilion paint-pot upset in the foreground, represents a cornfield with black-birds; and only on retreating to the far side of the gallery do you find that the vermilion splodges might – this is a suggestion, not a statement of fact – be roads. No wonder Dr. Gachet, next door, looks horribly seasick; this portrait must surely be a mean revenge on a cherished enemy.

Relief is afforded by the two fragrant and delightfully painted *Orchards in Provence*, also by van Gogh, who, although he did his best work in a lunatic asylum, seems at times to be the sanest of the lot.

Words are powerless to describe an epileptic landscape by Henri Matisse, quite without form, its kaleidoscopic colour scheme only bearable from the next room.

A revolution to be successful must presumably revolve; but, undeniably clever as they often are, the catherine-wheel antics of the Post-Impressionists are not likely to wake many responsive chords in British breasts. Right or wrong, however, they add to the gaiety of nations, and will form a favourite topic of dinner-table conversation for many months to come.

13. Charles Ricketts, 'Post-Impressionism'

Morning Post, 9 November 1910, 6

Ricketts (1866–1931) was trained as an illustrator, founded the Vale Press in 1896 and in 1904 turned to painting, modelling and theatrical design. He wrote at length about Post-Impressionism in his collected *Pages on Art* (1913) (see no. 37).

Sir, – May I be allowed to congratulate you on the admirable article on Post-Impressionism in yesterday's issue of the Morning Post? It is, therefore, in no spirit of hostility that I venture to challenge, not indeed the trend of the article, but one or two statements in it which discount its force and obscure its issue. Mr.

Ross tends to implicate Manet in the reactionary movement which Mr. Fry has described as Post-Impressionism, or as a species of Neo-Byzantinism, while he considers that it possesses 'the promise of a larger and fuller life than our own Pre-Raphaelite movement,' and has also 'a contrary tendency to the impressionist conception of art.' In the last statement Mr. Fry is correct. Manet must not be held responsible for the Grafton masters. Whatever limitation lies in the theory of impressionism itself the new 'decorative' art of M. Matisse, for instance, has turned over a new leaf; it is, in fact, on a different sheet of paper. In describing the works of the post-impressionists as decorative I take it that Mr. Fry considers that what is too silly to be called painting may be good enough for decoration. The impressionists were painters and some of them artists. Owing to the presence in England of work which is recognised abroad, post-impressionism has not touched us, and it is a novelty to our critics. The 'cult' has its organised headquarters in Paris, its prophets in America, and its cosmopolitan travellers in Germany. Mr. Roger Fry, the apostle of this new creed, this new message to the Gentiles of England, has pointed out that not Manet but Cézanne is the true begetter of the new school. This Mr. Ross forgets. Cézanne was not properly an impressionist, though he borrowed impressionist pictures to copy. He remained for a while only a compromising provincial satellite of the school, till he retired to the country, where he would leave his paintings in the fields, not deeming them worthy to take home, thereby proving that, if a poor painter, he was an astute critic.

Egotism, exasperated by opposition or resistance, has led to the works which Mr. Fry has praised at the expense of Pre-Raphaelitism and impressionism. Morbid and suffering egotism is the spring of all anarchy in conduct, in art, as in criticism, but do not let us blame the impressionists for those who were unable to follow them or accuse France of decay owing to a few bad and mad painters. All great capitals have those whom the stress of life has maimed; our hospitals can show even the victims of strange imported maladies from abroad; yet London is not leprous or plague-stricken. Mr. Ross is again wrong in suggesting that the "Grafton decorations" should be burnt; they might interest the doctors of the body and the students of the sickness of the soul. Let us rather be angry not with post-impressionist painters – they are doing, perhaps, their best – but with the curious sophistries that preface the catalogue. This, like other wastepaper, might be burnt. On the whole Mr. Ross treats these paintings too seriously. One would rather learn about the public reasons which have brought together three directors of our museums and three trustees of the National Gallery among the sponsors of this venture. I trust that the pathetic last words of the preface are an explanation that 'the ladies and gentlemen on the Honorary Committee, though not responsible for the choice of the pictures by lending their names have been kind enough to give the project their general support.' Were they responsible at all? There is to me, however, an element of tragedy in this Grafton comedy of 'Art killed by kindness' when we recognise the names of admirers of Watts, Burne-Jones, and Gustave Moreau among the friends of this 5th of November plot, as Mr. Ross has aptly described it.

If post-impressionism has achieved, with a section of the Parisian Press, what I would call 'a success of intimidation,' the fear of being 'out of the movement' may capture some of our less cautious critics and some experts. Yet there is hope. Under the elegant heading 'Shocks in Art' I have read in a paper that 'the worst work at the Grafton is stimulating and vastly preferable – not for permanent possession, but for the amusement of an hour – to the dull level of mediocrity that prevails at most of our exhibitions.' I breathe again – mediocrity is a safer investment than these 'Shocks in Art' which cause but a little laughter and are gone in an hour. I disagree, however. 'Neo-Impresso-Byzantinism' bores me far more than the 'honest work' of our British post-impressionists of the pavement, who have to make literary appeals under their still-life drawings in the best Cézanne manner, not to kind committees, but to the public by statements possibly true, such as 'I am starving' or 'This is done by 'and.' I long to see the Grafton decorations bought by the Honorary Committee lest Mr. Fry should stand ashamed before those Parisian dealers who would save us. If the wish is too cruel let the executive, all of whom cannot be wholly innocent, be made to purchase, but of these it is better perhaps only to say 'Non ragionism di lor, ma guarda a passa.'

14. Laurence Binyon, 'Post-Impressionists'

Saturday Review, 12 November 1910, 609–10

Binyon (1869–1943), poet and art critic, was employed in the British Museum Department of Prints and Drawings from 1894 to 1933, becoming Keeper in 1932. He published numerous volumes of poetry, and in 1910 was art critic of the Saturday Review.

The Goupil Gallery Salon has become a recognised centre for our younger painters. Talent and promise of the most varied kind gain admittance here; and though under the auspices of no society or committee, the exhibition is as well selected and arranged as any of the kind in London. In the present show Mr. Nicholson and Mr. Orpen must be accounted the protagonists. Mr. Nicholson's *Marie*, a half-length portrait of a lady, the delicate profile in shadow against a grey background, has a subtlety and distinction which he has not always attained. It is challenged by Mr. Orpen's *Bright Morning by the Sea*, a woman's figure bathed in the pearly clearness of sunshine and moist air with a dappled sky behind, a variation in light tones on similar themes which we have seen from the artist's hand. In the same room is Mr. Pryde's striking composition *The Flying Dutchman*, more fanciful than imaginative, but boldly planned and spaced. There is, of course, an abundance of landscape, aiming for the most part at brightness, freshness, breeziness; M. Blanche's and Mr. Nicholson's still-life painting is as brilliant as ever; there are pleasant interiors by Mr. Connard and others; and among modern idylls the *Full Summer* of Mrs. Knight stands out by its gaiety of colour and graceful vigour. It is all very attractive and accomplished, this exhibition, and as modern and youthful as we expect to find in London. But how sober, how staid it seems, if we cross Piccadilly and pay a visit to the Grafton Gallery! For there at last we belated islanders are privileged to make acquaintance with the masters of Post-Impressionism. By an admirably discreet arrangement, reminding one of a Turkish bath, the shock of the revelation is only administered by degrees. In the first room you need scarcely be uneasy; Manet reigns there, and Manet is already a classic; in the second room the temperature is more exciting, you are in face of Gauguin and Van Gogh; and only when sufficiently acclimatised need you venture yet further into the wild realms of Matisse and his compeers.

But what is Post-Impressionism? French art, at least since the Revolution, cannot get on without 'movements'. These are created by one or two real talents, joined almost invariably sooner or later by a crowd of followers who hope to make up for their want of talent, want of ideas, or want of both, by making a deal of noise and catching the new accent as well as they are able. My insular prejudice distrusts these 'ists' and 'isms'. But perhaps it is no fault of the artists themselves that they have these portentous labels hung round their necks. At any rate, this particular movement is quite easily accounted for, whatever the claims made for what it is supposed to have achieved. Impressionism in France developed from certain instinctive preferences and methods into a very conscious theory. Its efforts became concentrated on one particular problem, which it was hoped to solve by scientific inquiry and experiment – the rendering of the effects of sunlight by pigment. That is to say, it degenerated into a side-issue. Some canvases by Signac and Seurat exhibited at the Grafton illustrate the cold-blooded puerilities which some years ago were supposed in Paris to be the last word in painting. Anyone could foretell that reaction must come. Painters had strayed from the highroad and wanted to get back to the main concerns of art; they had given themselves up to a passive acceptance of the appearances of Nature, an attitude consciously divested of all intelligent interest, so far as that is possible for thinking man. Mere optical sensation had been the one aim in view. What was lost was, to put it briefly and roughly, design. The new movement was obviously a reaction. But it is not fashionable to be a reactionary; therefore the movement must be represented as a step forward. It must be carried to extremes. The 'new' aim was to get behind the appearances of things, to render the dynamic forces of life and nature, the emotional significance of things. Away with imitation, let us get to primal energies and realities, was the cry. Away with plausibility, let us have the extreme of directness and simplification! Well, at bottom, was not this a healthy reaction, a movement in the right direction? European art has become so cumbered with its complex endeavour to represent the complete effect of a scene, sculptural form, light and shadow, natural colour and atmosphere, that nine painters out of ten forget the first business of art, which is with rhythmical design. Having often reproached our young painters for being too contented with merely pleasant aspects, with their cosy interiors and nice furniture caressed by sunshine, their pretty Early Victorian antiquarianism, their picnic landscapes; having exhorted them to think of painting less as an amusing game and more as a serious expression of thought and feeling, I should have the greatest sympathy with a movement aiming at the recovery of profounder and more strenuous moods, daring greatly at whatever sacrifice. Had I only read the preface to the catalogue of the exhibition, I could keep my sympathy. But I confess that the exhibition itself leaves me sad. Not one of these Post-Impressionists seems to me strong enough to carry out the programme. Reluctantly I am driven to the conclusion that it is only one or two in a generation from whom we can hope for success in the attempt to communicate the real realities, the deep places of human emotion and experience. The majority are better left with the gracious and agreeable, the lighter impressions of existence. Here is the cardinal fallacy, the fallacy illustrated so conspicuously by so great an intellect as Francis Bacon, that it is possible to discover a method or an instrument which will make all gifts equal, or at least put all equally on the way to producing works of depth and truth. Is it that uncompromising French logic, of which we hear so much, or is it merely the bluff and bunkum which seem fated to surround every Parisian art movement, that must be accounted responsible for the childish rubbish so pompously presented to us on certain walls of the Grafton Gallery? It matters little. Childish is what they aim at being. The child, uncorrupted by the ambition to create a deceptive imitation, goes directly for the essential things in drawing. Yes, but there is nothing more tedious than affected naïveté; and the efforts of unpromising children, of which we are reminded more than once, may bore even the fondest parent. We do not get nearer to the roots of things by painting houses which would fall down before they were built in landscapes where foreground and distance jumble in chaos. We may have suppressed the superficial aspect, dear to the bourgeois eye, but this, after all, is only a negative achievement.

Let us recall a drawing by Rembrandt. Is anything in this exhibition more simplified in method than one of those sketches of landscape or figures, done to all appearance with clumsy rude strokes of reed-pen and sepia? And yet those are landscapes into which one can walk surely into far distances; those are figures which live and move, which are wrung with poignant emotion. Or let us take a nearer instance: the paintings of Daumier. Neither Gauguin nor any of this school approaches Daumier's power of summing up the animating passion of a figure in a gesture, nor surpasses him in audacity of simplification. But Daumier was trained and exercised in inventive design; he did not grow up among Impressionists. It is really rather ludicrous to pretend that these Post-Impressionists have discovered anything new. One would infer from the preface to the catalogue that all contemporary painters had lost their way in a fog, that masters like Daumier had been forgotten, and that Cézanne alone could deliver them. One cannot write the history of modern art from the standpoint of a Parisian coterie. The reputation of Cézanne is a mystery to me. The two later men, Gauguin and Van Gogh, are more remarkable. Van Gogh was a colourist, and had a personal way of seeing things; it is a pity that he had not more power in carrying out what he wanted to do. One can see what he was attempting in his *Blés d'Or* (No. 70); but think what Blake would have made of such a vision of the waving corn! Gauguin is a stronger man, with a sense of the sinister, the exotic, the bizarre – a gift for strange and sullen colour; and at times his simplified form has hints of primitive grandeur; you feel the interest of a personality behind the work. But here again there is more struggle than mastery. None of these paintings could hold a candle to the *Smiling Woman* of Augustus John, which the Contemporary Art Society has, I am delighted to see, acquired for the nation.

Besides the Gauguins and Van Goghs, the things of interest (apart from the fine show of Manets) are the examples of Maurice Denis, Jules Flandrin, and Picasso. The *Nude Girl with a Basket of Flowers*, by the last named, is painted and modelled with wonderful subtlety; and the same artist's fine drawings in the end room should not be missed. M. Flandrin works on somewhat similar lines to M. Denis, but seems a better artist. His *Danse des*

Vendanges has a great deal of charm; one feels in it the tradition of Ingres and Chassérian still flowering naturally in the atmosphere of newer influences.

---

15. Wilfrid Scawen Blunt, *My Diaries*

1920, entry for 15 November 1910

Blunt (1840–1922), poet, politician and landowner, represents one of the most extreme responses to the pictures at the first Post-Impressionist exhibition.

To the Grafton Gallery to look at what are called the Post-Impressionist pictures sent over from Paris. The exhibition is either an extremely bad joke or a swindle. I am inclined to think the latter, for there is no trace of humour in it. Still less is there a trace of sense or skill or taste, good or bad, or art or cleverness. Nothing but that gross puerility which scrawls indecencies on the walls of a privy. The drawing is on the level of that of an untaught child of seven or eight years old, the sense of colour that of a teatray painter, the method that of a schoolboy who wipes his fingers on a slate after spitting on them. There is nothing at all more humorous than that, at all more clever. In all the 300 or 400 pictures there was not one worthy of attention even by its singularity, or appealing to any feeling but of disgust. – I am wrong. There was one picture signed Gauguin which at a distance had a pleasing effect of colour. Examined closer I found it to represent three figures of brown people, probably South Sea Islanders, one of them a woman suckling a child, all repulsively ugly, but of a good general dark colouring, such as one sees in old pictures blackened by candle smoke. One of the figures wore a scarlet wrapper, and there was a patch of green sky in the corner of the picture. Seen from across the room the effect of colour was good. Apart from the frames, the whole collection should not be worth £5, and then only for the pleasure of making a bonfire of them. Yet two or three of our art critics have pronounced in their favour, Roger Fry, a critic of taste, has written an introduction to the catalogue, and Desmond MacCarthy acts as secretary to the show. I am old enough to remember the pre-Raphaelite pictures in the Royal Academy exhibitions of 1857 and 1858, and it is pretended now that the present Post-Impression case is a parallel to it, but I find no parallel. The pre-Raphaelite pictures were many of them extremely bad in colour, but all were carefully, laboriously drawn, and followed certain rules of art; but these are not works of art at all, unless throwing a handful of mud against a wall may be called one. They are the works of idleness and impotent stupidity, a pornographic show.

---

16. Sir William Blake Richmond, 'Post-Impressionists'

Morning Post, 16 November 1910, 5

The Royal Academican, Sir William Blake Richmond (1842–1921) was one of the most implacable and outspoken opponents of Post-Impressionism. He was Slade Professor of Fine Art in Oxford (1879–83), was elected RA in 1895 and had written strongly against Degas's *Absinthe* when it was shown in London in 1893. He was probably best known for his decorations of St Paul's Cathedral.

Sir, – The article published in the *Morning Post* on Monday last, signed 'Robert Ross,' excited my curiosity to visit the Grafton Galleries. At the same time I wondered if Mr. Ross had exaggerated his vituperation, and if his reasons for non-conversion to the dogma set forth in the catalogue were justifiable. The idea was attractive. Here, so near to us in sensible old London, we were to find the natural offspring of noble Byzantine artists, in a fashionable quarter. Rumours were afloat that Mr. R. Fry had discovered the real thing in Paris, and had commenced his mission in London. Prominent 'experts,' enlightened connoisseurs, directors of picture galleries served as credentials, the official being immaculate to the crowd. So it all promised well; a nobly-spent hour or two could not be wasted surely under such auspices, and a final victory for 'Synthetics' was in store.

A bitter disappointment! Mr. Ross exercised too much charity; he has been too kind to quite tell the truth that is in him. It has led him to see merits, or their possibility, where none exists, except relatively to the hopeless impotence of the majority of these productions, which maintains, increasing *ad nauseam* from room to room, till it culminates in daubs by living French experimentalists.

Poor Manet! It is scarcely fair to attribute the parentage of this rotten egotism to him. A disagreeable artist, a brutal painter, yet a man of genius as he must have been to have executed *The Lady with the Parrot*. In the gallery there is nothing to vindicate his name, the work shown is poor and common. But Cézanne might well be the father of the Post-Impressionists. Mr. Ross thinks he might have become an artist. I differ. Cézanne mistook his vocation; he should have been a butcher, not a brave but a harmless profession. The Octagonal Gallery contains the kind of thing to be expected later on. There are far worse productions to be seen presently.

The gospel which inspired the dogma of the preface is in evidence on the walls of the large gallery and onwards. This preface is a lamentable medley of sophistry, crudities, and, not to mince matters, absence of good sense. Read carefully, traces of insecurity emerge, melancholy suggestions, quite unintentional no doubt, that the writer was not convinced by his matter or of himself in relation to it. The paradox of *The Rocking Horse* is an instance, and his painfully needed apology for the Committee, which gives away the whole show.

Well, what can be said? One cannot reason where there is no reason to start with. The marvel is that these hysterical daubs should have been perpetrated at all, but various forms of disordered mentality are common to these times. That they should not have been exhibited is obvious. It was unkind to expose them. It is futile to attack the authors of these morbid excrescences, I mean those especially which inspired Mr. Fry's dogma, and are the substance of this mission of folly to England. These incompetents neither could nor can draw, paint, or design. Blind to the beauty of line, colour, or motive, they do not possess the first faculty of a painter, power to observe. Yet we are told these are the inheritors of the genius which created the Mosaics in Venice, the tomb of Galla Placidia, the Ravenna Moccus, the dalmatic of Constantine, and countless other glories. The simile is strangely unfortunate. The work of Byzantine masters at its worst bears a kind of reminiscence of greater things, while at its best it recalls the reserve and strength of the purest Greek work, from which it can claim descent. No, the relationship is not as Mr. Fry would wish; it is to be found in the criminal lunatic asylum at Broadmoor, and the sad Bethel on the Island of Buvano, with a difference, the one a furious, terrible, passionate scourge, these melting their vigour before the embers of their expired Ego. Childish, not childlike!

To be fair, two pictures relieve the mournful monotony of degenerate cynicism, *Orchard in Provence* and *Saint Tropez en Fête*; both are harmless and pretty essays in impression.

One will try to forget this joyless and melancholy exhibition. I hope that in the last years of a long life it will be the last time I shall feel ashamed of being a painter. It was a relief to breathe the petroleum laden air of Bond-street, even the chill of a November afternoon came invigorating as a kind message of health after the suffocating tomb containing scarcely even the ashes of intelligence. For a moment there came a fierce feeling of terror lest the youth of England, young promising fellows, might be contaminated here. On reflection I was reassured that the youth of England, being healthy, mind and body, is far too virile to be moved save in resentment against the providers of this unmanly show. They will not growl at the perpetration of these atrocities, but pity them; but they are, and justly, indignant at the unwisdom of the sorry show.

It would be nauseating to myself and to your readers to dwell on details; it would be almost as unpleasant to read as to see. There is no fear of permanent mischief, I hope; the things are too bad for that. There is no regeneration for deluded egoists, 'sons of noble France, alas!' They are lost morally in the inferno where Dante places the unfaithful to God and to his enemies. The greater the artist, the more intense the vision, the loftier the imagination, the more modestly he will subjugate himself to Nature. Seeing more in her than others see, more beauty, more character, more diversity, and more divinity than is given to ordinary folk, whatever he creates in colour or form he admits to have been primarily Nature's property. Nature is not only the stern mentor, but the confiding friend of the true artist, the friend that keeps him straight and sane, Should he in a moment of weakness seek to escape her chiding he will speedily cease to be an artist and become an egoist, perhaps finally to cut his throat or die impotent and purposeless. Self-restraint once lost, vanity once permitted to take charge, Art is doomed to temporary destruction.

I hope the Press will teem with resentment against the insult offered to the noble arts of Design, Sculpture, and Painting, an insult also to the taste of the English people implied by the suggestion or presupposition that they have became so degenerate as to accept even from 'experts' such an imbecile mission. This invasion of depressing rubbish is, thank Heaven, new to our island of sensible and, in their own fashion, poetic people, and we hope the reckless prophet and promoters of this rotten league will be hard pressed by indignant cries of 'Shame,' correct the directions of their wandering, and evince regret for the hysteria by which they may have been temporarily mastered. Egoistic exultation is one of the sure signs of approaching insanity. The dogma set forth by Mr. Fry is egoism of the most insidious kind. The terrible warnings, perhaps necessary, however regrettable, now hang upon the walls of the Grafton Galleries.

---

17. Ebenezer Wake Cook, 'The Post-Impressionists'

Morning Post, 19 November 1910, 4

Wake Cook (1843–1926) was a landscape watercolour painter who exhibited at the Royal Academy from 1875 until 1910. He was also the author of *Anarchism in Art* (1904) – an attack on modernism in art modelled on Max Nordau's *Degeneration* (1895). Together with Robert Ross, William Blake Richmond and Philip Burne-Jones, Cook saw in Post-Impressionism the symptoms of a moral decay in western civilisation, and his voice continued to be prominent for another two years condemning each new movement in art.

Sir, – While agreeing with the excellent letters of Sir W.B. Richmond and Sir Philip Burne-Jones, I seriously question whether this is a case for art criticism at all; it is more like a postmortem on the death of the 'Modernity' movement, which reached its logical climax in the 'Impression,' exhibited for a time this year at the Salon des Indépendants, 'painted' by the swishing of a donkey's tail with a paint-brush attached. The Post-Impressionists present a case for historical, psychological, and pathological analysis, rather than art criticism; the whole show being intentionally made to look like the output of a lunatic asylum, its aim being to shock the bourgeoisie, and make talk, to attain the only success they aimed at, the *succès de scandale*.

Historically we know that when once decadence in any branch of art sets in painters and critics seem to race to see who shall reach the bottom of the hill first; or who shall wade deepest into the mire. Every downward step is hailed as an achievement. Art reflects life, and, as I showed in my exhaustive analysis of these movements, in *Anarchism in Art*, they are the analogue of the anarchical movements in the political world, the aim being to reduce all institutions to chaos; to invert all accepted ideas on all subjects, and the most amazing spectacle is that, with a few honourable exceptions, the most conservative papers have had the most reckless anarchists for art critics. The first downward step was taken when brains were eliminated from art and criticism, when every manifestation of invention, imagination, poetic or dramatic, was denounced as 'literary,' as 'art in its anecdotage,' and nothing but tricks of execution appealed to the 'new' critics. They encouraged a vogue of vulgarity, and every new trick, every advance in blatant badness, was hailed as a trimph of art. Now, it is not the wrongness of all this which arouses my protest; it is the silliness of it which makes me blush for my countrymen. What are we to think of 'critics' who could mistake decadence for progress, as the new critics did?

In my writings I have shown the psychological factor in the case. We are suffering from a surfeit of fine art; and from the democratisation of art by reproductive processes; and overworked critics having so much to do with it, their delicate æsthetic faculties became jaded to their death; judgment was vitiated, and works were lauded in the ratio of their badness; and the new 'criticism' became an orgy of topsy-turvydom. Monet gave away one secret of these new movements. He said, 'One year men paint things purple and people scream; and next year men paint much more purple.' Lacking inspiration and the hope of fame, they seek its monetary equivalent, notoriety, and out-scream each other in an advertising frenzy. During the height of the frenzy the critics of the *Morning Post* were among the few who kept their heads.

As for the pathological aspect of the case, your own columns on Thursday gave some hints under the heading of 'Eccentricities of Diet' and 'Abnormal Tastes,' Dr. Soltau Fenwick showing that people in abnormal states eat paper, hair, thread, varnish, polish, mud, clay, soot, sand, glass, and live fish. There have been 'dirt-eating epidemics,' as described by Hunter. This is analogous to 'Modernity' movements in art. In France we have had *The Flowers of Evil*, Satanism, and the Black Mass; and many people enjoy a morning in the Morgue. This is the pathological aspect of the case, and the Grafton Galleries have been turned into a Morgue for 'Modernity' art. Some years ago at the International there was a mania for painting flesh with mud, making Eve's fair daughters look unwashed; while others painted it in ghastly greys and greens, as if in the last stages of decomposition. This 'Black Death in Art,' as I called it, was checked by one strong, outspoken article of mine in *Vanity Fair*, checked in all those who were not absolutely colour blind. These sickening aberrations could never have got a footing had it not been for the 'Modernity' critics, who received them with acclamation and defamed all our national art, depreciating it by millions of pounds in value, unsettling aims, causing widespread distress, making criticism ridiculous, and dragging the papers they misrepresented into an anti-patriotic campaign in favour of anarchism and ultimate chaos.

Mr. Cunninghame Graham errs in saying this movement has its analogue in literature. We should have its like in a movement to abolish grammar and orthography, to invert the meanings of words, reduce language to chaos, subvert beauty, represent Nature as suffering from measles, and libel and distort everything, and to outrage all artistic sensibilities. Will Mr. Graham venture to assert that there is insanity enough among editors and publishers to make such a movement possible?

The gravest aspect of this matter is that no less than seven directors of our national collections support by their names this last degradation of the art they are paid to guard; this will discredit their judgment for years to come.

18. Roger Fry, 'The Grafton Gallery – 1'

Nation, 19 November 1910, 331

Ignoring the journalistic brickbats which continued to be hurled at the collection of paintings at the Grafton Galleries, Fry gave a calm and reasoned exposition of what he considered to be the principal features of Post-Impressionist art – though he avoids the term. Characteristically his argument draws upon historical precedents. The second part of this article was published in December 1910 (see no. 21).

Having been asked to advise on the choice of the pictures and their arrangement in the present exhibition, my remarks on it must be taken rather as explanatory – in the nature of an apologia – than as the expression of entirely independent criticism. The reader, thus duly warned, may discount my judgment to whatever degree he thinks fit. I have been accused of a strange inconsistency in admiring, at one and the same time, the accredited masterpieces of ancient art and the works here collected together, which are supposed to typify the latest and most violent of all the many violent reactions against tradition which modern art has seen. Without being much interested in the question of my own consistency, I believe that it is not difficult to show that the group of painters whose work is on view at the Grafton Gallery are in reality the most traditional of any recent group of artists. That they are in revolt against the photographic vision of the nineteenth century, and even against the tempered realism of the last four hundred years, I freely admit. They represent, indeed, the latest, and, I believe, the most successful, attempt to go behind the too elaborate pictorial apparatus which the Renaissance established in painting. In short, they are true pre-Raphaelites. But whereas previous attempts – notably our own pre-Raphaelite movement – were made with a certain conscious archaism, these artists have, as it were, stumbled upon the principles of primitive design out of a perception of the sheer necessities of the actual situation. At once the question is likely to arise: Why should the artist wantonly throw away all the science with which the Renaissance and the succeeding centuries have endowed mankind? Why should he wilfully return to primitive or, as it is derisively called, barbaric art? The answer is that it is neither wilful nor wanton but simply necessary, if art is to be rescued from the hopeless encumbrance of its own accumulations of science; if art is to regain its power to express emotional ideas, and not to become an appeal to curiosity and wonder at the artist's perilous skill. The fundamental error that is usually made is that progress in art is the same thing as the much more easily measured and estimated progress in power of representing nature. All our histories of art are tainted with this error, and for the simple reason that progress in representation can be described and taught, whereas progress in art cannot so easily be handled. And so we think of Giotto as a preparation for a Titianesque climax, forgetting that with every piece of representational mechanism which the artist acquired, he both gained new possibilities of expression and lost other possibilities. When you can draw like Tintoretto, you can no longer draw like Giotto, or even like Piero della Francesca. You have lost the power of expression which the bare recital of elementary facts of mass, gesture, and movement gave, and you have gained whatever a more intricate linear system and chiaroscuro may provide in expressive power. But the more complex does not really subsume the simpler: it replaces it by a new unity, in which the old elements of unity are lost sight of. Even if they are there, they are no longer put forward so as to make their full appeal; they are muffled and shrouded in the new elaboration.

Now, modern art had arrived in Impressionism at a point where it could describe everything visible with unparalleled ease and precision, but where, having given to every part of the picture its precise visual value, it was powerless to say anything of human import about the things described. It could not materially alter the visual values of things, because the unity cohered in that and in that alone. To give to the rendering of nature its response to human passion and human need demanded a re-valuation of appearances, not according to pure vision, but according to the pre-established demands of the human senses. It is this re-valuation of the visual that Cézanne started (or, rather, it already began in Manet). He discovered distortions and ruthless simplifications (which are, of course, distortions, too) of natural form, which allowed the fundamental elements of design – the echo of human need – to reappear in his representations. And this has gone on ever since his day in the group of artists we are considering. More and more regardlessly they are cutting away the merely representative element in art to establish more and more firmly the fundamental laws of expressive form in its barest, most abstract elements. Like the Anarchists, with whom they are compared, they are not destructive and negative, but intensely constructive. This does not mean that they will not destroy much that we cling to still. Growth, and not decay, is the real destroyer, and the autumn leaf falls, not because the wind and frost attack it, but because next year's bud has undermined its base. And so if all the accumulated science of representation, all the aids to perspective, all the anatomical diagrams, all the lore about atmospheric values goes, it will, no doubt, be built up again one day, but with passionate zest and enthusiasm, as it was once before; only we to whom it has been transmitted as a corpus of dead fact, alien to the imagination, must press on to the discovery of that difficult science, the science of expressive design. No artist can profitably use a single fact beyond what his imagination can grasp; every fact that he uses must have been passionately apprehended. We have become so accustomed to accepting masses of dead, undigested facts, facts of observation, and not re-creation, that we must begin at the beginning, and learn once more the A.B.C. of abstract form. And it is just this that these French artists have set about, with that clear, logical intensity of purpose, that absence of all compromise, of all regard for side issues, which has so nobly distinguished the French genius.

And there is at least this consolation, if we must surrender that too complex language of complete naturalistic representation, which, to tell the truth, very few artists in all these centuries since the Renaissance have been able quite evidently to master – namely, that the biggest things demand the simplest language, and Cimabue could tell us more of divine, and Giotto more of human, love than Raphael or Rubens.

From another point of view, moreover, the effect of discarding the actual illusion of three-dimensional space – of losing chiaroscuro and atmospheric color – is not without its compensations. I believe that anyone who will look, without preconception, merely at the general effect of the walls of the Grafton Gallery must admit that in no previous exhibition of modern art has the purely decorative quality of painting been more apparent. If only the spectator will look without preconception as to what a picture ought to be and do, will allow his senses to speak to him instead of his common-sense, he will admit that there is a discretion and a harmony of color, a force and completeness of pattern, about these pictures, which creates a general sense of well-being. In fact, these pictures, like the works of the early primitives, and like the masterpieces of Oriental art, do not make holes in the wall, through which another vision is made evident. They form a part of the surface which they decorate, and suggest visions to the imagination, rather than impose them upon the senses.

I should like also to appeal against another misconception of the aim and purposes of art, which often hinders spectators from a true perception of beauty, namely, the idea that the artist takes refuge in certain formulæ, merely to avoid trouble, and that he thereby cheats us of his part of the contract.

This is surely importing moral considerations into a field where they do not apply. What, indeed, could be more desirable than that all the world should have the power to express themselves harmoniously and beautifully – in short, that everyone should be an artist. The beauty of the resulting work has nothing to do with the amount of effort it has cost. While no effort is vain that is needed to produce beauty, it is the beauty, and not the effort, that avails. If by some miracle beauty could be generated without effort, the whole world would be the richer. As a matter of fact, many of the artists whose work is shown at the Grafton have already proved themselves accomplished masters in what is supposed to be the more difficult task of representation. That they have abandoned the advantage which that professional skill affords is surely rather a sign of the sincerity of their effort in another direction. No doubt the acrobatic feats of virtuosity always will appeal to our sense of wonder; but they are better reserved for another stage. If only an artist has genuine conviction, he rarely lacks sufficient skill to give it expression. And as things are at present, with our gaping admiration for professional skill, we are in less danger of finding a prophet whose utterance is spoiled by imperfect articulation than of being drowned beneath floods of uninspired rhetoric.

19. H.M. Bateman, 'Post-Impressions of the Post-Impressionists'

Bystander, 23 November 1910, 375

Visitors to the Grafton Galleries in the first two weeks of the exhibition spoke of the laughter and embarrassment generated by the pictures. Cartoonists immediately saw this as a fertile source of pictorial amusement and exploited it to the full.

For the cartoon, see p. 126

20. Desmond MacCarthy, 'The Exhibition at the Grafton Galleries: Gauguin and Van Gogh'

Spectator, 26 November 1910, 902–3

For MacCarthy see headnote to no. 10. MacCarthy as secretary to the first Post-Impressionist exhibition was also guide and apologist for the new paintings while the exhibition remained open. This article summarizes what he must have repeatedly told the curious and the hostile who visited the Grafton Galleries. MacCarthy was also responsible for the business arrangements and, as he pointed out in his book *Memories* (1953), in spite of the scandal which surrounded the show it was a financial success.

Sir, – The Chinese, who bestow lasting fame for an apparent trifle, for one trait of character exhibited, have cherished during hundreds of years the memory of a statesman, Yiu Hao. Of him it is recorded that when disgraced he took his punishment without complaint, except that he spent his days in writing with his finger in the air the four words: 'Oh! Oh! Strange business.' He should be a patron saint to all who find themselves in the public pillory. The promoters of the Exhibition of the Post-Impressionists at the Grafton Gallery have been abused as baffled and unhappy egotists, as inoculators of a new noisome disease; but retorts to such personal charges are better written upon air than paper.

To read many of the letters which have appeared in the Press, one might suppose that, with the exception of Manet's pictures, the whole Exhibition contained nothing but artistic impostures. It is strange that so few critics should have recognised, for instance, the decorative beauties of the Gauguins which hang upon the right wall of the large gallery, – their sombre and rich colour, their gravity and expressiveness of line, the large execution of their design, the restfulness of the general effect. There is a classical quality, too, in some of them, notably in the group of figures lent by M. Alphonse Kann, which has escaped the notice of many; perhaps because the Exhibition contains some headlong experiments they never thought to find such a quality; perhaps they did not see it because this classical quality is combined with barbaric sentiment. But there it is, nevertheless, plain to the eye for any one who is not, for one reason or another, too angry to look at the pictures. The epithet 'barbarous' often resounds through the Grafton Galleries, and there is a sense in which the term is applicable to Gauguin's best work. It has the barbaric quality which, as Baudelaire pointed out, is often visible at a completed stage of art, whether Egyptian or Assyrian. This quality springs from 'the impulse to see things largely, to consider them in their total effect, to synthesise and abbreviate all detail.' Gauguin's work is the work of a man whose eye and memory have absorbed the colour and silhouette of the human figure, who refuses to be distracted by the clamouring crowd of detailed impressions which compete to be recorded, – the work of one to whom nudity is a simple and familiar fact. There is an imaginative side to these pictures, too, which must appeal to any one who has recovered from the surprise of such a simplification of appearances. What an imaginative grasp of the savage's constant fear of the other world is shown, for instance, in *L'Esprit des Morts qui Veille!* A naked girl lies flat upon a couch, her palms open on the pillow. She is still with terror, while behind her stands a strange wooden-faced figure, cowled like a monk, with a long, white eye, and one hand outstretched. The yellow gleam upon the girl's limbs suggests the light of a dim lamp; the background is a disquieting violet-purple, sown with phosphorescent flowers, flowers which shine in the dark and remind the natives that the dead are thinking of them. It is a composition of horizontal and undulating lines, of harmonies of orange and blue, united by derivative violets and yellows, a superb piece of decorative painting.

There is a story of Napoleon meeting on a narrow path at St. Helena a man staggering under a load of hay; the *aide-de-camp* swore at him to get out, whereupon Napoleon uttered his finest *mot*, 'Respect the burden.' Though the counter-ideal – so different from Gauguin's – which Van Gogh pitched against contemporary civilisation is not set forth in parable-pictures on the walls of the Gallery, no one of any imagination can look at the portraits hanging there without feeling how profound is Van Gogh's sense and respect for the human burden. It may be said: 'And what has that to do with art, whose concern is with beauty? I wish to look upon beautiful, unmarred faces, not upon portraits like those of Dr. Gachet or the postman.' But that sense of the burden is necessary to the appreciation of much of the beauty of life, and poetry does enter into painting. How much of Rembrandt would we miss if that sense had been denied him! A lady stops in front of *La Berceuse* and exclaims with comic exaggeration of horror at the idea of hanging such a picture in her room. She is quite right. It would never do as a background to her life. It was not meant to beautify some rosy, cosy interior, but a sailors' cabaret in Marseilles or Sainte-Marie. That yellow, stolid, unchangeable woman does not represent a mother to her, whose ideal of maternity is much more graceful and intimately tender; but look at it with some imagination of the experience of those for whom it was meant, and it is full of significance; look at the colour and design of the picture, keeping in mind its proper surroundings, and the colour and design are grand and appropriate. It would be a great disaster if artists only painted pictures suitable for the homes of those who can buy their pictures at high prices; not because such pictures cannot be of the finest beauty, but because such a restriction bars out so much beauty of another kind that might find expression in art. Van Gogh's still-life pictures ('still-life' is almost a misnomer for studies so vibrant with vitality), his sunflowers and irises, would never do in the modern home. Such a shout of colour and such vitality of form would destroy any scheme of decoration, the charm of which depended upon preserving the *chez soi*; but in a bare workshop full of the stir of men and common work, what a magnificent, exhilarating decoration they would be!

---

21. Roger Fry, 'The Post-Impressionists – 2'

Nation, 3 December 1910, 402–3

This article is a continuation of no. 18 in which Fry deals with the individual artists in the first Post-Impressionist exhibition. He focuses most strongly on the work of Cézanne, which, as he admits, had risen greatly in his admiration as a direct consequence of the exhibition.

In my first article I tried to urge one or two points of general consideration about the group of painters shown at the Grafton Galleries. I will now try to discuss the artists separately. And first let me admit, in reply to the flamboyant diatribes of those who wish to see me burned together with the pictures which I arranged with such effrontery to insult the British public, that the collection is far from being perfect as an expression of this movement in art. Anyone who has tried to collect so large a body of pictures in a short time will know how many accidental obstacles occur to prevent one's getting just those pictures on which one has most set one's heart, and how often in despair one has to accept a less perfect example of such and such an artist. One kindly critic is quite right in saying that there are too many Gauguins, and that there are Van Gogh's which it would have been most desirable to add. Then again, Matisse, owing to the absence of a well-known collector, is quite inadequately represented, and Picasso should have been seen in bigger and more ambitious works. But at least the exhibition has given an opportunity to the British public to judge of a great movement of which it had hitherto remained in almost total ignorance, and it has given Sir W.B. Richmond the opportunity to express publicly his shame at bearing the designation of artist. That is perhaps even more than one had ventured to hope.

Another confession – the Manets are not, on the whole, good examples, and perhaps establish an unfair comparison with Cézanne. I always admired Cézanne, but since I have had the opportunity to examine his pictures here at leisure, I feel that he is incomparably greater than I had supposed. His work has the baffling mysterious quality of the greatest originators in art. It has that supreme spontaneity as though he had almost made himself the passive, half-conscious instrument of some directing power. So little seems implied at first sight in his apparently accidental collocation of form and color, so much reveals itself gradually to the fascinated gaze. And he was the great genius of the whole movement; he it was who discovered by some mysterious process the way out of the cul-de-sac into which the pursuit of naturalism *à outrance* had led art. As I understand his art, and I admit it is exceedingly subtle and difficult to analyse – what happened was that Cézanne, inheriting from the Impressionists the general notion of accepting the purely visual patchwork of appearance, concentrated his imagination so intensely upon certain oppositions of tone and color that he became able to build up and, as it were, re-create form from within; and at the same time that he re-created form he re-created it clothed with color, light, and atmosphere all at once. It is this astonishing synthetic power that amazes me in his work. His composition at first sight looks accidental, as though he had sat down before any odd corner of nature and portrayed it; and yet the longer one looks the more satisfactory are the correspondences one discovers, the more certainly felt beneath its subtlety, is the architectural plan; the more absolute, in spite of their astounding novelty, do we find the color harmonies. In a picture like *L'Estaque* it is difficult to know whether one admires more the imaginative grasp which has rebuilt so clearly for the answering mind the splendid structure of the bay, or the intellectualised sensual power which has given to the shimmering atmosphere so definite a value. He sees the face of Nature as though it were cut in some incredibly precious crystalline substance, each of its facets different, yet each dependent on the rest. When Cézanne turns to the human form he becomes, being of a supremely classic temperament, not indeed a deeply psychological painter, but one who seizes individual character in its broad, static outlines. His portrait of his wife has, to my mind, the great monumental quality of early art, of Piero della Francesca or Mantegna. It has that self-contained inner life, that resistance and assurance that belong to a real image, not to a mere reflection of some more insistent reality. Of his still life it is hardly necessary to speak, so widespread is the recognition of his supremacy in this. Since Chardin no one has treated the casual things of daily life with such reverent and penetrating imagination, or has found as he has, in the statement of their material qualities, a language that passes altogether beyond their actual associations with common use and wont.

If Cézanne is the great classic of our time, Van Gogh represents as completely the romantic temperament. His imagination responds to the call of the wildest adventures of the spirit. Those who have laughed at this great visionary because he became insane, can know but little of the awful adventures of the imagination. That Rembrandt saw as far into the heart of pity and yet remained sane is true, but that should rather be imputed to Rembrandt as his supreme greatness and good fortune. To laugh at a less fortunate adventurer is to ignore the perilous equilibrium of such genius, to forget how rare it is to see God and yet live. To Van Gogh's tortured and morbid sensibility there came revelations fierce, terrible, and yet at times consoling, of realities behind the veil of things seen. Claiming his kinship with Rembrandt, Van Gogh became a portrayer of souls; souls of broken, rugged, ungainly old women like the *Berceuse*, whose greatness yet shines in the tender resignation of her folded hands; souls of girls brutalised by the associations of utter poverty, and yet blazing with an unconscious defiance of fate. And souls of things – the soul of modern industrialism seen in the hard splendor of mid-day sun upon the devouring monsters of a manufacturing suburb; the soul of the wind in the autumn corn, and, above all, the soul of flowers. Surely no one has painted flowers like Van Gogh. We know how deeply Van Gogh's own predecessors of the seventeenth century sinned in their thick-skinned cleverness and self-assurance, using flowers as a kind of animate furniture. But modern European art has almost always maltreated flowers, dealing with them at best but as aids to sentimentality until Van Gogh saw, with a vision that reminds one of Blake's, the arrogant spirit that inhabits the sun-flower, or the proud and delicate soul of the iris. The gibe of insolent egotism was never more misapplied than to so profound, so deeply-enduring a genius as Van Gogh's; for his distortions and exaggerations of the thing seen are only the measure of his deep submission to their essence.

Of Gauguin I find it harder to speak. With him one must make excuses and concessions if one is to be perfectly honest. Of his astonishing talent as a designer, his creation of new possibilities in pattern, and his unrivalled power of complex color harmony, these pictures tell plainly enough, and to that I must add a real sense of nobility and elemental simplicity of gesture, and at times a rare poetic insight. But I do not always feel sure of the inner compulsion towards the particular form he chooses. I cannot shake off an occasional hint of self-consciousness, of the desire to impress and impose; in fact, of a certain rhetorical element. The mere statement of this seems to exaggerate it; perhaps it only means that he is a Parisian, and that certain turns of his whimsical wit strike us as having a tinge of perversity. Yet all this must be unsaid before his greatest designs, before the touching and entirely sincere *Agony in the Garden*, before his *L'Esprit veille*, with its sympathy with primitive instincts of supernatural fear and its astounding physical beauty, before landscapes of such fresh and rare beauty as No. 44a, and perhaps, above all, before his splendid flower-piece, No. 31.

I know that to dismiss Gauguin thus is unfair, but space is wanting to deal with so much new material fully. Henri Matisse is, as I have said, but poorly represented. As I understand him, he is an artist not unlike Manet, gifted with a quite exceptional sense of pure beauty – beauty of rhythm, of color harmony, of pure design; but at the same time perhaps a little wanting in temperament, without any very strong and personal reaction to life itself, almost too purely and entirely an artist. The *Femme aux yeux verts* strikes me as a more convincing and assured creation every time I see it. To my eye, it appears singularly perfect in design, and at once original and completely successful in the novelty, frankness, and bravery of its color harmony. In his drawings, of which a considerable number are shown, he proves I think, beyond doubt his masterly sense of rhythmic design and the rare beauty of a handwriting which, in its directness and immediacy, reminds one more of Oriental than European draughtsmanship. That the plastic feeling in painting is by no means dependent upon light and shade, but may be aroused quite as surely by line and color, might be guessed indeed from his paintings, but is made evident by the examples of his bronze statuettes. Whatever one may think of his figure, *Le Serf*, as an interpretation, it cannot be denied that it shows a singular mastery of the language of plastic form.

Picasso is strongly contrasted to Matisse in the vehemence and singularity of his temperament. In his etching of *Salome* he proves his technical mastery beyond cavil, but it shows more, a strange and disquieting imaginative power, which comes at times perilously near to the sentimental, without, I think, ever passing the line. Certainly, in the drawing of the *Two Women* one cannot accuse him of such a failing, though its intimacy of feeling is hardly suspected at first beneath the severity of its form. Of late years Picasso's style has undergone a remarkable change, he has become possessed of the strangest passion for geometrical abstraction, and is carrying out hints that are already seen in Cézanne with an almost desperate logical consistency. Signs of this experimental attitude are apparent in the *Portrait of M. Sagot*, but they have not gone far enough to disturb the vivid impression of reality, the humorous and searching interpretation of character.

One or two of the younger artists must just be mentioned here: Othon Friesz appears in the three canvases here shown as inclining towards Impressionism, but he has carried over much that he has learned in his more synthetical designs; his color has an extraordinary gaiety and force, and he shows how much more vivid to the senses and imagination are interpretations of sunlight like these than anything achievable by direct observation.

Vlaminck is a little disconcerting at first sight, by reason of the strangely melancholy harmonies he affects, but he has the power of inventing admirably constructed and lucid designs, a power which is perhaps even more clearly seen in his paintings upon *faïence*. I would call special attention to these, since, if the group of artists here exhibited had done nothing else, their contribution to modern art would be sufficiently striking, in that they have shown the way to the creation of entirely fresh and vital pattern designs, a feat which has seemed, after so many years of vain endeavor, to be almost beyond the compass of the modern spirit.

---

22. Jacob Tonson [Arnold Bennett], 'Books and Persons'

New Age, December 1910, viii, 135

Bennett (1867–1931) had given up the editorship of *Woman* in 1896 to write novels full-time. In 1910 he published *Clayhanger* but contributed a regular column of literary criticism to the *New Age*. In Virginia Woolf's essay 'Mr Bennett and Mrs Brown', she associated his style with conservatism and old-fashionedness but he spoke out strongly in favour of Post-Impressionism. Bennett had spent much time in Paris and had been familiar with the painting of Cézanne and Gauguin for some years before 1910.

The exhibition of the so-called 'Neo-Impressionists' over which the culture of London is now laughing, has an interest which is perhaps not confined to the art of painting. For me, personally, it has a slight, vague repercussion upon literature. The attitude of the culture of London towards it is of course merely humiliating to any Englishman who has made an effort to cure himself of insularity. It is one more proof that the negligent disdain of Continental artists for English artistic opinion is fairly well founded. The mild tragedy of the thing is that London is infinitely too self-complacent even to suspect that it is London and not the exhibition which is making itself ridiculous. The laughter of London in this connection is just as silly, just as provincial, just as obtuse, as would be the laughter of a small provincial town were

Strauss's *Salome*, or Debussy's *Pelléas et Mélisande* offered for its judgement. One can imagine the shocked, contemptuous resentment of a London musical amateur (one of those that arrived at Covent Garden box-office at 6 a.m. the other day to secure a seat for *Salome*) at the guffaw of a provincial town confronted by the spectacle and the noise of the famous *Salome* osculation. But the amusement of that same amateur confronted by an uncompromising 'Neo-Impressionist' picture amounts to exactly the same guffaw. The guffaw is legal. You may guffaw before Rembrandt (people do!), but in so doing you only add to the sum of human stupidity. London may be unaware that the value of the best work of this new school is permanently and definitely settled – outside London. So much the worse for London. For the movement has not only got past the guffaw stage; it has got past the arguing stage. Its authenticity is admitted by all those who have kept themselves fully awake. And in twenty years London will be signing an apology for its guffaw. It will be writing itself down an ass. The writing will consist of large cheques payable for Neo-Impressionist pictures to Messrs. Christie, Manson and Woods. London is already familiar with this experience, and doesn't mind.

Who am I that I should take exception to the guffaw? Ten years ago I too guffawed, though I hope with not quite the Kensingtonian twang. The first Cézannes I ever saw seemed to me to be very funny. They did not disturb my dreams, because I was not in the business. But my notion about Cézanne was that he was a fond old man who distracted himself by daubing. I could not say how my conversion to Cézanne began. When one is not a practising expert in an art, a single word, a single intonation, uttered by an expert whom one esteems, may commence a process of change which afterwards seems to go on by itself. But I remember being very much impressed by a still-life – some fruit in a bowl – and on approaching it I saw Cézanne's clumsy signature in the corner. From that moment the revelation was swift. And before I had seen any Gauguins at all, I was prepared to consider him with sympathy. The others followed naturally. I now surround myself with large photographs of these pictures of which a dozen years ago I was certainly quite incapable of perceiving the beauty. The best still-life studies of Cézanne seem to me to have the grandiose quality of epics. And that picture by Gauguin, showing the back of a Tahitian young man with a Tahitian girl on either side of him, is an affair which I regard with acute pleasure every morning. There are compositions by Roussel which equally enchant me. Naturally I cannot accept the whole school – no more than the whole of any school. I have derived very little pleasure from Matisse, and the later developments of Felix Vallotton leave me in the main unmoved. But one of the very latest phenomena of the school – the water-colours of Pierre Laprade – I have found ravishing.

---

23. A.J. Finberg, 'Art and Artists'

Star, 14 December 1910, 2

A.J. Finberg (1866–1939) was an art student at Lambeth and Paris. He exhibited at both the Paris Salon and the New English Art Club. He drew illustrations for the *Graphic* and the *Illustrated London News* and was art critic for the *Morning Leader*, the *Manchester Guardian*, the *Saturday Review* and the *Star*. He is best remembered for his work on Turner whose drawings he catalogued and about whom he wrote a number of influential books.

Finberg records the enormous success of the first Post-Impressionist exhibition and gives the sense of a genuine debate growing up around the pictures in the British Press.

When the exhibition of the Post-Impressionists was first opened at the Grafton Galleries it was feared by its organisers that it would be treated with silent contempt; that the public would not go to see it, and that it would prove a financial failure. These fears have proved groundless. Sir William Richmond's flamboyant letter in *The Times* gave the exhibition a splendid advertisement. Sir William Richmond and the other old fogies who went about foaming at the mouth had so often been in the wrong that people began to say that there was probably something in Post-Impressionism after all. Men like Sir William Richmond cannot apparently understand that loud and immoderate abuse is the best advertisement that any artistic movement can have. As soon as the attention of the public is drawn to such matters they immediately want to investigate for themselves. So the Grafton Galleries gradually filled with spectators bent on seeing and judging for themselves. For the last three or four weeks the show has been doing excellent business. The takings, I have been told, have averaged over £40 a day. The exhibition has turned out a brilliant success – a success of scandal – and the organisers have to thank those who have abused them immoderately for this happy result.

I went for the second time to the galleries on Saturday afternoon, and found them uncomfortably crowded with a horde of giggling and laughing women. Like Mr. Hind, though for different reasons, I find it hard to understand how anyone can laugh at the Post-Impressionists. But there are some women, I suppose, who find a visit to a lunatic asylum comic and amusing. The antics and gestures of the unfortunate inmates must be so very funny. I confess I prefer other forms of amusement. To walk round the rooms of the Grafton Gallery now is as painful and depressing to me as a visit to a lunatic asylum or hospital would be. The solemn pretensions, the uncouth and sterile productions of these unfortunate creatures fill me with pity. Most of the work is undoubtedly sincere. The artists have put themselves into their pictures. The whole background of their pitiable mental life hangs round their awkward and forbidding daubs. I came away from the Grafton Galleries with the smell of the lazar house clinging to me and with the gibbering of idiots ringing in my ears.

Yet there is 'something' in these strange works, I am constantly assured by well-meaning friends. Of course there is 'something' in them. But people have got into such a fixed habit of thinking that painting is merely an art of imitation that the discovery of its direct and innate powers of expression takes them completely by surprise. The power of representation is only a small part of the function of painting. The greater part is its power of direct presentation – the presentation of what is rather loosely called 'the personality of the artist.' The quality of the paint, of the color scheme, the character of the artist's brushwork and his design – these are direct presentational elements in painting which reveal the texture and quality of the artist's mind and character as clearly as the pitch and tone of his voice does in speaking. A man or a woman may utter some commonplace words to you, but the nervous tension of the voice may tell you very much more than the words. The words uttered are equivalent to the subject of the picture, but the way the picture is painted is equivalent to the tone of the speaker's voice. And this is the 'something' in the works of Cézanne, Gauguin, Van Gogh, and others which their admirers have discovered. The few apples on a plate in Cézanne's *Nature Morte* (lent by M. Alphonse Kann), the rocks and water in his *L'Estaque* (8), the man's head and badly drawn shoulders in his *Portrait de l'Artiste* (10), the ungainly woman in his *Portrait de Mme. Cézanne* (11) – these are the dull and feeble and insignificant words the artist utters. But through them all rings the voice which tells of the artist's unstable and badly balanced emotive life. As Mr. Roger Fry, the most eloquent and thoughtful champion of this kind of art, says of Cézanne, 'So little seems implied at first sight in his apparently accidental collocation of form and color, so much reveals itself gradually to the fascinated gaze.' The remark is perfectly true. The abnormal character of the artist's mind reveals itself at once to anyone who understands the language of pictorial art. It is the same with Gauguin's *Le Christ au Jardin des Oliviers* (85) and Van Gogh's *La Berceuse* (76). The gibber of the artists half-formed thoughts, vague aspirations, and pitiful self-satisfaction is expressed by the sweep of their hands, and 'reveals' itself in spite of the grotesque and banal forms in which their ideas are represented.

But what appears so strange to me is to find the works of these men praised without moderation because of this expressive quality in the painter's language. It is like saying a man is a great orator because you can hear his words. This directly expressive power of painting is as common as the quality of audibility which spoken words possess. It is not only common but it is innate and inevitable. There are no pictures that do not possess it. It is a mere function of all pictorial art. But what makes the difference between a masterpiece of art – like Leonardo's *Monna Lisa*, Rembrandt's *Jewish Rabbi*, Turner's *Frosty Morning*, or any drawing by Jean Francois Millet – and abortions like the pictures I have named at the Grafton Galleries is simply that this common function of painting brings us into direct contact, in the one case, with a fuller and finer personality than our own, and, in the other case, with a poorer and less normal personality. Who would not gladly hear the tone of Plato's or Shakespeare's voice? Who can find the same pleasure in listening to the shrieks and mumbling of unfortunate beings whose minds have escaped from the control of reason?

That it is the character of the background of thought and feeling revealed by the expressive qualities of painting which matters is proved by the different way the works of the different artists represented at the Grafton Galleries affect you. One does not feel the same about the works of Maurice Denis and Jules Flandrin as about those of Cézanne and Van Gogh. Pictures like Flandrin's *La Danse des Vendanges* (27) and Denis's *Orphée* (33), *Madone au Jardin Fleuri* (124), *Nausicaa* (100), and *Calypso* (82) are certainly sufficiently strange and extravagant. But the strange forms and extravagant colors are clearly adopted for the artists' own purposes. Maurice Denis, one knows – his work tells us so – is a capable and courageous artist. He does what he wants to. The pathos of the works of Cézanne and the others is the grotesque break between intention and performance, between the gorgeousness and splendor of the demented man's dreams and the importance of his actions. Let us not laugh at these works. For pity's sake, let them be decently removed from the gaze of the public.

24. Spencer Frederick Gore, 'Cézanne, Gauguin, Van Gogh &c., at the Grafton Galleries'

Art News, 15 December 1910, 19–20

Gore (1878–1914) was a Slade student between 1896 and 1899. In 1904 he met Walter Sickert in Dieppe and was co-founder and first President of the Camden Town Group in 1911.

A criticism of this exhibition presents some difficulty. The paintings have been so ill received that one feels more ready to pour a stream of violent abuse on the heads of the critics and public than attempt a piece of more or less intelligent criticism. Cézanne, Gauguin, Van Gogh, Seurat, and Signac have their accepted places on the Continent, just as Joseph Israels, Mauve, and the brothers Maris have in England. France has as much right to have its Mattisse, Piccasso, Marquet, Denis, and other painters with various aims, as England its Steer, Ricketts, Shannon, John, Orpen, & c.

If there was an exhibition in Paris of Rossetti and Burne Jones might not the French critic point out that none but madmen could have produced such curious creatures, such quaint angularities, such boneless monstrosities? I only suggest how it might strike him, and how wrong he would be!

There seems very little need for the inclusion of Manet in this exhibition. Cézanne and Gauguin have very little connexion with him, and there is probably no other painter represented here would admit any at all. Besides, as M. Blanche points out in his letter to *The Morning Post* – a letter that must have appealed to every Englishman, fearful that he might entrust his daughter's portrait to some budding Van Gogh – the Manets are inferior Manets. Perhaps it is only this which accounts for the poor show they make beside the Cézanne's.

Cézanne was a painter who sought above everything else an exact harmony of colour, as exact as Whistler's relations of tone. In attaining this he often lost the drawing, which he would then recover with a line. Hence incompleteness. The incompleteness not of the shirker, but of the man who has pursued a thing as far as he can go, and still finds the end just out of reach. And it is the intense sincerity with which he pursues his object which gives to his painting that wonderful gravity, and makes such paintings as Manet's *Bar at the Folies-Bergère* or *The Café* appear, as they really are, pictures 'faked' in the studio by a virtuoso of great skill, but without any particular object.

Of all the painters represented here Gauguin seems to be the least disliked. He is certainly the best represented. Those interested in lineage might find amusement in tracing the influence of Camille Pissarro, not only in the Breton pictures, but also in the Tahitan.

The attempt to separate the decorative side of painting from the naturalistic seems to me to be a mistake. Durer is supposed to have said just before he died, that he had begun to see how simple nature was. Simplification of nature necessitates an exact knowledge of the complications of the forms simplified. This may be done to produce a greater truth to nature as well as for decorative effect. I should like to know, for instance, into which category any one is going to put *Martinique* (34), or *Les Laveuses* (83), or *Paysage Bretonne* (87).

Every picture has its origin, in something seen either at first hand in nature or second hand in some other picture – something that has filtered through one brain, through the brains of a generation, or many generations. Gauguin gives his idea of Tahiti just as Goya gave his of Spain. It is equally untrue to say of Pissarro, Sisly, Signac, or Seurat that they cared for nothing except the momentary effects of light on objects as it is to say of Cézanne or Gauguin that they simplified objects to express the emotional significance which lies in things. All of them were equally interested in the character of the thing painted, and if the emotional significance which lies in things can be expressed in painting the way to it must lie through the outward character of the object painted.

Again Cézanne's and Gauguin's method was just as subtle and complicated as the methods of the others. Cézanne differed in using various tints of the same colour instead of the broken colour used by the others in a greater or lesser degree as it suited their purpose at different times, Seurat and Signac, of course, carrying the broken colour method to its furthest point.

Signac more than either Gauguin or Cézanne simplified nature, reducing it to a series of silhouettes filled with colour. A future generation, forgetting their quarrels, forgetting the names they were given, and the names they gave themselves, will certainly find them much more closely linked together than we are able to. It is possible to imagine them seated at a round table. Gauguin between Degas and Pissarro, Cézanne, Van Gogh, Seurat, Signac, Monet, and so on round to Manet, Renoir, and Degas again.

Of the present-day division of the exhibition it is not possible to say very much, none of them being very well represented. Piccasso, and Matisse have nothing there of any great interest, or that gives any clue to their work as a whole. Frietz's *Balcony*, with the river behind, is extremely interesting. Marquet (108–114) Serusier (93, 95, and 96) and Puy (147) are artists of whom we would like to see more. Such painting as Herbin's *Maison au Quai vert* (108) arouses our curiosity, as also do the paintings of Deraud. Let us hope next time for an entirely modern and representative exhibition of French painting.

---

25. Henry Holiday, 'Post-Impressionism'

Nation, 24 December 1910, 539

Holiday (1839–1927) began exhibiting at the Royal Academy in 1858. He illustrated Lewis Carroll's *Hunting of the Snark* but was best known for his *Dante and Beatrice* (1883) now in the Walker Art Gallery, Liverpool.

Sir, – I was moved by Mr. Roger Fry's article in your issue of December 3rd, and by my interest in new developments in painting or music (as in many other things), to go and see the work of the Post-Impressionists. I have heard in my time so many things abused which to me seemed full of interest, that I went with an open mind, wondering if I should find signs of a new perception underlying some imperfections and eccentricities such as Mr. Fry himself admitted.

Mr. Fry's enthusiasm is aroused to the highest pitch by Cézanne, who, he observes, 'is incomparably greater than I had supposed.' In the 'human form' he is 'supremely classic,' and 'he seizes individual character in its broad static outlines.'

A picture of some bathers seemed to promise a good example. It was a small picture with three or four small figures. I am bound to admit that these were recognisable as human figures, and some objects behind as trees, but, short of that, they were as nearly formless as possible – feeble and flabby, painted with patchy color, expressing nothing. The man on the right has a black eye and a great blobby nose, suggesting that he has just come out second-best at a prize-fight. But in one respect Mr. Fry's words are literally true; we have here 'broad static outlines' and no mistake; it is true that the outlines might be dynamic for all that I could tell, but though the figures are small, the outlines are in parts a quarter of an inch broad and black. I have sometimes seen bathers, but not being a Post-Impressionist, I failed to see thick black lines round their limbs. I suppose the perception of these lines is not given to ordinary intelligence; only those who possess 'intellectualised sensual power,' or who have become 'able to build up and, as it were, re-create form from within' can be expected to recognise such beauties. They appear to be outside the figure, but no doubt really belong to the 'inner life.'

Let anyone who seeks for real beauty look at this picture and recall the memory of Fred Walker's *Bathers* – fresh, sunny, and exquisite in form and color.

And yet this is, if anything, rather a favorable specimen of the works hung at the Grafton Gallery. After going over the whole exhibition with some care, one feeling predominated – that the thing is an impudent sham. Mr. Sadleir's remarks on the absurdity of men imitating children in order to ape primitivism reminded me of my own exposure of the sham medievalism in tradesmen's stained glass. 'If a man were to fulfill the command to "become as little children" by wearing a child's frock, short socks and shoes, and by imitating a child's toddling walk, lisp, and language, he would be the intellectual counterpart of the glass-painting trades-men, who thought that they were following the principles of medieval artists by making childish caricatures of their manner-isms. The mimicry of children by grown men is only practised in real life at the pantomime. The cathedral is considered the proper place for the corresponding antics where tradesmen's art is concerned.'[1] We must now add a Post-Impressionist Exhibition as another stage for the display of similar absurdities. If space permitted, it would be pleasant to speak of a few pictures in the gallery which do possess charm of color and atmosphere, but these are evidently regarded as oldfashioned by the true Post-Impressionists.

No young painter who has fallen in love with Nature will be seduced from his pure affection by such stuff as this. It is satisfactory to note that many of the pictures are from twenty to thirty years old, and yet the 'school' is still practically negligible.

NOTE

1 Michael Sadleir, *Stained Glass as an Art*. [Holiday's note]

26. Holbrook Jackson, 'Pop Goes the Past'

T.P.'s Weekly, 16 December 1910, 829

Jackson (1874–1948) became joint editor of the *New Age* in 1907, was acting editor of *T.P.'s Weekly* between 1911 and 1914, becoming editor in 1914. Like Frank Rutter (but unlike Fry) he stressed the revolutionary aspect of Post-Impressionism which he saw in a larger European context than most of his contemporaries. He was also aware of the Futurist enterprise. The Futurist Manifesto had been published in the *Tramp* (August 1910) but the pictures were not seen in London until 1912.

Quite recently I was crossing the Place du Carrousel with a well-known British artist; it was his first visit to Paris, and I felt that he was somewhat in my charge. My sense of responsibility ventured a little further than mere guidance from one point to another, and every now and then I felt called upon to indicate some historic building or other mark of the past. But my enthusiasm for such things – never very great, let me say – was not encouraged. As we crossed the Place du Carrousel, however, it was inevitable, and perhaps natural, that I should indicate the Louvre beyond the pleasant greenswards, and the ugly statues on our left. My artist friend was unmoved. I thought he did not quite realise what the word Louvre actually meant, so I annotated my information. 'It is the place,' I said, 'where they keep the great art collections of the past – the temple of the Venus de Milo and of the Olympia de Manet' 'To hell with the Louvre!' he said, earnestly, but cheerfully. 'Let's get on to the Salon d'Automne, and see the Art of the Future.'

I reproduce his words in all their limpid brutality, because they represent a modern point of view which cannot be otherwise expressed. The attitude of mind has, of course, existed in past ages; it has, indeed, existed in every age that stands out with any sort of distinction, but never before has this attitude been deliberate and self-conscious. The expression of my friend is but one hint of a spreading and deep-rooted revolt against the past. The most lively spirits of to-day are up in arms against what is traditional, what is established, what is classical. The tiny upheavals in modern politics, which fill the newspapers of the hour with consternation and dismay, are but the far-away echo of a movement beginning deep down in the soul of European ideas. The leaders of the revolt know little and care less about politics. They are not organised, they laugh at organisation. They do not even know each other. Their movement is not a movement, it is an impulsion towards an altogether new interpretation of life. It believes in nothing that was, in little that is, and it is even suspicious of that which is to be. It has no leaders, but here and there you may come across dark and forbidding names, like Bakounin and Max Stirner, which may be taken as keynotes of the new intellectual cacophony. Beside the fierce gospels of such men, Bernard Shaw sounds like Samuel Smiles and Friedrich Nietzsche like Isaac Watts. But Bakounin and Max Stirner are dead, and their books almost unknown; still, in no other books do you get the fervid hatred of the past which is more than beginning to dominate European art. In Warsaw and Munich, in Paris and Brussels, and Milan there are workers in all the arts who defy tradition in every sweep of the brush, in every stroke of the chisel, in every variation of tone and word. England is as yet untouched by these new ideas; intellectually we are still at the tail-end of the realist movement which produced the great Impressionists: Rodin in sculpture, Ibsen in drama, Whitman in poetry, Manet and Degas in painting, Zola in the novel, Richard Strauss in music, and Nietzsche in philosophy. In Germany, in France, in Italy the discussions about these creators of new values are over. It is recognised that the Impressionists have done their work – it is established safely in the opera houses of the people, the popular libraries, the public museums, and the Louvre.... The Objectors hold the forums of Europe to-day, and their texts are not Nietzsche and Ibsen, Strauss and Cézanne, but Post-Impressionism and Futurism.

We in England are beginning to hear something of both these subjects. The exhibition of Post-Impressionist paintings at the Grafton Galleries has attracted a considerable amount of attention; people are still going to the show to have a good laugh, just as they used to go to the Impressionist Exhibitions in Paris. How they did laugh at Manet! – and Manet is now in the Louvre! Still, some who go to the Grafton to laugh remain to pray, and at least one English art critic has publicly announced that he is on the eve of conversion. Post-Impressionism is not, however, mere repudiation of the past; it is something more than that, something greater. Post-Impressionism is the affirmation of the present; it is the hot portrayal of feeling following perception; not the thing seen, mind you – that would be Impressionism. The Impressionists painted what they saw, and they are of the past. 'We've got far ahead of them,' they say on Mont Parnasse. The Post-Impressionist paints what he feels about the thing seen; he cares not whether you like what he feels, or whether he communicates his feeling to others. His work is the result of a half-frenzied desire for individual expression; it considers nothing but itself and its need. All the old conceptions of art are ignored; his work is quite traditionless. And he has as little reverence for beauty as he has for the antique. 'Beauty is all very well in its way,' he will tell you, 'but what has it got to do with art?' These Modernists are trying to do for Europe what Walt Whitman tried to chant his own fellow-countryman into doing for America. 'The cleanest expression,' said the poet, 'is that which finds no sphere worthy of itself, and makes one.' That is what the new movement is doing. In the immediate past the artist has looked with suspicion upon science, and in so doing he has forgotten his vocation; for it is one of the functions of art to anticipate life, and he could not do that in a scientific age by ignoring the moving spirit of that age. Supremely great artists like Michael Angelo and Leonardo da Vinci did not ignore science; they used it. And it is in the use of science that the most modern painters are distinguishable from the back numbers. The Post-Impressionist movement is as scientific as it is artistic; many of the most astounding results of its leaders are due to the scientific recognition of the laws of vibration in regard to light and colour and form.

27. Roger Fry, 'A Postscript on Post-Impressionism'

Nation, 24 December 1910, 536–7

In this article Fry responded to some of the criticisms of Post-Impressionism voiced in the press in November and December 1910, and at the same time gave some idea of the scope and intensity of the feeling which the exhibition roused in artists and critics.

I am asked by the Editor to reply to some of the critics of the Post-Impressionist Exhibition and of my remarks about it. Mr. Robert Morley says that all the abuse of the Post-Impressionist which has come from certain quarters is more than justified. Having thus thrown the great weight of his name into the scale against these unfortunate artists, all that I can do is to pile on to the other scale such names as Degas (who owns several Gauguins), Dr. von Tschudi, Mr. John S. Sargent, and Mr. Claude Phillips (at least as regards Cézanne), Mr. Berenson, Mr. Herbert Horne, and Mr. Charles Loeser, in the faint hope that the balance may be redressed. I ought, in fairness, to allow Mr. Morley the German Emperor as a fellow-passenger. Having done this, I must leave posterity to read the verdict. Such weighing of names does not appeal to me as a very useful proceeding, but there is, alas, no other way of meeting Mr. Robert Morley's *ipse dixit*. Had he condescended to argument, I might have opposed him more profitably. He is quite right, by-the-by, when he compares their pictures to the work of the Benin artists, but then I must differ quite as strongly from his contemptuous view of these.

Next comes Mr. Sadler, who is temperate and reasonable, and whom it would require much space to answer completely. I fear the difference between us lies quite as much in our estimate of Cimabue as it does in our reaction to the Grafton Gallery. I cannot find Cimabue's technique 'defective'; on the contrary, it seem to me masterly to the highest degree. I do not think his art 'struggles into defective expression.' I think it is complete and perfect expression of something which no one else, either before or since, has ever said. In fact, his pictures (among which I agree with Mr. Sadler in counting the Rucellai Madonna) are works of art. That is to say, they are final and complete expressions of certain spiritual experiences. What is defective in Cimabue is not technique, but representative science. He has enough of this to say what *he* wants, but not enough to say what Mr. Sadler, with Titian and Velasquez impertinently intruding from the back of his mind, demands. Our habitual way of looking at early art historically, with a table of dates in our mind, tends, unless we take special precautions, to make us unable to see the work of art itself. If only our art historians would look at the old masters as though they were contemporaries, we should probably get some very instructive criticisms and some very frank admissions of a kind to horrify the conventionally cultured.

When Mr. Sadler goes on to ask whether Van Gogh 'burned with the same passion as Cimabue?' I must certainly answer no. No two artists burn with the *same* passion, their passions are aroused by different things, and colored by their different personalities. If he means, Is Van Gogh's passion as pure and as intense as

Cimabue's? – the question becomes one of great delicacy and scarcely to be answered off-hand. I should be inclined to say that it was as intense, but much less simple, much less serene, more troubled by the conflicts and ironies of modern life, more tortured and less healthy. This may amount to admitting either that the conditions of Van Gogh's life were less felicitous than Cimabue's, or that his temperament was less harmoniously composed, or, perhaps, both. This admission may even make me go further and say that I think Cimabue's achievement the more noble of the two, but that does not prevent me from being grateful that, in an age of vulgar commercialism in art, so passionate a spirit as Van Gogh's did arrive at beautiful expression.

Mr. Sadler finally contends that Mr. Augustus John has succeeded in doing all that the Post-Impressionists did without any open breach with tradition. Now I have always been an enthusiastic admirer of Mr. John's work. In criticising the very first exhibition which he held in London I said that he had undeniable genius, and I have never wavered in that belief, but I do recognise that Mr. John, working to some extent in isolation, without all the fortunate elements of comradeship and rivalry that exist in Paris, has not as yet pushed his mode of expression to the same logical completeness, has not as yet attained the same perfect subordination of all the means of expression to the idea that some of these artists have. He may be more gifted, and he may, one believes and hopes, go much further than they have done; but I fail to see that his work in any way refutes the attainments of artists whom he himself openly admires.

Next let me take Mr. Henry Holiday's letter, though I should prefer to leave it unanswered.

Mr. Holiday criticises Cézanne's *Bathers* entirely from the point of view of representation. He thinks nothing can have justification in a picture which does not happen in nature. From this point of view Cézanne's *Bathers* deserves all that he says, but unfortunately Mr. Holiday proves too much. Almost all that he says would apply equally to a drawing of the *Virgin and Child* by Raphael. Who ever saw, Mr. Holiday would say, if he were to maintain his position with unfailing consistency – who ever saw a woman with two or three lines round the oval of her face, who ever saw the line of the skull under the hair, who ever saw a number of parallel black scratches across her cheek? Such or such like would surely have to

be Mr. Holiday's criticism. He forgets that Art uses the representation of nature as a means to expression, but that representation is
not its end, and cannot be made a canon of criticism. I believe that
there is no symbol for natural appearance employed by the painters
at the Grafton Gallery which is one whit more disconcerting than
the symbols of the shaded pen drawing with which we have
fortunately become so familiar that they cause us no inconve-
nience. We have learned to read them with perfect ease, and I feel
sure that the same will happen as regards the paint symbols used
by Van Gogh, though I am not the least surprised that their
unfamiliarity makes them a stumbling-block for a time.

Mr. C.J. Holmes, the Director of the National Portrait Gallery,
has put out a little pamphlet on the exhibition which I believe is on
sale at the Grafton Gallery. Of this, thanks to his courtesy in
sending me an advance copy, I can speak here. It is welcome as
being a quite sincere and open-minded effort at a just appreciation
of the works on view, and of the artists, so far as they are
represented at the Grafton Gallery. But Mr. Holmes seems to me
too much of the schoolmaster. He goes round with a set of
principles, applies them in turn to each of the pictures, and reads off
the result. He finds Cézanne clumsy, frequently incoherent, and,
though a sincere artist, of modest rank. I confess the idea that
Cézanne is clumsy surprises me. His feeling for facets of color
prevents him from using a flowing or sinuous brush stroke which
would inevitably break up the severe architecture of his planes, but
I see no evidence of clumsiness, given the particular feeling for
form which is personal to him. On the contrary, the quality of his
pigment seems to me singularly beautiful. As the director of one of
the largest Continental picture galleries once said, 'You know, we
really like El Greco's handling, because it reminds us of Cézanne's.'
I cannot withhold admiration for the courage of Mr. Holmes's
patronising estimate of Cézanne, in view of the almost complete
unanimity of opinion among foreign critics in giving him a much
more exalted position.

Towards Van Gogh Mr. Holmes relents a little. He is even now
and again carried away with real enthusiasm, and forgets to correct
his errors. But what does he mean by saying that the yellow
background of the irises 'renders the picture useless for the
decoration of any ordinary room'? It would be so much easier and
more desirable to alter the room than such a picture as this. All

through Mr. Holmes's criticism runs the question whether a picture is 'serviceable' or not, a question which suggests an odd idea of the artist's function, as purveyor to the conveniences of the middle classes. It was certainly not so that these artists worked, nor so, I believe, that any noble or lasting creations were made.

Gauguin naturally gets rather good marks from Mr. Holmes. His qualities are, indeed, more easily estimated by the critic's measuring line than those of more elusive and spontaneous artists. Vallotton is, I think, much overpraised. To me he appears altogether too 'serviceable,' while the treatment of Matisse seems, to me, to show a misunderstanding of his aims. Mr. Holmes compares him somewhat contemptuously to Holbein, a comparison which is not really illuminating because of the complete difference of aim of the two artists. Nor has Mr. Holmes felt the importance of Derain's and Vlaminck's work. These seem to me to be among the most remarkable of all the contemporary men. Derain, in particular, shows a strange and quite new power of discovering those elements in a scene which appeal to the imagination with an immediacy comparable to that of music. His *Deserted Garden*, No. 118, expresses the essential emotion of such a scene, which the presence of any particularised or actual forms would inevitably weaken. It is here that I think we may find the main achievement of the Post-Impressionist artists, namely, that they have recognised that the forms which are most impressive to the imagination are not necessarily those which recall the objects of actual life most clearly to the mind.

The first Post-Impressionist exhibition had been an enormous success (as nos 31 and 32 record). It closed on 14 January 1911 by which time the hysterical reaction to the works on show had given way to a more measured response. But the spring and summer of 1911 were marked by a significant polarisation of attitudes to modern art amongst critics. What had previously been dismissed as a passing phase or a Continental freak by the detractors now began to pose a real threat to English art – especially when, almost immediately, the influence of Post-Impressionist techniques was observable in exhibitions of modern British painting. Those who supported the modern view, on the other hand, also had time to gather their forces and to publish longer and more considered critiques of Post-Impressionist painting. Throughout the year much greater British interest was shown than ever before in current exhibitions in Paris, and in the summer of 1911 John Middleton Murry and Katherine Mansfield produced the first issue of the magazine *Rhythm* with the intention of bringing before the British public the latest movements, developments and ideas from the French capital.

28. John Singer Sargent, 'Post-Impressionism'

Nation, 7 January 1911, 610

Sargent (1856–1925) had, since his settling in London in 1884, become internationally famous as a society portrait painter. He became an RA in 1897 and disliked his name being associated with those who supported the Post-Impressionist painters.

Sir, – My attention has been called to an article by Mr. Roger Fry, called 'A Postscript on Post-Impressionism,' in your issue of December 24th, in which he mentions me as being among the champions of the group of painters now being shown at the Grafton Gallery. I should be obliged if you allow me space in your columns for these few words of rectification.

Mr. Fry has been entirely misinformed, and if I had been inclined to join in the controversy, he would have known that my sympathies were in the exactly opposite direction as far as the novelties are concerned that have been most discussed and that this show has been my first opportunity of seeing.

I had declined Mr. Fry's request to place my name on the initial list of promoters of the Exhibition on the ground of not knowing the work of the painters to whom the name of Post-Impressionists can be applied; it certainly does not apply to Manet or to Cézanne. Mr. Fry may have been told – and have believed – that the sight of those paintings had made me a convert to his faith in them.

The fact is that I am absolutely sceptical as to their having any claim whatever to being works of art, with the exception of some of the pictures by Gauguin that strike me as admirable in color, and in color only.

But one wonders what will Mr. Fry not believe, and, one is tempted to say, what will he not print?

29. Walter Richard Sickert, 'Post-Impressionists'

Fortnightly Review, January 1911, n.s. xcv, 79–89

Sickert (1860–1942) had worked with Degas in Paris in 1883 and after a period of painting in Dieppe between 1900 and 1905 founded the Fitzroy Street Group which later formed the nucleus of the Camden Town School. He had long been familiar with French painting of the nineteenth century, and viewed Fry's championship of Post-Impressionism as a form of opportunism. For Sickert the differences between the artists who made up the so-called Post-Impressionist movement were more significant than their similarities. This article is the text of a lecture which Sickert gave in the Grafton Galleries just before the exhibition closed in January.

Mr. Roger Fry and his committee have earned the gratitude of all painters, students, and lovers of art in this country by the illuminating and interesting collection they have formed at the Grafton Gallery. That they have entitled it 'Manet and the Post-Impressionists' is a detail of advertisement. Only those who have never had to decide on what I may call poster-editing will quarrel very seriously with him on this score. That he has included in a collection of many great works a few nonsense-canvases and a few nonsense-bronzes, leaves me undisturbed. None of us, not straight out of the egg, take the leaders or the 'tendencious' arrangements of news in the daily papers seriously. This does not prevent almost any copy of any daily paper from containing enough of mere news to guide us sufficiently towards the judgements we desire. Perhaps the very nonsense element in this exhibition, and in the claims gravely put forward for it with some ingenuity of undergraduate ratiocination, has its utility. It has caused a rumpus. The rumpus has collected a crowd, and the crowd is quite ready to listen to reason and to learn. Who knows whether, without the rumpus in question, and the consequent crowd, I should at present have the honour of addressing a cultivated audience of my countrymen again on the subject nearest my heart? Monsieur Matisse obligingly parades before the Grafton Street booth with a string of property sausages trailing from the pocket of his baggy trousers. John Bull and his lady, who love a joke, walk up, and learn a few things, some of which have been known in Europe for a decade, and some for a quarter of a century.

We are citizens, and nothing is gained by denying it, of a country where painting forms no living part of national life. Painting here is kept alive, a dim little flickering flame, by tiny groups of devoted fanatics mostly under the age of thirty. The national taste either breaks these fanatics, or compels them to toe the line. The young English painter, who loves his art, ends by major force, in producing the chocolate-box in demand.

I have never indulged in gibes, and I never will, at the popular academician and his annual 'Picture of the Year.' I had occasion the other day to see a spirited and intelligent copy of *The Weavers* of Velasquez. Guessing here and there as to the authorship among the possible men whose work I respect, I was told at last that it was by Edwin Long, the author of the works we know. No example could be more striking and complete. Successful shade, accept my hand in fraternal contrition! We are druv' to it. John Bull will have it so. *Tu l' as voulu John Dandin!* And his lady still more! Let us toe the line, my brothers, and invest with care. *Londres vaut bien une messe.*

What is it I say has been known in Europe for a decade or for a quarter of a century? Manet, to begin with, and the strange importance of Cézanne, partly intrinsic, and partly relative. The Salon d'Automne of some years ago revealed to the stupefied admiration of the world the life-work of Gauguin. Van Gogh we appraised in the early 'eighties. M. Blanche, in a letter to the *Morning Post*, thinks it of critical relevance to emphasise the theory that Van Gogh was a Jew, and, what appears to make the matter worse in the eyes of my brilliant and talented friend, an apparently intolerable aggravation, a Jew from Holland at that! Truly it is difficult for the fashionable portrait painter to be a just critic! A lifetime spent in *tête-à-tête* with the *femme du monde*, his customer, cannot but tinge his views of life and art. Mr. Ross struck in the *Morning Post* with no uncertain trumpet, the protectionist note on this exhibition. Mr. Ricketts, I suspect, is merely naughty and knows better, a delightful and witty *advocatus angelorum*. I said to myself, 'We shall have Sir William Richmond and Sir Philip

Burne-Jones.' '*Pan!*' as they say in France, '*ça y était.*' 'There
remains now,' I said to myself, 'only the regularly recurring Mr.
Wake Cook. '*Tac!*' which I must explain is the Venetian for '*Pan!*' I
have lived so much abroad.

Mr. Fry is nothing if not an educationist and an *impresario*. I offer
him the following scheme for consideration. Before the end of the
exhibition now open at the Grafton, let him bare one of the best
walls, and hang, in their order as correspondents, works by Mr.
Ricketts, Sir William, Sir Philip, Mr. Wake Cook, and M. Jacques
Blanche. If this arrangement will make it appear puzzling that Mr.
Ricketts, brilliant and spoilt child of Delacroix and Daumier as he
is, should be so anti-Gallic, the other antipathies will, perhaps,
seem natural, and the paintings will constitute a useful and
complementary appendix to the letters in the *Morning Post*.

Mr. Fry, and the writer of the preface to the Grafton Gallery
catalogue, and later, Mr. Lewis Hind, in the *English Review*, with
the freedom that belongs to literature, affect to take Monsieur
Matisse and Matissism in bronze and in paint seriously. I gather
that Mr. Berenson has done the same. In my younger and less
composite days, before I had soiled my fingers with ink, I
remember how those of us who were purely painters were shocked
at the levity and irresponsibility of most writers' steps on the
flower-beds that to us are sacred. I remember, with another earnest
inquirer, a painter of talent, addressing to a brilliant novelist and
critic *d'avant garde*, some question like the following: 'But do you
really think so-and-so?' 'I don't know that I do, but I thought that
that point of view would make rather a nice article.' Naïf children
of the palette, my friend and I had forgotten that his article was his
picture, and that he had the same right, on paper, of distortion, of
deformation of fact, as we claim on canvas. If a critic did not
arrange and touch-up, underline and suppress, as it suits his prose,
the subject in hand, it would make dull reading. A critical pontiff,
who dines out for Art with a capital A, must prime the bomb of
paradox he has prepared for the dowager he takes in to dinner with
fairly strong, and above all, unexpected stuff, or it will miss fire. M.
Blanche says that M. Matisse should be grateful to Mr. Fry for
essaying to transplant a carnival reputation on to English soil. The
obligation is all the other way. Art stuff is devilish dull copy. M.
Matisse was sent by a merciful providence to enable us poor critics,
gravelled for lack of matter, by gravely professing admiration for

patent nonsense, to *épater* more successfully than we could by discovering a new Rembrandt.

I am, unfortunately, debarred from the piquant satisfaction of making these effects. Several times a week I have to face a critical audience of some fifty-odd students, who look to me for guidance over a field that extends from the elements of drawing, to the finest and most debateable points in style and execution. I have to send them to the museums to look at Rowlandson and Leech and Cruikshank, at Keene and Beltraffio, at Ingres and Millet and Degas, while suggesting that, for the present, they might do worse than avert their eyes from the too numerous reproductions of drawings by Alfred Stevens, with which there is just now a fashion to over-nourish our academic walls. These counsels of mine I have to back up with comprehensible reasons, or I should hear of it.

We are to suppose then, that one of my students, a young man who is making considerable sacrifices to work as many hours a day as he can at drawing, asks me, 'What about Matisse and Picasso?' say. I cannot pull his leg, as a diner-out for Art would a dowager's, nor can I reply in a bright article without serious consequences, as Mr. Hind or Mr. Berenson might.

I should be obliged to say, 'If you will look at M. Matisse's drawings, you will see that he has acquired the most fluent school-facility, just the kind of school-facility that you do not find in good drawings or great drawings. The great artist is humbler, and a shade clumsier, *un tantinet plus gauche*. Matisse has all the worst art-school tricks. Just a dashing hint of anatomy is obtruded; and you will find a line separating the light from the shade. You know what we think of that trick. The instinct of self-preservation, conscious or unconscious, must have dictated to him that this slickness of empty perfection, of a poor order, would never make its mark. So we have wilful deformations, wilful distortions, either the glutei maximi or the abdomen inflated like a balloon, or pectorals like hat-pegs. These distortions arrest if they do not please, and the, consciously or unconsciously, desired end is attained in the bronzes as in the drawings. In the paintings, unrelated colours tell us no more than the empty drawings do.

'If you look at Picasso's little nude girl with the bowl of geraniums, you will see a quite accomplished sort of minor international painter. Like all Whistler's followers, he has annexed Whistler's empty background, without annexing the one quality by which Whistler made his empty backgrounds interesting, the relation of colour and tone. The child looks a little like a sawdust doll, but a very animated and very *chic* little sawdust doll. Why a nude child is carrying a bowl of flowers in front of grey vacancy were, perhaps, an old-fashioned question to ask. I understand the tip has gone round that pictures need have no sense. In any case, here we have quite presentable gallery accomplishment. Turn from this to the nonsense-portrait of M. Saget by the same hand, and we see a superficial and very feeble caricature of Cézanne's failings. We used to call the room of these things in the Salon d'Automne, *la salle des fauves*. Here we have a *faux fauve*, a sham wild beast. I have heard the defence put forward for this stuff that these painters are so gifted that they have done everything that accomplishment can achieve, and that therefore only monstrosity is left to interest them. There is one thing they have not done that is work of fine quality. If they had, they would not have left it off so soon, and they would have found it takes a lifetime to develop a tiny talent to its utmost. An instructive sidelight has been thrown on the whole Matisse movement by one of the group, who having attracted attention for a few years by being a *fauve*, is now doing a quiet, but steady, little trade in portraits in the manner of Sargent for Brough. Look at *Le château* (149), by Maurice de Flamineck, after examining a landscape by Cézanne. I shall have occasion to speak later of Cézanne's violent and persistent restraint of his own exuberance. Examine a patch of sky or foliage by Cézanne, and you will see the meticulous labour, the Benedictine application of fitting strip to rectangular strip, in the search for infinitesimal variation of what he called accords of colour. In the Flamineck we have what has been well called in French *la blague extérieure de la chose*. There also we find rectangular strips, gradations which assume in a sky, as in Cézanne, a disconcerting suddenness, as of a blot of ink on a page; only there are no accords, there is no observed and subtle variety, there is no travail, no sweat and no groans. There is a jaunty and superficial imitation of a style which chances of association, and be it said, Stock Exchange speculation, have placed in a fashionable position on the European and American markets.'

I should regret that Marquet, a real painter, allows his work to be compromised by such company, if it were not that nothing can compromise beauty. His *Notre Dame* stands out clearly as serious painting by a born colourist. Valotton we know and respect, and I am not sure that Laprade is not a sketcher of talent like our Brabazon.

I notice that when an English writer is faced with a question about Cézanne, he waves the inquirer gracefully forward to another department, and begs Monsieur Maurice Denis to serve the gentleman. Mr. Fry even, called upon for explanations, clears his throat and – translates Maurice Denis, admirably, I need not say, quite admirably.

Before discussing Cézanne it seems to me convenient to lay down for reference the law as to deformation or distortion. I have never found this done by any writer on art – I should be grateful if anyone who has would forward me a note on the subject. Not only have I not found it done well, I cannot find that is has been done at all; and yet it is of the highest importance. A right understanding of this law must govern all critical consideration of painting or sculpture, and what is still more important, must affect teaching, for good, or for most nonsensical ill.

Deformation or distortion in drawing is a necessary quality in hand-made art. Not only is this deformation or distortion not a defect. It is one of the sources of pleasure and interest. But it is so on one condition: that it result from the effort for accuracy of an accomplished hand, and the inevitable degree of human error in the result.

The departures from the geometrical forms intended in an Eastern carpet are not intentional. The design is strictly adhered to, but instead of being dead, as are machine-made designs, it may be said to breathe. It may be said to have the variation of individual gesture that a regiment standing at ease displays. The ranks remain unbroken in their frame.

Cézanne was fated, as his passion was immense, to be immensely neglected, immensely misunderstood, and now, I think, immensely overrated. Two causes, I suspect, have been at work in the reputation his work now enjoys. I mean two causes, after all acknowledgment made of a certain greatness in his talent. The moral weight of his single-hearted and unceasing effort, of his sublime love for his art, has made itself felt. In some mysterious way, indeed, this gigantic sincerity impresses, and holds even those who have not the slightest knowledge of what were his qualities, of what he was driving at, of what he achieved, or of where he failed.

Then we must remember that, if dealers cannot easily impose on the world as fine work, work which has not some qualities, dealers, and those critics who directly or indirectly depend on them, can to a great extent hold back or unleash a boom to suit themselves. In Cézanne there were all the conditions most ideal for the practice of great 'operations,' as they are called in Paris by the able 'brewers of affairs' who control the winds from their caves full of paintings.

'*Ah, Mademoiselle, je n'arrive jamais à faire quelque chose de complet,*' Cézanne said to someone I know. I can hear her imitation of his particular accent. Canvas after canvas was begun, worked on eternally, redrawn, worked on again, and abandoned anywhere, while the fury-driven painter pursued the perfection he had in his mind on new versions of the same problem. Cézanne was a rich man, these essays had no market value. He left them anywhere, as one leaves the shell of a walnut or a half-eaten apple. We know that twenty years ago le père Tanguy sold them, retail, at forty francs. Let us exaggerate and suppose their value, wholesale, to have been twenty francs. Operations which will turn a louis into four hundred louis are worth considering. What is that? Forty thousand per cent.? Decidedly, for a dealer, Cézanne was a great painter! The greatest of all perhaps. And if, of two unhappy apples standing by a shaky saucer, one is without the blue authenticating contour, would it be a very great crime to employ a talented youth to surround it – oh! for Germany or America? *C'est si loin tous ces pays là.* The greatest living painter, now an old man, was looking regretfully the other day at canvases that he knows he is not destined to finish. Thus, at least, a sympathetic visitor thought to interpret his sigh. 'No. It is not that. It is that they will take care to finish them for me when I am gone.'

For the exaggerated homage of a *cénacle* to a given master we have to look to other reasons. The Frenchman is nothing if not thoroughgoing, if not *entier*. A reaction throws a group of students as a protest from one set of errors into the arms of another. Claude Monet's too-much-advertised haystack series, Rouen cathedral series, and pond and water-lily series, reduced one of the theories of Impressionism, if not to the absurd, at least to the banal. Heard through the megaphones of the Rue Laffitte and the Press, the theory took on something of the patter of a syndicate. Monet was not the first to discover that, as the sun declines, new arrangements of light and shade arise. Older masters than he knew this as well as he did. But they also knew that, of the possible arrangements of light and shade on an object, one arrangement brings out the form and character, reveals the essential soul of the object, more than another. It seems to me that Monet's demonstrations required, for their full utility, that the series should be kept together and hung in sequence. If you come, as you do at the Rouen museum, upon an isolated canvas of the cathedral series (framed, by the way, in a terrible chocolate-coloured frame), the very limited information it gives as to colour is not sufficient to compensate for the vagueness of form in a subject which cries aloud for drawing above all things. Turn, at Rouen, from this to the exquisite early Sisley, with a long, low building standing by an inundated road, and we get the pure painter's poetry of Impressionism, before the cult had become Americanised.

The notoriety of these series had one effect. They were supposed to epitomise the impressionist doctrine. The part was taken for the whole. Two persistent errors may now be found in nearly all writers on the Impressionists. The first, that Impressionism was limited to effects of colour, and implied the absence of composition or pattern in a design. The second, that Impressionist pictures were all painted on the spot, or, as people say, 'from nature.' I am here using the word Impressionist, which I detest, exactly in the same precise and strictly limited sense as I would use the phrase 'member of Boodle's.' '*Comme nous avons mal fait de nous laisser appeler Impressionistes,*' everyone has heard Degas say. But the harm is done, and a critic writing in 1910 cannot but use the term. Pissarro, whose intrinsic and relative importance is not yet at all understood, even in France, executed many works in the studio, composed from studies in pencil or pastel or *gouache*. His etchings and his tempera paintings were done in the same way. The bulk of the work of Degas is founded, like the work of the old masters, on drawings.

It is absurd to call Cézanne a Post-Impressionist, embedded as he was in the Impressionist movement. Influenced at first by Delacroix, Daumier, and Courbet, he was drawn into the Impressionist group by his sympathies. In number 16 we can still see the influence of Daumier. In number 48, of the Grafton exhibition, we can see some traces of painting with the knife, the legacy of Courbet. Indeed, at one time Cézanne had some enormous palette knives made to order, so as to cover large spaces of canvas with paint. It is impossible to disentangle influence and counter-influence in the work of a group so closely connected as were the Impressionists. They spoke one language. They modified it daily in replies, and echoes, and quotations, tossed forwards and backwards in the heat of their eloquence. A fluid style was modified by daily and passionate reference to, and inspiration from, nature. If Pissarro influenced Cézanne, certain canvases of Pissarro, certain preferences of Pissarro, would not have existed except for his association with Cézanne. Is not Gauguin's art distilled from essences gathered shrewdly in the gardens of Degas, of Cézanne, and of Pissarro?

History must needs describe Cézanne as *un grand raté*, an incomplete giant. But nothing can prevent his masterpieces from taking rank. I fancy that is the sober truth. I remember one year, when there was a real unpacking, a *déballage* of, it seemed, all the studies by Cézanne in creation, catching a fragment of conversation between two men who passed me in the Salon d'Automne. '*Ils vont réussir à tuer Cézanne.*' The *mot* had point, but it was a half-truth. It is useful that we should see the hundred failures on which is built the hundred and first success. No doubt such exhibition is trying to the supergoose at whose next tea-party, where she is not, happiness awaits her. 'Dear Mr. Fry, or Berenson, or McColl, do tell me quickly, ought I to like these things or not, because I have got to go on to Lady So-and-so's?' 'Transient Madam, we can tell you nothing quickly. Toss up one of the sovereigns we see dangling in your purse of golden mail, and go on to the next party, or to the palmist.'

Of masterpieces by Cézanne, I recall a black marble clock in the Salon d'Automne, a still-life of fruit at the New Gallery some years ago, that made its pendent by Monet look thin. The landscape numbered eight at the Grafton is a marvel of tones in mother-of-pearl. The delicate but abrupt transitions lead the eye down to the shore and away round the magic bay without *trompe l'œil*. It is the painter who wields the bâton. It is he who conducts, and compels us to accept the time and the rhythm he chooses to impose.

I have never forgotten some words I heard Degas let fall in 1885 about the direction that was being taken in painting at the time, and his attitude towards that direction. 'They are all exploiting the possibilities of colour. And I am always begging them to exploit the possibilities of drawing. It is the richer field.'

That same summer I had speech, in a street in Dieppe, of a sturdy man with a black moustache and a bowler, who was destined to carry out the master's advice to some effect. My friend, Horace Mélicourt asked me to come and see a comrade of his, no longer a youth, who was thinking of throwing up a good berth in some administration in order to give himself up entirely to the practice of painting. I am ashamed to say that the sketch I saw him doing left no very distinct impression on me, and that I expressed the opinion that the step he contemplated was rather imprudent than otherwise. I lived to be the obscure person who saw, collected at the Salon d'Automne, three or four years ago, a posthumous collection of the work of the man in the bowler, and his name was Gauguin.

So intelligent a French painter as M. Blanche is surely making himself the mouthpiece of less-informed English prejudice, when he speaks of Gauguin as if he had abdicated the civilised French tradition of hundreds of years, and had imitated barbaric methods. I would invite M. Blanche to look again. To look first at the pre-Tahitian works, and their lessons exquisitely learnt from Degas and Pissarro. I would invite M. Blanche to reflect that Degas inherits the tradition of Ingres and Poussin, and Pissarro those of Millet and Corot. Does M. Blanche think it likely that the painter of *Bretagne* (26) and *Les coiffes blanches* (84) would throw away, for nothing, qualities that place him with the masters? Qualities of composition, of *esprit*, of drawing, of colour, and of execution? If Gauguin's profound and real penetration into the wonders of a strange civilisation, so different from ours, was a re-birth of inspiration, was the luxury of coming to life again in a new planet, is it probable that he would lay down the arms of the art which he had so magnificently acquired? If, in his Maori subjects, he simplified and transmuted his technique, taken largely from Cézanne, it was to become, in his young maturity, one of the finest flowers of modern European painting, or of modern French painting, which is the same thing: since all modern painting is founded on the French school. 'When I say "religion,"' as the lady said, 'I mean Christian. And when I say "Christian," I mean the Church of England.'

A strange grandeur has crept into Gauguin's figures, a grandeur that recalls Michaelangelo. Look at the poise of the figure holding a dish behind the little nun in the picture numbered 40. It will be a crime if number 42, *L'esprit veille*, or number 83, *Les laveuses*, or both, are not acquired for the nation. Here is National Gallery quality at its highest level. Was ever painted figure more sculpturesque than the awe-stricken Vahina prone on the couch, not daring to move in the haunted room? Has paint ever expressed perfect form more surely and with more fulness? Look at the muscle of the right forearm, and at the perfect hands laid out, plump and *élancé* at the same time. What an atmosphere of forest is created in *Les laveuses*! And with what sober means! Here is no war with the medium, no exasperation, no rebellion. The paint obeys the inspiration with the suavity of a mastered thing. All of the old weapons of the old masters are here. Gauguin disdains to use none of them, while he enriches them with a modern vision. Cunningly chosen and expressive silhouette is here. Learned chiaroscuro is here, no longer as in the later old masters in shades of brown, but again, as in the primitives, in the subtlest harmonies of lovely colour. And all these the painter accomplishes without distress, without raising his voice, without loading or embroiling his paint. This picture is sonorous because it is not shrill. The tones are brilliant because they are never colourless. Degas once said to me, 'If you want to sell your pictures nowadays, you must paddle in flake white,' *'patauger dans le blanc d'argent.'* Look at the figure in blue looking off on the prompt side of the picture, and holding a basket. Is not all freshness, all youth, all innocence, and all woodland poetry expressed in that one figure? The proper study of mankind is man; and no country can have a great school of painting when the unfortunate artist is confined by a puritan standard to the choice between the noble site as displayed in the picture-postcard, or the quite nice young person, in what Henry James has called a wilderness of chintz.

Which is more like the idea of the outlawed and persecuted Jew in his anguish in the garden, Gauguin's picture of the redhaired man in his strange setting, or the well-groomed image of our own Holman Hunt? Is the profound force and feeling of the former impious, and the levity of the latter conceit reverent? I am afraid Mr. Ross has no argument but the protectionist flag which he has nailed so bravely to the *Morning Post*. 'Burn the pictures of these triple aliens, and give the poor Academicians a chance!'

'And Van Gogh; why, he went mad, and we are actually insulted by being asked to subscription dances where such works hang on the walls!' Have not some quite dull painters gone mad also? It is like the expression 'the artistic temperament.' Haberdashers have been known to be regrettably irregular in their domestic and financial relations. Yet I have never heard invectives against the 'haberdashing temperament.'

I think Voltaire somewhere thus defines madness. If I remember him rightly, he says that madness is to think of too many things, too quickly, one after another, or to think too exclusively of one thing. I must say I wish Van Gogh would bite some of our exhibitors, who think, not of too many things, nor too insistently of any one thing, but as the yokel said, of 'maistly nowt.'

I have always disliked Van Gogh's execution most cordially. But that implies a mere personal preference for which I claim no hearing. I execrate his treatment of the instrument I love, these strips of metallic paint that catch the light like so many dyed straws; and when those strips make convolutions that follow the form of ploughed furrows in a field, my teeth are set on edge. But he said what he had to say with fury and sincerity, and he was a colourist. *Les Aliscamps* is undeniably a great picture, and the landscape of rain does really rain with *furia*. Blonde dashes of water at an angle of 45 from right to left, and suddenly, across these, a black squirt. The discomfort, the misery, the hopelessness of rain are there. Such intensity is perhaps madness, but the result is interesting and stimulating.

This exhibition, a very *frittura mista*, opens up more questions than can be even touched upon in an article. I have no space to speak of Girieud, who inspires interest; of Puy, a real painter, a true colourist.

Maurice Denis is altogether too large a question to be disposed of summarily. It would be necessary to make a study of Puvis de Chavannes, which would carry us too far afield. I must reserve these things for another occasion. As we have been considering the Impressionists, it may be well to remind ourselves that Puvis de Chavannes would not have been at all what he was but for the enormous influence that the Impressionist movement had on him. As it would be impossible to establish any logical thread running through the exhibition at the Grafton, any complaints as to absentees would be beside the point. Why Manet? Why skip the other Impressionists? Since 'post' is the Latin for 'after,' where are Vuillard and Bonnard? Where is Albert André? We must always remember that, if the innocent and none too discriminating enthusiasm of an English committee proposes exhibitions of this kind, it is the French dealer and the state of his stock which disposes. But not all the remainder biscuit of Manet's great studio can induce us to swallow Matisse as next-of-kin.

30. Roger Fry, 'Post Impressionism'

Fortnightly Review, May 1911, n.s. xcv, 856–67

This is the text of the lecture which Fry gave at the Grafton Galleries at the close of the first Post-Impressionist exhibition in January 1911.

I find a certain difficulty in knowing to whom I am to address myself in this lecture. The cultivated public is sharply divided by the question of Post Impressionism. I think that I may assume that there will be representatives here of three classes. First, those who, like myself, admire it with enthusiasm. Secondly, those who think that the exhibition is a colossal farce got up for the deception and exploitation of a gullible public. Some of these have expressed the desire that all the pictures now in the Grafton Gallery should be burned, and that I myself should be offered up upon the holocaust as a propitiation of the outraged feelings of the British public. Such expressions of opinion appear to me to be somewhat exaggerated, I might even have said hysterical, were it not that that is the word which has been applied to myself with a view to explaining the aberrations of a mind hitherto supposed to be fairly well balanced. Finally, there is the class of those who are frankly puzzled, and yet inclined to doubt the explanations of fraud or self-deception which are put forward by the second group. In the main, I mean to address myself to this last class, to the intelligent but doubting inquirer. But here let me put in a plea for tolerance. Suppose even, as is roundly declared by the adversaries, that there are in the exhibition paintings tinged with the sin of charlatanism, paintings executed by the artist, not from sincere conviction, but from a desire to flout and irritate the public. Let us suppose this for the sake of argument, even though personally I should be inclined to deny it. Now charlatanism is undoubtedly a sin on the part of an artist, it is a departure, and a lamentable departure, from the strict course of artistic probity, but even if it be a sin, it is one which is promptly punished; while there is another sin, another departure from the straight and narrow way, which is not only not punished, but constantly rewarded with titles, money, and social prestige, I mean the sin of compromising with the public demand for pictures which arouse curiosity or gratify sentimental longings. This sin is so frequently and so openly committed by the artists of modern times that we scarcely feel indignant at it, we certainly do not rush to denounce it in the papers in the way that has been done of late with regard to the work in this Gallery. Honestly, it appears to me a much more dangerous and insidious sin for the artist, and I would far rather be responsible for the hanging of works intended to flout the public than of any single painting in which I detected any desire to flatter it. In the main, then, I mean to address myself to the doubting inquirer, but first a word of consolation and encouragement to the adversaries. These think that an entirely fictitious, degenerate, and irrational mode of artistic expression is being foisted upon the public by means of advertisement and all those subtle arts of corruption which modern journalism has discovered, and that this iniquitous campaign is organised by a few speculative dealers. Now I do not believe that any such adventure can succeed, I believe that even in art, Abraham Lincoln's dictum holds, that you cannot fool all the people all the time – and we have an excellent example of this in the fate of that abortive movement known as *art nouveau*, which certainly had all the advantages which lavish advertisement in the more popular and less scholarly artistic reviews could give, and which has, nevertheless, so evidently and undeniably fizzled out. Let the adversary, therefore, take heart, if this movement is of the same kind I not only believe, but sincerely hope, that it will meet with a similar fate.

But naturally I do not think that it is of the same nature, that it is merely a new way of startling the public into attention and frightening it into purchasing these ingenious painters' catchpenny wares. I believe that even those works which seem to be extravagant or grotesque are serious experiments – of course, not always successful experiments – but still serious experiments, made in perfectly good faith towards the discovery of an art which in recent times we have almost entirely forgotten.

My object in this lecture is to try to explain what this problem is and how these artists are, more or less consciously, attempting its solution. It is to discover the visual language of the imagination. To discover, that is, what arrangements of form and colour are calculated to stir the imagination most deeply through the stimulus given to the sense of sight. This is exactly analogous to the problem of music, which is to find what arrangements of sound will have the greatest evocative power. But whereas in music the world of natural sound is so vague, so limited, and takes, on the whole, so small a part of our imaginative life, that it needs no special attention or study on the part of the musician; in painting and sculpture, on the contrary, the actual world of nature is so full of sights which appeal vividly to our imagination – so large a part of our inner and contemplative life is carried on by means of visual images, that this natural world of sight calls for a constant and vivid apprehension on the part of the artist. And with that actual visual world, and his relation to it, comes in much of the painter's joy, and the chief though not the only fount of his inspiration, but also much of his trouble and a large part of his quarrel with the public. For instance, from that ancient connection of the painter's with the visual world it comes about that it is far harder to him to get anyone, even among cultivated people, to *look* at his pictures with the same tense passivity and alert receptiveness which the musician can count on from his auditors. Before ever they have in any real sense *seen* a picture, people are calling to mind their memories of objects similar to those which they see represented, and are measuring the picture by these, and generally – almost inevitably if the artist is original and has seen something with new intensity and emotion – condemning the artist's images for being different from their own preconceived mental images. That is an illustration of the difficulties which beset the understanding of the graphic arts, and I put it forward because to understand the pictures here exhibited it is peculiarly necessary that you should look at them exactly as you would listen to music or poetry, and give up for once the exhibition attitude of mind which is so often one of querulous self-importance. We must return to the question of the painter's relation to the actual visible world.

I am going to assume that you will all agree with me in saying that the artist's business is not merely the reproduction and literal copying of things seen: – that he is expected in some way or other to *mis*represent and distort the visual world. If you boggle at the words misrepresent and distort, you may substitute mentally whenever I use them, the consoling word idealise, which comes to exactly the same thing. Now it is when we come to consider how far this distortion ought to go that our difficulties begin. Mr. Sickert, in his lecture here, suggested that the distortion should be entirely unconscious, due mainly to the incapacity of the artist to reproduce visible things exactly, though he characteristically suggested that the artist ought to draw in such a manner as would inevitably produce the greatest amount of distortion – the method of drawing from point to point. At the same time, Mr. Sickert pointed out that this distortion was precisely what characterised any work of art as opposed to any machine-made object. To me there seems something of Jesuitical casuistry about this: 'Distortion is inevitable, and it is even desirable as a characteristic of a work of art, but it must be always unwilling and unsuspected. Therefore, put yourself in such a position that you cannot possibly avoid it, and then try your utmost to prevent its occurrence.' Is not this rather like the Quaker's advice to his son: Thee must not marry money, but thee had better marry where money is.

The question of how much distortion – how much unlikeness to the totality of appearance – is allowable to the artist has always been a difficult one, and has been answered very differently at different times. I can remember a very sensitive judge of art, himself an artist, who belonged to the generation of Leighton and Frederick Walker, and who knew Italian art well. To him Titian and Raphael represented the minimum of naturalism possible. Being a very humble-minded man he used sadly to admit that there must be something in Botticelli and Mantegna, but for him the incompleteness of their representation was a fatal bar to accepting their revelations.

Our generation has moved on a step further in appreciation. It no longer finds any difficulty in understanding the symbols of the Italians of the fifteenth century. Rather it fully and freely enjoys them. This much distortion of nature – and do not forget that it is already a very great distortion – is perfectly allowable. We have even got, by means of a kind of archæological imagination, to give lip-service to the real primitives, to Cimabue and the Byzantines, and to the French sculptors of the twelfth century. I say lip-service because I notice almost always a kind of saving clause in people's admiration of these things – a way of saying how wonderful they are considering the time when they were produced – how interesting as a foretaste of the great art that was to follow, and so forth. Now this way of looking at a work of art, this evolutionary method, is, I think, entirely fallacious. It is the result of false analogies taken over unconsciously from our habits of thought when dealing with science. The work of a physicist of fifty years ago – say of a man like Joule – in so far as it is true, is completely absorbed in the works of more recent physicists. It is completely subsumed in the later work, and the later work replaces it entirely. It would make no difference now to mankind if every word Joule wrote were completely effaced. The law of the conservation of energy would none the less be the accepted basis of our thoughts about physics. But if the works of Giotto were destroyed, the fact that we still possessed the works of Raphael and Titian would afford us no sort of consolation or recompense. The human inheritance would be forever definitely impoverished. A work of art can never rightly be regarded as a means to something else, it is only rightly seen when regarded as an end in itself.

It so happens that the period of art that students generally concern themselves with – the period from 1300 to 1500 – is one which, as well as producing many great masterpieces, shows a continual and fairly steady progression towards the more complete science of representation. To those who undertake the paradoxical task of teaching art, this progression appears as a godsend. Among all those terribly elusive realities of human passion and feeling of which art is the triumphant but hardly decipherable record, here at least is a thing capable of easy and lucid demonstration, one upon which one may even set examination papers and give strictly judicial marks. And so art is conceived as a progressive triumph over the difficult feat of representing nature; a theory, which, if it were really believed, would put Meissonier above Raphael and Alma Tadema above Giotto. The fact is that changes in representative science are merely changes in the artist's organs of expression.

These are not the changes that matter most. The changes on which we ought to fix our attention are the changes in the feelings and sentiments of humanity, and I firmly believe that if perspective had never been invented, the art of the eighteenth century would have differed as profoundly from the art of the thirteenth as it actually does; and I am confirmed in this by the fact that Utamaro's prints, with their rudimentary perspective, belong just as decisively to the eighteenth century as the paintings of Boucher or Fragonard.

For the sake of argument I have perhaps exaggerated a little the indifference to the essential purposes of art of representative science, the naturalism which the artist makes use of. In reality it is not quite so simple as that – first, a certain amount of naturalism, of likeness to the actual appearances of things is necessary, in order to evoke in the spectator's mind the appropriate associated ideas. If I have to express the idea of a tiger attacking a man, it is essential that the spectator should realise the animal to be a tiger and not a hippopotamus – if, however, the given idea is merely a wild animal attacking a man, such doubts are different, and all that is necessary is the expression of ferocity and wildness. There is, therefore, varying according to the idea to be expressed, a real minimum of naturalness allowable – though I believe it is a very low one and corresponds fairly closely with the amount of natural appearance called up to the mind by words. But there is also a limit in the other direction. One may have too much naturalism for the expression of a particular idea. Let me take an example. Suppose the artist wishes to describe an armed crowd attacking a palace from which an emperor escapes in a carriage, dressed as a woman. Supposing he adopts a method of more or less complete naturalism which we are familiar with in modern art, he will be troubled by the fact that the policeman in the foreground, even if he does not obscure the principal actors in the drama, will occupy a quite disproportionate area of the composition. And this difficulty has made the composition of all ceremonial painting either impossible or ridiculous in modern times. But if the artist supposes himself to be suspended above the scene, and represents all the figures as seen from above, he may be able to get a composition expressing coherently the whole effect of the action; but then each in ividual will be seen from a very unusual and frequently ridiculous point of view, certainly one which will take away from the expressiveness of the figure. But if the artist frankly gives up any strict following of the laws of appearance, and groups the figures as though they were seen from above, and yet draws them as though seen more or less on a level, he will have a really adequate method of narrative composition. And such was the method employed by Japanese artists, and frequently by the miniature painters of mediæval times. We see here, then, the case of a dramatic idea which is much better and more lucidly conveyed by disregarding the laws of appearance than it could be by following them.

We come, then, to this, that it entirely depends upon the nature and the character of the sentiment which he wishes to convey, how much or how little naturalism the artist should employ. And I think we may say this, that those sentiments and emotions which centre round the trivialities of ordinary life – that kind of art which corresponds to the comedy of manners in literature – will require a large dose of actuality, will have to be very precise and detailed in its naturalism: but those feelings which belong to the deepest and most universal parts of our nature are likely to be actually disturbed and put off by anything like literal exactitude to actual appearance. It is not really the absence of naturalism which disturbs us, when we are disturbed, in an artist like Blake, but the introduction of a false and unfelt realism. But this is not all. When a high degree of completeness in the representation of things seen is demanded of the artist, his energies are usually exhausted in the mere process of representation. This becomes, indeed, a feat so difficult that its mere accomplishment rouses wonder and admiration at the artist's skill, and public and artist alike, left gaping at this wonderful tight-rope performance, forget that it is only a means to some quite other end, that art ought to rouse deeper and far other emotions than those with which we greet the acrobat or billiard player.

And this worship of skilful representation has had several bad effects upon art. It has caused the artist to abandon technical skill in the strict sense of the word. Technique, is usually now applied simply to skill in representation, but I mean here the actual skill in the handling of the material, the perfection of quality and finish. And if in that point the artists whose works are here exhibited compare unfavourably with the artists of early ages, the fault must be set down, at least in part, to the exigencies of that representative science which has resulted in the loss of the tradition of craftsmanship.

However, in every century a few men actually do come through the ordeal which our rage for representation has imposed; these men do succeed in actually saying something. Hence the worship of genius. Genius alone has the right to exist in the conditions of modern art, since genius alone succeeds in expressing itself through the cumbrous and round-about method of complete representation. The rest remain, not what they should be, definite minor artists, but often in spite of much talent and individuality, entirely ineffectual and worthless. They do not produce beautiful objects, but only more or less successful imitations.

But supposing the artist to be freed from the incubus of this complete representation – suppose him to be allowed to address himself directly to the imagination – we should get a genuine art of minor personalities, we might even attain to what distinguishes some of the greatest periods of artistic production, an anonymous art.

Now it is precisely this inestimable boon that, if I am right, these artists, however unconsciously they may work, are gaining for future imaginations, the right to speak directly to the imagination through images created, not because of their likeness to external nature, but because of their fitness to appeal to the imaginative and contemplative life.

And now I must try to explain what I understand by this idea of art addressing itself directly to the imagination through the senses. There is no immediately obvious reason why the artist should represent actual things at all, why he should not have a music of line and colour. Such a music he undoubtedly has, and it forms the most essential part of his appeal. We may get, in fact, from a mere pattern, if it be really noble in design and vital in execution, intense æsthetic pleasure. And I would instance as a proof of the direction in which the post impressionists are working, the excellences of their pure design as shown in the pottery at the present exhibition. In these there is often scarcely any appeal made through representation, just a hint at a bird or an animal here and there, and yet they will arouse a definite feeling. Particular rhythms of line and particular harmonies of colour have their spiritual correspondences, and tend to arouse now one set of feelings, now another. The artist plays upon us by the rhythm of line, by colour, by abstract form, and by the quality of the matter he employs. But we must admit that for most people such play upon their emotions, through pure effects of line, colour, and form, are weak compared with the effect of pure sound. But the artist has a second string to his bow. Like the poet he can call up at will from out of the whole visible world, reminiscences and remembered images of any visible or visually conceivable thing. But in calling up these images, with all the enrichment of emotional effect which they bring, he must be careful that they do not set up a demand independent of the need of his musical phrasing, his rhythm of line, colour, and plane. He must be just as careful of this as the poet is not to allow some word which, perhaps, the sense may demand to destroy the *ictus* of his rhythm. Rhythm is the fundamental and vital quality of painting, as of all the arts – representation is secondary to that, and must never encroach on the more ultimate and fundamental demands of rhythm. The moment that an artist puts down any fact about appearance because it is a fact, and not because he has apprehended its imaginative necessity, he is breaking the laws of artistic expression. And it is these laws, however difficult and undiscoverable they may be, which are the final standard to which a work of art must conform.

Now these post impressionist artists have discovered empirically that to make the allusion to a natural object of any kind vivid to the imagination, it is not only not necessary to give it illusive likeness, but that such illusion of actuality really spoils its imaginative reality.

To take a single instance. In the first room of the Gallery there hangs a picture by Manet, the bar at the Folies Bergères, in which there is a marvellous rendering of still life – marvellous in the completeness and the directness of its illusive power. In that there is a circular dish of fruit. Now the top of a circle seen in perspective appears as an ellipse, and as such Manet has rendered it. In a *nature morte* by Cézanne, hanging close by, there is also a dish of fruit, but Cézanne has rendered the top as a parallelogram with rounded corners. This is quite false to appearances, but a comparison of the two paintings shows one how much more vivid is the sense of reality in the Cézanne. I do not pretend to explain this fact, but it would seem that Cézanne has stumbled upon a discovery which was already the common property of early artists. Both in Europe and the East you will find the wheels of chariots, seen in perspective, drawn exactly in this way. It occurs in Japanese paintings of the thirteenth century. And you will find St.

Catherine's wheel drawn in the same way by Siennese painters of the fourteenth.

Or compare the girl in the Folies Bergères with Cézanne's portrait of his wife. In the first the modelling is elaborately realistic, however brilliant the short-hand in which it is expressed. In the second there is very little attempt to use light and shade, to give illusion of plastic relief. But none the less, I find that Cézanne's portrait arouses in my imagination the idea of reality, of solidity, mass and resistance, in a way which is altogether wanting in Manet's picture.

I do not pretend altogether to explain these facts; we have to find out empirically what does impress the imagination, the laws of that language that speaks directly to the spirit.

But there is perhaps one fairly obvious reason why the imagination is not readily impressed by anything approaching visual illusion, namely, that the illusion is never quite complete, never, indeed, can be complete, for the imaged reality has not the same proofs of coherence and continuity which appertain to actual life; it follows that recognising how near to actuality the illusive vision is, the mind inevitably compares the picture with actuality and judges it to be less complete, less real. It has for us only the reality of a reflection or echo. Now the world of the imagination is essentially more real than the actual world, because it has a coherence and unity which the actual world lacks. The world of the imagination, though more real, is much less actual, and the intrusion of actuality into that world of imagination tends to disturb the completeness of our acquiescence in it.

A great part of illusive representation is concerned with creating the illusion of a third dimension by means of light and shade, and it is through the relief thus given to the image that we get the sensual illusion of a third dimension. The intrusion of light and shade into the picture has always presented serious difficulties to the artist; it has been the enemy of two great organs of artistic expression – linear design and colour; for though, no doubt, colour of a kind is consistent with chiaroscuro, its appeal is of quite a different order from that made when we have harmonies of positive flat colour in frank opposition to one another. Colour in a Rembrandt, admirable though it is, does not make the same appeal to the imagination as colour in a stained-glass window. Now if it should turn out that the most vivid and direct appeal that the artist can make to the imagination is through linear design and frank oppositions of colour, the artist may purchase the illusion of third dimensional space at too great a cost. Personally I think he has done so, and that the work of the post impressionists shows conclusively the immense gain to the artist in the suppression or re-intepretation of light and shade. One gain will be obvious at once, namely, that all the relations which make up the unity of the picture are perceived as inhering in the picture surface, whereas with chiaroscuro and atmospheric perspective the illusion created prevents our relating a tone in the extreme distance with one in the near foreground in the same way that we can relate two tones in the same plane. It follows, therefore, that the pictures gain immensely in decorative unity. This fact has always been more or less present to the minds of artists when the decoration of a given space of wall has been demanded of them; in such cases they have always tended to feel the need for keeping the relations upon the flat surface, and have excused the want of illusion, which was supposed to be necessary for a painting, by making a distinction between decorative painting and painting a picture, a distinction which I believe to be entirely fallacious; a painting of any kind is bound to be decorative, since by decorative we really mean conforming to the principles of artistic unity.

But in regard to this question of three dimensional space in the picture, another curious fact becomes apparent when we look at the pictures in the Grafton Gallery. We find, for instance, that a painter like Herbin, who goes to the utmost extreme in the denial of light and shade and modelling, who makes all his tones in perfectly frank flat geometrical masses, actually arouses in the imagination the idea of space more completely than those pictures – of which we may take Valtat's and Marquet's as examples – in which the gradations of tone, due to atmosphere, are taken as the basis for the design. I confess that this is a result which I should have never anticipated, but which seemed to me undeniable in front of the pictures themselves.

It appears then that the imagination is ready to construct for itself the ideas of space in a picture from indications even more vividly than it accepts the idea when given by means of sensual illusion. And the same fact appears to be true of plastic relief. We do not find, as a matter of empirical fact, that the outlines with which some of these artists surround their figures, in any way interfere with our imaginative grasp of their plastic qualities – particularly is this the case in Cézanne, in whom the feeling for plastic form and strict correlation of planes appears in its highest degree. His work becomes in this respect singularly near to that of certain primitive Italian artists, such as Piero della Francesca, who also relied almost entirely upon linear design for producing this effect.

Many advantages result to art from thus accepting linear design and pure colour as the main organs of expression. The line itself, its qualities as handwriting, its immediate communication to the mind of gesture, becomes immensely enhanced, and I do not think it is possible to deny to these artists the practice of a particularly vigorous and expressive style of handwriting. It is from this point of view that Matisse's curiously abstract and impassive work can be most readily approached. In his *Femme aux Yeux Verts* we have a good example of this. Regarded as a representation pure and simple, the figure seems almost ridiculous, but the rhythm of the linear design seems to me entirely satisfactory; and the fact that he is not concerned with light and shade has enabled him to build up a colour harmony of quite extraordinary splendour and intensity. There is not in this picture a single brush stroke in which the colour is indeterminate, neutral, or merely used as a transition from one tone to another.

Again, this use of line and colour as the basis of expression is seen to advantage in the drawing of the figure. As Leonardo da Vinci so clearly expressed it, the most essential thing in drawing the figure is the rendering of movement, the rhythm of the figure as a whole by which we determine its general character as well as the particular mood of the moment. Now anything like detailed modelling or minute anatomical structure tends to destroy the ease and vividness with which we apprehend this general movement; indeed, in the history of painting there are comparatively few examples of painters who have managed to give these without losing hold of the general movement. We may say, indeed, that Michelangelo's claim to a supreme place is based largely upon this fact, that he was able actually to hold and to render clear to the imagination the general movement of his figures in spite of the complexity of their anatomical relief; but as a rule if we wish to obtain the most vivid sense of movement we must go to primitive artists, to the sculptors of the twelfth century, or the painters of the early fourteenth.

Now here, again, the Post Impressionists have recovered for us our lost inheritance, and if the extreme simplification of the figure which we find in Gauguin or Cézanne needed justification, it could be found in this immensely heightened sense of rhythmic movement. Perfect balance of contrasting directions in the limbs is of such infinite importance in estimating the significance of the figure that we need not repine at the loss which it entails of numberless statements of anatomical fact.

I must say a few words on their relation to the Impressionists. In essentials the principles of these artists are diametrically opposed to those of Impressionism. The tendency of Impressionism was to break up the object as a unity, and to regard the flux of sensation in its totality; thus, for instance, for them the local colour was sacrificed at the expense of those accidents which atmosphere and illumination from different sources bring about. The Impressionists discovered a new world of colour by emphasising just those aspects of the visual whole which the habits of practical life had caused us to under-estimate. The result of their work was to break down the tyranny of representation as it had been understood before. Their aim was still purely representative, but it was representation of things at such a different and unexpected angle, with such a new focus of attention, that its very novelty prepared the way for the Post Impressionist view of design.

How the Post Impressionists derived from the Impressionists is indeed a curious history. They have taken over a great deal of Impressionist technique, and not a little of Impressionist colour, but exactly how they came to make the transition from an entirely representative to a non-representative and expressive art must always be something of a mystery, and the mystery lies in the strange and unaccountable originality of a man of genius, namely Cézanne. What he did seems to have been done almost unconsciously. Working along the lines of Impressionist investigation with unexampled fervour and intensity, he seems, as it were, to have touched a hidden spring whereby the whole structure of Impressionist design broke down, and a new world of significant and expressive form became apparent. It is that discovery of Cézanne's that has recovered for modern art a whole lost language of form and colour. Again and again attempts have been made by artists to regain this freedom of imaginative appeal, but the attempts have been hitherto tainted by archaism. Now at last artists can use with perfect sincerity means of expression which have been denied them ever since the Renaissance. And this is no isolated phenomenon confined to the little world of professional painters; it is one of many expressions of a great change in our attitude to life. We have passed in our generation through what looks like the crest of a long progression in human thought, one in which the scientific or mechanical view of the universe was exploited for all its possibilities. How vast, and on the whole how desirable those possibilities are is undeniable, but this effort has tended to blind our eyes to other realities; the realities of our spiritual nature and the justice of our demand for its gratification. Art has suffered in this process, since art, like religion, appeals to the non-mechanical parts of our nature, to what in us is rhythmic and vital. It seems to me, therefore, impossible to exaggerate the importance of this movement in art, which is destined to make the sculptor's and painter's endeavour once more conterminous with the whole range of human inspiration and desire.

31. G.R.H., 'Art Notes: "The Revolution in Art"'

Pall Mall Gazette, 3 January 1911, 4

This reviewer refers to the publications of C.J. Holmes (no. 33), Clutton–Brock (no. 36) and Frank Rutter (no. 35).

The exhibition of the so-called 'Post-Impressionists' was the most disturbing feature in the history of British Art in 1910. The public refused to take it seriously; but it had a disastrous effect on certain artists and critics of insular view. They accepted it as if it were the completed performance of a new school grown up all 'unbeknown' to them, and refused to regard it as the mere passing phase of protest which it certainly was. In Paris and elsewhere the movement – if there be an organized movement – has long since been understood and properly valued. It is recognised by a public familiar with the Salon d'Automne that to be a Post-Impressionist is not necessarily to be a genius. The inaccuracy of the nickname accounts for a good deal of the disturbance here.

THE EARLIER REVOLUTIONARIES

Cézanne, Van Gogh, and Gauguin, the three most prominent men at the Grafton, are, of course, 'Post-Impressionists' as to date, but are not correctly included in any group or movement. They are the legitimate descendants of Daumier and Courbet. The contemporary 'Post-Impressionism' or 'Neo-Impressionism' of Paris has only a fortuitous relation to them in the fact that it is founded upon an individual expression of nature without any agreed principles beyond a perfect freedom. Each of the three dead painters, now acclaimed as the leaders of a living school, was a man of strong individuality which refused to be pigeon-holed. Cézanne worshipped Manet, but did not imitate him, and Van Gogh painted – and quarrelled – with Gauguin, but there is no idea common to the three, and no bond, except protest against convention, to link them with the far more accomplished moderns like Henri-Matisse, Maurice Denis, and Picasso.

PAUL CÉZANNE

But the subject is very fruitful to those who are eager to translate the difficulties of the painter's craft into words and phrases, and much ink has been spilt. It matters nothing that these gratuitous translations remain incomprehensible to the artists concerned. The past week has produced quite a heavy crop of magazine articles and pamphlets from writers who may be sometimes trying to find their own bearings by instructing a public which is still further out at sea. There is Professor C.J. Holmes, for example (something of a 'synthesist' himself, by the way), who publishes his *Notes on the Post-Impressionist Painters* (London: Mr. Lee Warner), marked by a fine impartiality, which will allow him at a later period to descend quite comfortably on either side of the fence. With his view of Cézanne I disagree altogether, imagining, as, indeed, he hints, that he has had inadequate opportunity of studying that master's work. Everywhere he finds 'clumsiness' and 'heaviness of touch,' and regrets that Cézanne did not paint like Rubens or Conder! Here speaks the painter, interested more in some absent charm of surface quality than in the structural permanence of the idea. Cézanne's conceptions and methods were like Rodin's. He carved his subjects out of nature in blocks, with a unity so complete that the smaller accidents of structure and mass were each an epitome of the whole, and so take their places absolutely in the central conception. As Mr. Clutton Brock points out in the January *Burlington Magazine*, 'He revolted from the Impressionist insistence on the momentary aspects of reality.' His view was so large and so simple, in which the elementary relations were so perfectly expressed, that he had no room for the fripperies and elegances of paint. Is it essential that the painter should appeal to the painter? What Professor Holmes calls 'want of competence' is only Cézanne's nonconformity with his critic's idea of pictorial analysis. But it was unlikely that Paul Cézanne's methods should be appreciated by a writer who, in this and other essays, persistently judges the completed work of art by its suitability or otherwise for the decoration of a wall! Certainly in a 'modern boudoir or drawing-room' Cézanne's primitive solidity and seriousness might be out of place, but that would be the fault of the boudoir or drawing-room – hardly of Cézanne. Where, in his villa, would Professor Holmes hang a masterpiece by Tintoretto or an acre of paint by Henri Matisse?

GAUGUIN

Professor Holmes's appreciation of Gauguin is more generous. He hails him as the wall decorator in perfection, the artist whose pictures group so effectively because they are not realistic, 'but broadly and superbly conventionalised.' Yet here, again, he brings out his infallible test, 'A rich collector might with some pains furnish a Gauguin room . . . but in more ordinary surroundings the placing of a single Gauguin picture might prove difficult.' I remember hearing Whistler maintain with whimsical persistence that no room should hold more than one picture in colour (preferably, of course, one of his own); that it should be the key to which the rest of the decorations were subordinated. Perhaps Gauguin thought likewise, and never anticipated that medley of a modern room which Professor Holmes uses as his test of excellence. Like Cézanne, Gauguin revolted against irrelevance, and the chief irrelevance to him was the art of illusion contained in literary detail or local colour. Even in his earlier still-life studies, when he carried representation up to a certain point, it is not the imitation of nature, but the dominant expression of the man, that persists. Later he conceived his Tahitian subjects in a monumental splendour of large forms and barbaric colours that admitted no realism and yet gave the abstraction of life with inevitable logic. Mr. Clutton Brock thinks him a great artist 'because he created a world of his own,' which may mean only that the content of his art was novel. One would prefer to think that he brought new light and a new interpretation to a world that was old and misunderstood.

VINCENT VAN GOGH

As regards Van Gogh, one notices that everybody drops theory before Van Gogh's realistic work, and turns with relief to pictures easily understood, like *The Ochard in Provence* and the *Effet de Pluie*, which are not 'Post-Impressionist' in any sense, and much nearer Claude Monet than Gauguin or Matisse. Mr. Clutton Brock's estimate of Van Gogh is very high, and is worth quoting: 'No modern pictures are so near to poetry as Van Gogh's; and yet he is never literary or sentimental. There is no need of words to explain what he means, and he never tries to make beauty by the imitation of beautiful things.' In his pictures of flowers 'he shows us living things, and not mere instruments of pleasure, and they have in them the mystery of life, not the material attraction of prettiness.' This is, of course, the secret of beauty in painting, the secret especially of the early Eastern painters and of all times and schools of great art.

M. Henri-Matisse remains the insoluble problem round which all the writers skate warily. Professor Holmes compares him with Holbein to note that in all that makes art valuable the modern Frenchman is nowhere in comparison with the great master. A more inept comparison would be impossible. Mr. Clutton Brock omits him altogether, and it remains for Mr. Frank Rutter, whose *Revolution in Art* has just been published, to give him some measure of justice. The *Femme aux yeux verts* at the Grafton was certainly a disconcerting introduction to the study of Henri-Matisse. His search for 'an abstract harmony of line'; for rhythm, has taken him further and further from any conventionalised view of nature. His theory of painting is directly opposed to the idea that any representation of objects, animate or inanimate, necessarily constitutes a work of art. He is entitled to carry his theory to its logical conclusion, and you need not admire him unless you are 'so dispoged.' But in making the protest, and in evoking along with it a new spirit in wall decoration, he has done enormous service to the cause of art.

32. Unsigned review, 'An Art Victory: Triumphant Exit of the Post-Impressionists'

Daily Graphic, 16 January 1911, 15

On Saturday afternoon the Grafton Galleries, where for three months the Post-Impressionist pictures have sustained the onslaughts of the critics and the jeers of the Philistine, presented a curious spectacle. It was the last day of the exhibition, and the galleries were thronged. Never has there been such a crowd since the Whistler Exhibitions at the New Gallery, and the fact seems to show that public taste in pictures is advancing faster than the critics. Moreover, to one who, like the representative of the *Daily Graphic*, has visited the gallery many times during the exhibition, the demeanour of the spectators was very instructive.

During the first week or two of the exhibition a considerable proportion of the spectators used to shout with laughter in front of Van Gogh's *Girl with the Cornflower,* or Gauguin's *Tahitians*, or the allegories of Maurice Denis, the architecture of Otho Friesz, or the stippled landscapes of Seurat and Signac, the pointillistes, or the archaic figures of Flandrin. But on Saturday the general attitude was one of admiration and of regret that an exhibition which has furnished so much food for discussion must close. A brisk trade was being done at the last in photographs of the pictures.

Mr. Desmond McCarthy, the secretary, to whose energy a good deal of the success of the show has been due, expressed much satisfaction at the way in which the public had shown their appreciation of it.

---

33. C.J. Holmes, Introduction to *Notes on the Post-Impressionist Painters*

1910, pp. 12–14

Holmes (1868–1936) was the editor of the *Burlington Magazine* between 1903 and 1909 (when Fry took over); he was Slade Professor of Fine Art at Oxford (1904–10) and Director of the National Portrait Gallery (1909–16). At Fry's request he lent the weight of his name to the first Post-Impressionist exhibition by allowing it to be added to the list which made up the organising committee. His *Notes* were published late in 1910 and express a bemused tolerance of Post-Impressionist painting.

We must begin therefore with Cézanne. In his lifetime Cézanne was a humble follower of the more famous Impressionists, and a review of his work at the Grafton Galleries does not compel any drastic reconsideration of his place. Wherever he seeks to render delicacy, the coarse handling of his school is accentuated by consistent personal clumsiness of touch. This clumsiness is not relieved by any supreme intellectual power; indeed, Cézanne frequently fails in grouping his matter into a coherent design – sometimes in expressing his intentions at all. Nevertheless, he was a genuine artist. His landscapes in particular have sincerity and sometimes real force, while he often shows distinction as a colourist. Honesty is his paramount virtue, and sometimes this quality enables Cézanne to impress us more than men of infinitely finer gifts. The verdant landscapes of Daubigny, the exquisite fruit-pieces of Fantin, might often appear to those who admire Cézanne's work in the same field as having, in comparison, ever so slight a savour of the potboiler about them. That, however, does not prove Cézanne to be more than a very sincere artist. His modest rank will at once be made clear if we compare his life's work with that of Courbet, whom he not infrequently recalls.

With Gauguin no such Laodicean verdict seems natural. Though the surface of his paint is less obtrusive than that of the Impressionists and of many English painters of the present day, his frank use of thick dark outlines, no less than his barbaric subject matter, his unusual schemes of design and colour, constitute him an innovator still more daring than Claude Monet. Even those who are not in sympathy with his work do not deny him occasional grandeur of planning and occasional success as a colourist, though they would still regard the result as more suitable for a hut on a Pacific island than for any conceivable edifice in Europe. To argue that the same results could have been achieved without all this deliberate simplification of forms, without those rude outlines, is to argue on uncertain ground. The colour in Gauguin's work, which makes his pictures group so effectively, is not realistic but broadly and superbly conventionalised. To have combined this conventional colouring with wholly realistic forms would have led to an incongruity. Moreover Gauguin's simplification of the human figure undeniably emphasises the idea of large primitive dignity which is at the root of his most characteristic designs. Indeed had these very compositions been executed on the scale of an Indian miniature, or even of the size preferred by M. Vallotton, their magnificent colour and monumental design would have won them universal admiration. As it is they appear to be a stumbling block to the many and a stimulus only to the few. They are original, powerful, and the work of a master, but they fail of their full effectiveness, because they are not serviceable also. A rich collector might with some pains furnish a Gauguin room, and a wonderful room it would be, but in more ordinary surroundings the placing of a single Gauguin picture might prove difficult. 'Less valuable as an example than as a stimulus' might be no unjust summary for him.

The insanity of Van Gogh's last years has furnished the enemies of the Post-Impressionists with a cheap cudgel, but taints only a very small proportion of his work. The remainder is of wonderful quality and variety. His design is large and original, his colour powerful or refined as the occasion demands, his intensity of vision almost disquieting. Only in his handling does he frequently appear to pass beyond the confines of oil-paint and encroach upon the province of mosaic. But the haunting power and beauty of his best work is so great as to annihilate all minor objections, and among the masters with whom he is associated not one is more deserving of study. Beside Van Gogh, Gauguin appears narrow and sometimes heavy handed, and all the others appear like men who have very little of their own to say. If I may judge by my own feelings, Van Gogh is destined to become a considerable figure, perhaps the most considerable figure of the movement, and if some millionaire admirer would only buy up his occasional failures and burn them, Van Gogh would not have to wait long for universal fame.

34. C. Lewis Hind,
The Post Impressionists
1911, pp. 1–7

Hind (1862–1927) wrote regularly for the *Art Journal* and the *Daily Chronicle*. He had seen the collection of paintings by Matisse at the Stein house in Paris early in 1910 and had already began to publish accounts of the 'new' art before the first Post-Impressionist exhibition. *The Post Impressionists* was the first full-length account of Post-Impressionist painting in English. It contained twenty-four illustrations and was largely made up of articles which Hind had published in the previous year.

Post Impressionism has been called the heart of painting: it has also been described as an insult to the intelligence. To some it is a re-birth of vision and feeling, to others the foul fruit of a horrid egotism. In a word it is a novelty – to England.

As the essence is sincerity, and as even sincerity is contagious, the idea of Post Impressionism inclines to make people sincere in their utterances about the idea. Calling it a spiritual movement, I have been charged with hypnosis and accused of imagining that which it never contained. I accept the reproach. St. John saw more in Patmos than was contained in the landscape, and some can read infinity into the words – 'Raise the stone and you will find me. Cleave the wood and there am I.' I submit that when one is in harmony with the spirit that informs the movement, one has a clearer vision of the vital things in life, and becomes impatient of rhetoric and the rhodomontade of unessentials. The danger is that if we rid our souls of the glamour of unessentials and of rhetoric it may happen that there is no glimmer of a soul to be seen.

If a child were to ask – 'What is Post Impressionism?' I think I should tell that child about the Sermon on The Mount, and say – 'If the spirit that gives life to the movement we call Post-Impressionism is in your heart you will always be trying to express yourself, in your life and in your work, with the simple and profound simplicity of the Sermon on The Mount. You will say what you have to say as if there were nobody else but you and Nature or God.'

Expression, not beauty, is the aim of art. Beauty occurs. Expression happens – must happen. Art is not beauty. It is expression; it is always decorative and emotional. And it can be greatly intellectual too. There is as much intellect as emotion in the Parthenon and the Sistine Vault. Art is more than the Emotional Utterance of Life. It is the Expression of Personality in all its littleness, in all its immensity. A man who expresses himself sincerely can extract beauty from anything. There is a beauty of significance lurking within all ugliness. For ugliness does not really exist. We see what we bring. He who expresses his emotion rhythmically, decoratively, seeking the inner meaning of things, is artist. He who represents the mere externals is illustrator. Frith was illustrator – delightful and competent; Cézanne, the true parent of Post Impressionism, was artist. To him the spiritual meaning was everything. Few are the artists. Many are the illustrators. The founders of Post Impressionism, Cézanne, Van Gogh and Gauguin, were artists.

II

Obviously, Expressionism is a better term than Post Impressionism, that avenue of Freedom, opening out, inviting the pilgrim who is casting off the burdens of mere representation, and of tradition when it has become sapless. Degas once said – 'If you were to show Raphael a Daumier he would admire it; he would take off his hat; but if you were to show him a Cabanel, he would say with a sigh, "That is my fault." Post Impressionism or Expressionism seeks synthesis in the soul of man, and in the substance of things; it lifts mere craftsmanship into the region of mysticism, and proclaims that art may be a stimulation as well as a solace. It tries to state the sensation or the effect. It is cheering, and it is as old as ecstasy. It has been called by many names. It informed the work of Botticelli when he expressed the gaiety of spring, Rembrandt when he expressed the solemnity of a Mill, Cozens when he expressed the serenity of Nature, Swan when he expressed forsakenness in his *Prodigal Son*. It is in the work of all who are artists not illustrators, and it would have glided on, coming unconsciously to the initiate, uncatalogued, unrecorded, had not three men – Cézanne, Van Gogh, and Gauguin – flamed its principles abroad, and by the very intensity of their genius forced the world to label their performances. The label chosen, when a very incomplete exhibition of such pictures was opened at the Grafton Gallery in November of 1910, was that of Post Impressionism. Little did the originators of the exhibition anticipate the storm those pictures would arouse in torpid art England – abuse and praise, gratitude and groans, jeremiads, and joy that a new avenue of expression had been opened to all who are strong enough to draw life from the idea behind the movement, and not merely foolishly to copy the pictures, the failures as well as the successes.

III

When the exhibition of Post-Impressionist pictures at the Grafton Gallery closed I sighed, supposing that the stimulus to thought, talk and writing of the pictures was ended – for the present. But no. A week later my morning paper contained a report of three different meetings in different parts of the country raging round Post Impressionism. This whirl of argument had continued, week by week, ever since the opening of the exhibition. What does this mean? Why should a mere picture exhibition stir phlegmatic art England? Moreover, it stirred people who are not particularly interested in art. Why should this be? There can be but one answer. Behind the movement there is a purpose, an idea partaking more of the spiritual than the material. Mere picture-making was lifted into a larger region. A few perceived the goal of the search, and wondered.

But the differences of opinion were bewildering. To some, the pictures were a crime, sprung from the devil; to others, they were a revelation, God-inspired.

To me they opened avenues: beginnings, yes, often imperfectly realised, often leading to regions where there is more reality than in the visible world. But the pictures themselves vary as widely as the personalities of the painters. What could be further apart in vision and technique than Van Gogh's lovely, spring-illumined *Orchard* in *Provence*, and the angry realism of his *Mad Girl*; than the faded, pathetic *Woman with the Beads*, by Cézanne, and that brilliant expression of artificial modernity, Matisse's *Woman with the Green Eyes*; than the fierce intensity of Van Gogh's *Self Portrait*, everything stated with a kind of scorching force, and the massive eloquence of Cézanne's *Portrait of Himself*, fatigued yet eager, as of a soul struggling for release, the paint a means not an end, the idea everything, the real man grown old searching for that which can never be wholly found? Like them or loathe them, but admit that these men at their best are themselves, naked souls before the living God, searching eye and eager heart, seeking the soul-meaning behind the bodily forms, dutiful to tradition, but violently aware of being alive – a Van Gogh actively, a Cézanne passively.

IV

Suppose one who had been devoted all his life to art merely for æsthetic enjoyment, found slowly, after long years, a new meaning in art – would not that be strange?

Suppose one found spiritual stimulus in the patient insight shown by Cézanne in a half articulate portrait of the glimmering soul of an old woman, and in what Catholic theologians call the 'substance' of a mere bowl of fruit; in Van Gogh's agony of creation and in the patterned gold of his harvest fields; in the weight of sorrow that bears Gauguin's *Christ* to earth, and in the solemn forms of his Tahiti women; in the intellectual striving of a Picasso, and in the search for simplification and synthesis in peace and in bustle of a Derain, a Vlaminck, and a Friesz – would not that be strange?

Art is but an episode of life. The artist's life is but a part of the whole. To be effective he must express himself. Having expressed himself his business ends. He has thrown his piece of creation into the pond of Time. The ever-widening ripples and circles are his communication to us, the diary of his adventures. We read the diary. We are comforted, stimulated, consoled, edified, helped to live according to the degree of life-force in the diary, and the idea behind it.

The Idea behind it! So I come again to my original statement as to the idea, the force, at the back of the movement. But first the ground must be cleared if we mean to understand the Idea behind Post Impressionism, if we mean to peer out, uplifted, through the opening avenue, noting that art can be great even when dealing with what the world calls little things, and that profound vision can be clothed in cheerfulness and gaiety.

35. Frank Rutter,
Revolution in Art
1910, pp. 14–18

Rutter (see headnote to no. 1) had championed Post-Impressionism in the columns of the *Sunday Times* (for which he had been art editor since 1903) and *Art News* (of which he was editor). His attitude, however, was far less conciliatory than that of Fry and where Fry emphasised the historical continuities of the new art Rutter stressed the iconoclastic, revolutionary aspects of Post-Impressionist painting. Rutter also has warm praise for pointillist painters, praise which was rare in English criticism at this time.

Dedication: To the rebels of either sex all the world over who in any way are fighting for freedom of any kind, I dedicate this study of their painter comrades.

For the last four or five centuries painting has been steadily growing more and more complex. Towards the latter end of the nineteenth century it had become an overstuffed portmanteau in which nothing fresh could be put except by taking out something already there. The army of art had so encumbered itself with baggage that, staggering under its self-imposed load, it was hardly able to bear, and almost unable to wield, its most effective weapon. That weapon always has been, still is, and ever will be design. The younger recruits to the army at the end of the last century set themselves resolutely to lightening the baggage train. One by one they threw away orthodox garments and accessories found of little use and even a hindrance to the fighting line. To the scandal of the staff college they went naked into action, and as franc-tireurs saw no necessity for adopting any recognized uniform.

Such combatants are apt to be harshly dealt with when caught by the enemy, and the rebels of art were shot down metaphorically as ruthlessly as the Communists were literally. Both had against them not only the organized forces of existing authority, but also the general opinion of a majority lazily content with existing conditions. But no one denies that the Communists of Paris were serious, and most people give them credit for being activated by ideals. The painter rebels, less fortunate, are rarely allowed to be in earnest, or to possess ideals, and the insult added to the injuries offered them are probably due to the fact that the general public knows far less of an artist's than of a social revolutionary's relation to itself. A lucid exposition of this relation has already been given in an introduction to the catalogue of the exhibition which has stirred this monograph into being.

[Here Rutter quotes from Desmond MacCarthy's introduction to the catalogue of the first Post-Impressionist exhibition (no. 10): 'In the earliest ages of art...hampers him as well']

It is, perhaps a national characteristic of the French to be intense on all they undertake, and if there is one quality common to the generation of painters who followed the earlier impressionists it is intensity. This earnest passionateness has produced developments in two main directions, towards more intense luminosity and towards more intense simplification. The first is exemplified in the work of the pointillists who carried to its logical conclusion the division of tones, and built up their pictures with points or square touches of pure colour. Paul Signac, for example, is dazzling in his scientific presentment of the power of light. It is difficult to believe that luminosity can be carried further than in his radiant canvasses whose force make the most brilliant Turner appear pale and weak in comparison. Signac's method, it may be noticed in passing, is a square touch of pure colour as opposed to the circular spots of Seurat, the inventor of pointillism Theo van Rysselberg, and the late Henri-Esmond Cross.

If Signac has reached the limit in intense luminosity, Henri-

Matisse, Othon Friesz and André Derain, among others, stand for intense simplification. But it is still a little too early to deal with their astonishing works, and any one sincerely desirous of comprehending the aims of these revolutionary painters may be recommended to commence their course of initiation by a serious study of Cézanne and Gauguin. These two deceased painters are to their younger comrades what Marx and Kropotkin are to the young social reformers of to-day.

36. A. Clutton-Brock, 'The Post-Impressionists'

Burlington Magazine, January 1911, xviii, 216–19

Arthur Clutton-Brock (1868–1924) was literary editor for the *Speaker* (1904–6) art critic for the *Tribune* and briefly for the *Morning Post*. He was art critic of *The Times* from 1908 and published his *Essays on Art* in 1919. His interpretation of Post-Impressionism was a highly literary one which stressed the personal, expressive qualities of the paintings.

The art of painting nowadays is as confused and uncertain in its aims as the art of poetry was in the eighteenth century. We have learned that the proper end of poetry is the expression of emotion, to which all reasoning and statement of fact should be subsidiary; but we have not learned that painting should have the same end, using representation only as a means to that end, and representing only those facts of reality which have emotional associations for the painter. In primitive pictures, it is true, we look for the expression of emotion rather than for illusion, and that is the reason why so many people get a real pleasure from primitive art. They judge it by the right standard, and ask of it what it offers to them. But from modern pictures they demand illusion – that is to say, the kind of representation they are used to; and when they do not get it they accuse the artist of incompetence. There is a prevalent notion, based upon the history of art in the fourteenth and fifteenth centuries, that painting must always advance in fulness of representation, that when it does not it is deliberately retrograde. But that notion is refuted by the facts of the modern history of painting. Ever since Turner, at least, the best painters have constantly sacrificed representation more and more to expression, and have been reviled for doing so. The Impressionists themselves, though they represented some new facts, discarded many old ones. Their very name is a token of their refusal to be bound by any laws of representation. But their aesthetics were confused by their scientific interest in the new facts which they chose to represent. Their advocates often talk as if the whole business of a painter were to produce the illusion of sunlight, as if sunlight were the essential fact without which no picture can be a work of art. Thus they ignored one set of facts for the sake of another set, not for the sake of expression. They did not give the artist freedom, but imposed a new bondage on him. They never arrived at the truth that to painting, as to poetry, no facts are essential, that all are subsidiary to expression, and therefore may be represented or ignored as the artist chooses. This is the truth upon which the art of the Post-Impressionists is based; and to distinguish them from the Impressionists we might, perhaps, call them Expressionists, which is an ugly word, but less ugly than Post-Impressionists.

Their art, of course, is not necessarily good because it is based upon a truth, but we must grasp that truth before we can judge their art fairly, and if once we grasp it we shall see that the task they set themselves is not an easy one. When Wordsworth tried to rid his poetry of all that rhetoric which had been employed by the poets of the eighteenth century to hide the prosaic nature of their subject matter, there were many critics who said that he was trying to conceal his own incapacity by pretending to despise skill in versification. They also reviled him for the meanness and ugliness of his work, and the most popular poet of the day sided with them. There is a passage about Wordsworth in one of Byron's published letters in which an important word is left blank, but anyone can guess what it is, Byron hated the art of Wordsworth as he loved rhetoric. He could not see that Wordsworth was at least trying to do something far more difficult than anything which he himself attempted, and he noticed only Wordsworth's failures, which were obvious just because he would not conceal them with rhetoric.

So the failures of the Post-Impressionists are obvious because they do not attempt to conceal them with irrelevant feats of representation. Their only end is expression, and when they do not accomplish that they accomplish nothing. If Cézanne, Gauguin, and Van Gogh were charlatans, they were like no other charlatans that ever lived. If their aim was notoriety, it is strange that they should have spent solitary lives of penury and toil. If they were incompetents, they were curiously intent upon the most difficult problems of their art. The kind of simplification which they attempted is not easy, nor, if accomplished, does it make a picture look better than it is. The better their pictures are, the more they look as if anyone could have painted them who had had the luck to conceive them; in fact, they look just as easy as the lyrical poems of Wordsworth or Blake.

But we must not suppose that a painter, if his aim is expression, need have no power of representation. On the contrary, a much more thorough mastery of facts is needed to make them expressive than to represent them without expression. We all know that it is easier to describe persons or things when we have just seen them for the first time than when we know them well. For at first sight they arouse in us the interest of curiosity which is easily expressed in mere description.

But when we know them well this interest wears off and changes either into mere boredom or into an emotional interest which is difficult to express in description. Now the interest expressed in much impressionist painting is only an interest of curiosity. The painter represents facts that he has only just noticed. He is like a clever journalist who makes an article out of his first observation of a new country. But the aim of the Post Impressionists is to substitute the deeper and more lasting emotional interest for the interest of curiosity. Like the great Chinese artists, they have tried to know thoroughly what they paint before they begin to paint it, and out of the fulness of their knowledge to choose only what has an emotional interest for them. Their representations have the brevity and concentrated force of the poet's descriptions. He does not go out into the country with a note-book and then versify all that he has observed. His descriptions are often empty of fact, just because he only tells us what is of emotional interest to himself and relevant to the subject of his poem; and they are justified, not by the information they convey, but by the emotion they communicate through the rhythm and sound of words. The Post-Impressionists try to represent as the poet describes. They try to give to every picture an emotional subject matter and to make all representation relevant to it. Hence the formal or decorative qualities of their pictures, which, like the formal qualities of poetry, are the result of expression. We hear a great deal about decorative painting in England; but our decorative painting is often a mere imitation of the Italian primitives. People like it, as they like modern imitations of Elizabethan verse, because it reminds them of beauties with which they are familiar. But the decoration of Gauguin and Van Gogh and Cézanne does not remind us of familiar beauties; rather it startles us at first sight because it does not express a mere admiration of Italian primitives, but the artists' own emotional experiences. Nothing could be more unjust than to accuse them of imitating any kind of primitive art. If they have qualities in common with the great primitive artists it is because they have the same aims. But the very accusation is based upon a misunderstanding of primitive art. The ordinary admirer of Giotto or Fra Angelico or Piero della Francesca professes to like them because they tried so hard to do what they could not do. He sees in them, not a magnificent artistic success in expression, but a morally creditable failure in representation. For this failure, he thinks, they are to be excused because they were born so early. But the Post-Impressionists are not to be excused for a like failure, which must be mere affectation or incompetence in them since they were born so late. We have not this view of Chaucer, though Dryden had it. We know that Chaucer was a master of poetry and that his means were perfectly fitted to express the emotional experience of this age. When Dryden rewrote the *Knight's Tale* he proved that he could not do better what Chaucer tried to do; and also that Chaucer had not tried to do what he, Dryden, could do so well, but something quite different which Dryden did not understand. So it was with the great Italian primitives. They also expressed the emotional experience of their time with splendid success, and Titian could not have repainted their pictures and made them better. Understand this and you will see that the Post-Impressionists have a right to express the emotional experience of their time in an art which is quite different from the art of Titian. The only question is whether they have expressed it.

Now the emotional experience of our time, as it is expressed in literature and art, is not rich or full or confident or joyous. The artist nowadays, partly because he is bewildered by the multitude of new ideas and the increase of new knowledge, partly because he is not at ease in our modern society, suffers from a great insecurity of emotion. Intellectual scepticism and exasperation produce in him emotional scepticism and exasperation. In a painter this state of mind is very unfavourable to fulness of representation. The sincere artist finds himself hating almost as many things as he loves for their emotional associations, and he finds that many things in the mechanical bustle of modern life have no emotional association for him whatever. Thus even the finest art of our time is inferior in richness, in serenity, and in accomplishment to the finest art of the past. But it has at least this merit, without which art is only a shadow of itself, that it does express the artist's own emotional experience of life.

Cézanne, Gauguin and Van Gogh were men of very different minds; but they were alike in this, that they all attempted to subordinate representation to expression, and were all determined to express only their own emotional experience. Cézanne, a friend of the Impressionists and of Zola, could not content himself with impressionist triumphs of representation. Above all, he revolted from the Impressionist insistence on the momentary aspects of reality. He was, so to speak, a kind of Plato among the artists of his time, believing that in reality there is a permanent order, a design which reveals itself to the eye and mind of the artist, and which it is his business to expose in his work. But this design he was determined to discover in reality itself, not in the works of other artists. His task was enormously difficult because he would take nothing whatever at second hand. Nature must tell him all her own secrets, and he would not listen even to her when she told him commonplaces. He was not interested, so to speak, in her caprices, in her chance effects of beauty that anyone can see. He painted landscape as Titian or Rembrandt painted portraits; searching always for the permanent character of the place, for that which, independent of weather or time, distinguished it from other places. This permanent element he found in structure and mass; but, like Titian and Rembrandt, he would not abstract these from colour.

For him, as for these masters, structure and mass revealed themselves in colour, and all three must be verified by incessant observation. He felt that there was a proper colour belonging to every representation of structure and mass, with the same kind of permanence based upon its relation to the permanent character of the scene represented; and in his pictures we can see the result of these convictions. There is no brilliant illusion in them, nor is there any sacrifice of colour to form. They do not represent some ideal state of being which he desires, nor do they express lyrically some passing emotion of his mind. He is a classical painter by nature and not because he admires the classics. Without help from the past he tries, by a classical balance of representation, to express his own permanent relation to reality by insisting upon its permanent elements. In his landscapes, although the colour has an independent force and freshness like the colour of the Impressionists, yet it never confuses the design, and that is always based upon structure and mass. He represents a world in which heavy things have not lost their weight, for that seems to him an essential part of their character. For him, a hill is not a screen for the play of light; it is built up of earth and rocks. Nor is a tree a mere rippling surface, but a living thing with the structure of its growth. Everywhere he looks for character; yet he subordinates the character of details to the character of the whole. And the character of the whole means for him its permanent character, which he expresses in a design not imposed upon it but discovered in it, as Michelangelo discovered the statue in the block of marble. It is the perfect balance of Cézanne's art that makes it difficult. At first sight he seems a clumsy painter, because he neither produces a vivid illusion nor insists eagerly upon his design. You might think from the character of his execution that he had no grip of form. The fact is that he does not need to grip form with his paint, his mind has so firm a grasp of it. The design of his pictures is like a conviction that a man holds so surely and deeply that he does not state it but only reveals it in his conduct; and it is a design based upon all the elements of reality. Sometimes, as in the portrait of his wife, he is so intent upon the revelation of character that he forgets all about illusion. You may say that that portrait is not like a woman; but it conveys to you through the eye Cézanne's idea of his wife; and it is the idea of a profound and noble mind. What more can you ask except beauty? And you will find beauty in it when once you have allowed the picture to express itself to you, when you have ceased to demand of it an illusion which you do not demand of the great Primitives.

Van Gogh is a more lyrical artist than Cézanne; and in many of his pictures, he expresses the immediate mood with which a particular scene inspired him. But this is a very different thing from merely representing the momentary aspect of a scene. He went mad, and that is enough in the opinion of some critics to condemn his art, and indeed all the art of the Post-Impressionists. They were all mad, no doubt, but only Van Gogh had the honesty to betray his madness. This he sometimes does even in his pictures. There are signs of madness, for instance, in his *Cornfield with Blackbirds*. The very forms of things seem distorted by the trouble of his mind; yet there is a firm grasp of fact behind this distortion. The picture is evidently inspired by reality. It does not tell us only that the painter is mad, but speaks rather of an effort to express something beyond his madness; and the fierceness of that effort gives it the tragic beauty of conflict. But in most of Van Gogh's pictures we see not madness, but only an extreme sensibility controlled by a sincerity no less rational.

Lyrical artists are always apt to grow impatient of reality unless they have a sincerity which makes them test all their emotions by the precise expression of them. Van Gogh, filled with the desire for precise expression, drew all his inspiration from reality. He was like a poet who makes music of his own experience. He used the impressionist technique, but in his hands it was never a mechanism. Compare, for instance, his landscape *Arles* with the landscapes of Signac and Cross and you will see at once the difference between poetic and photographic design, between an imitative and an expressive handling of nature. The execution with Van Gogh is a means of giving unity to the picture, the unity of his own mood; with the other two it is a means of producing illusion. And Van Gogh always fits his execution to his theme, usually with perfect success. In his finest works, such as the *Evening Landscape* and the *Rain*, the expression is so keen that it produces the effect of illusion, the more surely because the artist has not aimed at that effect. We believe in the world which he shows us because he convinces us that it is a world of his own experience and not merely one that he has glanced at. He paints the essence and character of rain, as a poet might describe them, with all irrelevant facts eliminated and with an expressive power in his execution which, like the rhythm and music of a poet's words, communicates some of his own sensibility to us, so that his representation means to us what the reality meant to him. In fact no modern pictures are so near to poetry as Van Gogh's; and yet he is never literary or sentimental. There is no need of words to explain what he means, and he never tries to make beauty by the imitation of beautiful things. Even when he paints flowers he does not, like nearly all other painters, try to imitate their beauty. His pictures of sunflowers are like Rembrandt's portraits of men and women. He shows us living things and not mere instruments of pleasure, and they have in them the mystery of life, not the material attraction of prettiness. In fact their beauty is the expressive beauty of art other than the imitated beauty of nature. This land of beauty is difficult to recognise when it is new and original. There is some excuse for critics who fail to see it, but none for those who call Van Gogh a mere incompetent. A glance at his drawings will prove that he was not that; and if he went mad, so did Cowper, who remained the most moving of poets in his madness.

Gauguin is a painter much easier to enjoy than Cézanne or Van Gogh; indeed his weakness is that he is too much afraid of dulness or irrelevance. There seem to be no accidental beauties in his pictures. Everything is aimed at and achieved with a French lucidity. And yet he is a great artist because he creates a world of his own. Most of us have never been to Tahiti, yet we seem to know it from his pictures as we know Russia from the great Russian novelists. He does not paint local colour for us, but life itself, making it seem not strange, but familiar. He again makes no literary appeal whatever, yet his *Esprit Veille*, his *Esprit du Mal* and his *Religieuse* interest like a story, and in each case all the story is in the picture. If there were more effort at illusion there would be irrelevance, like the irrelevance of elaborate scenery in a play. Those who think that Gauguin leaves out what he cannot represent should notice the expressive power of what he does represent, his command of gesture and especially his indication, so rare in pictures, of the psychological relation between his figures. He is so amusing a painter that it is easy to overlook the nobility of his forms and to forget that he, like Van Gogh, is a poet. He condescends to be witty, but it is a condescension for which we should be grateful.

Of the other Post-Impressionists, some, perhaps, have mistaken their vocation, but none try to conceal their want of artistic power with irrelevant skill. They never tell us a multitude of things that we do not want to know. If they have nothing to say, at least they say it briefly, and often they have more to say and say it with more precision than we should suppose at a first glance. Some of their pictures have been called insults to the British public; but that public seems to find them amusing, and after all it is better to be amused than to be bored.

37. Charles Ricketts, 'Post-Impressionism at the Grafton Gallery'

Pages on Art, 1913, pp. 149–64

This piece was orginally written for the *Nineteenth Century* in January 1911 but it was not published until 1913. Ricketts responded positively to some of the decorative qualities of Post-Impressionist painting, but in common with many of his contemporaries reacted against the unbridled individualism or 'egotism' which he saw as a symptom of degeneracy in the work of these painters.

Every art movement has to face not only the hostility of its opponents, it has to pass through a more dangerous ordeal when ridicule is brought upon it by its belated following, from which often emerges the reaction against it. Classicism failed in France in the hands of the pupils of Ingres; a compromise between their practice and a melodramatic view of Romanticism produced Delaroche, with whom both aims were combined. To-day Impressionism has to face its parody and the opposition fostered within its ranks, and Post-Impressionism stands to-day against its parent. Art is to become young again, but the slate must be cleaned for the new message to appear in the form of a subconscious ecstasy, from which our past and present must be erased, so that we may attain to the synthetic outlook of childhood, in which lies the only possibility of the future. Such at least is the programme of its advocates.

Not only has Post-Impressionism been championed in the *Burlington Magazine*, where it is promised a larger and fuller life than pre-Raphaelitism; it has been hailed elsewhere as a novelty, against which it would be injudicious to slam the door. Post-Impressionist works may some day be deemed masterpieces!

Caliban worshipped the god Setibos, and Novelty may also be a god! Alas, that criticism should doubt itself and forgo all choice when Caliban did not doubt; he only feared. These critics do both.

Novelty in itself is valueless. The spirit of beauty and power, of which art is the expression, has centuries behind it; it is as old as thought. From the day that the first flint flake chanced to resemble the shell-like hollow which might become a cup, art has aimed at permanence, not at novelty, which is a laterday fiction. The new has ever had to prove its value against the experience of the past before it could be considered admirable. To revert in the name of 'novelty' to the aims of the savage and the child – out of lassitude of the present – is to act as the anarchist, who would destroy where he cannot change. To wish to blot out the page upon which our knowledge is written, in the hope of a new thrill of expectancy, is an old form of petulance. It is as old as the spirit that denies. Bad in art, it is deplorable in criticism. Its advocacy here should be laid at the door of the impressarios of the movement, which claims Cézanne as its root and Monsieur Matisse as its flower. Against the painters themselves it is less urgent to be hostile; they may believe – Caliban believed – and among them are men of varying attainments. All these 'experimentalists' are united in one fault, they are over-confident; they forget that the place for the experiment is the studio; it is not an aim but a means. I would also accuse them of lacking in tenderness towards their craft, and of a lack of humour, were these qualities not rarer still in the apostles and advocates of the movement. Mr. Fry, who is now their champion, is without pity for his contemporaries, who are not Post-Impressionists; he is as merciless as Herr Meier Graefe in

Germany, who scoffs at Reynolds and Turner, and will none of our modern art – who does so, however, in a spirit of rapid and half-playful cynicism, which is in total contrast to the austerity of his British rival, who bids us 'Hope no more' save in the art where the spirit of the 'rocking horse' is one with the 'treeness' of the trees which flourish in the meadows of the mind of Monsieur Matisse.

I should not insist on the advocacy of Post-Impressionism, since it has come mainly from the sponsors or organisers of the Grafton Gallery Exhibition. Like the painting shown there, it may be sincere and, in its degree, interesting or curious, had it not been the cause, or at least so I imagine, of bringing together an imposing honorary committee, on which we find the names of three Museum directors and trustees of the National Gallery. Should these gentlemen have been familiar with Post-Impressionism and admirers of its achievements, we might feel nervous about the future character of our institutions; should they have been ignorant of this movement, which is twenty years old, one's astonishment is not less. Under no circumstances could we imagine their standing sponsors to a similar venture devoted to modern English art; there are public reasons which would even render this inadvisable.

The group of dealers who have loaned these pictures, and the organised propaganda in Germany, might afford the stuff for an article. The painters themselves merely labour under the blinding weight of their egotism when against them is placed the barrier beyond which the loser and the lunatic cannot stray, on which is written, 'Know thy Self.'

Was this reactionary temper, this singular emphasis of personal limitations, contained and foredoomed in the theory of Impressionism itself, in its advocacy of side issues and contempt for its foes? Who shall say? The triumph of Impressionism is denied by these laggard followers, who would also be crowned. To-day is not to-morrow, and beyond lie countless other 'to-days.' The inexorable power of selection, which often stumbles in its need, may choose amiss. Impressionism to-day or something else – of this who knows?

Success in art consists in the power of concentrating the result of countless experiences and emotions within the restricted surface of canvas. With this effort the outlook of the child upon life has nothing in common. To trust chance personal intuitions only, and never to doubt, is allied to the strange persistent egotism of animals; and against this stands the intellect of man, 'the paragon of animals.'

There is the fruit for small talk, the possibilities of self-assertion in the acceptance of Post-Impressionism; there is also the stuff for journalistic copy; but can we imagine this manner of painting upon the walls of our homes, or indeed anywhere else, than upon a hoarding, where the stridency of its appeal might arrest one for a while, till our streets resembled the rooms in the Grafton Gallery?

Post-Impressionism or Proto-Byzantinism, as it has been fatuously described, claims Cézanne as its half-conscious founder or pioneer. Cézanne's paintings are laboured in effect; a suffering sense of 'values' made him plaster his canvases with pigment in some sort of parody of the pictures of Manet. For a while he plodded on, affording Zola 'copy' for his novel L'Œuvre; to Manet and Degas he seemed but a provincial satellite, a compromising flower of Impressionism. He left Paris doubtful of himself, doubtful also of the school whose novelty and notoriety had cast a spell upon him. He is one of those countless failures who have set out 'to conquer Paris,' to use the 'romantic' phrase of M. Zola's.

Cézanne was failure, and believed that he had failed, yet Sir Claude Phillips recognises a savage grandeur in some of his still-life paintings of jugs and pumpkins. To me his pictures seem mere accumulations of thick pigment, applied to hack subjects designed in the style of our 'Proto-Byzantine' pavement artists; they lack only the written appeal, 'Please remember the artist.' This is done by his advocates and by the cosmopolitan dealer, who sees that he, at least, is not forgotten.

Monsieur Maurice Denis confesses, 'Cézanne seems to bring us health and promises us a renaissance by bringing before us an ideal akin to that of the Venetian decadence.' He also adds, 'I have never heard an admirer of Cézanne give me a clear or precise reason for his admiration.... Now of Delacroix or Manet one could formulate a reasoned appreciation which would be clearly intelligible, but how hard it is to be precise about Cézanne.' Cézanne will help us; he was quite explicit in his judgment of his paintings; he left them in the fields, not deeming them worthy to take home.

Monsieur Gaugin in his earliest canvases shows the influence of Cézanne, the 'Timon of Impressionism'; his Breton pictures are in part influenced by him. Dare I confess that I do not always dislike his pictures of Tahitian life. Their technical shortcomings have a left-handed affinity with Degas's later and 'less studied' works. The strangeness of his subject-matter attracts me, not his painting.

I have a childish liking for savage art, idols of feathers, amulets of wood; a mere shell-tipped arrow-head conjures up the magic of distances, or the aromatic gloom of forests where a savage might crouch, watching for his prey, silent, imovable, primæval, whilst at his side some large flower opens, fades, and sinks unnoticed upon its stem. I can imagine this in the space between the musical cries of the tide in the coral reef beyond and the boom, like distant cannon, of the sea on its return. In Gaugin's pictures I like the snakelike trees, the grape-coloured women; I do not hate their impassiveness and lack of 'significant gesture' or something at once animal and vegetal in the life he paints. His art is a by-path of painting. He is not Proto-Byzantine, but, like a sailor with shells and parrot feathers in his trunk, he would bring us something from those distant lands where the tree-roots clasp at the sea-weed on the water's edge, or bend beneath the hurricanes that send crashing through the trees strange wreckage from the seas. Of his painting it is possible to say, this I like, this I like less.

Of Cézanne there is little to be said, of his influence upon Van Gogh there is more; of Monsieur Matisse, the latest of the Proto-Byzantines, there is nothing to say; yet one and all are still related to the parent 'Impressionist' movement which they would deny. Most of these canvases are devoted to the old familiar subjects which contented Courbet and Manet fifty years ago, such as *Femme en bleu*, or the chance names of women or places; were they capably painted we would describe them as later-day Impressionist pictures. Even Gaugin's aim is in the main subjective, and it is only the left-handed workmanship which differentiates Post-Impressionism from other mere hack transcripts from nature done within the last thirty years. Monsieur M. Denis alone has brought a decorative or symbolic element to this 'agony of Impressionism,' and with him we are on familiar ground. Unlike the poor, mad Van Gogh and the other hermits of individuality, his work is allied to current tendencies in literature and music. His art is the fashion; it has the qualities and faults of fashion.

Ever since the rumour in France of English pre-Raphaelitism and the Anglo-Belgian influence upon the crafts, something which might be termed 'the cult of the Lily' has been intermittent in Paris, and spread with the importation of Liberty fabrics. This tendency has other roots in Symbolist literature; it can be traced back even to the success of *Parsifal* and what was written about it in Paris. On the whole, many of these elements are admirable in themselves, and revert to past noble aims and achievements. A trace of neo-classicism is never long absent from French thought, and Pan and his nymphs have been invited to stray in the vales of the new Avalon.

Monsieur Denis has caught something of all this; he is 'neo-pagan,' he is also a neo-Catholic; he remembers the great Puvis de Chavannes, but without the nobility of vision, breadth of design, and largeness of emotion of that master. I am irresistibly tempted to describe him as the new 'Paris de Chavannes.' He exhibits in the spring and autumn salons, and has 'arrived' unsigned from the realistic gardens of Klingsor, having conquered the Kundry of Impressionism. His art is bland, ingenious, not without charm; it is only in his use of colour that he forces a pleasant talent of compromise. Some day he may pass within the dome of the Institute of France.

Germany, who is omnivorous of anything Parisian, is enthusiastic about Maurice Denis; and if to-day we can hear energetic Wagnerian singers cry '*Ach! ach! ich bin nicht beglückt*' to the pale harmonies of *Pelléas et Mélisande*, so we shall find the decorative canvases of M. Denis enthroned in provincial museums, where the more local Boecklin ruled but yesterday, ranked in estimation only below the masterpieces of Van Gogh, in whose work lies the future, with which we have been threatened by some fervent critics, since he is at once the savage, the madman, and the child.

Has failure a magnetic attraction of its own which brings failure to failure, or is it merely united by a common effort to reach success which belongs to others? Lands without art have welcomed Cézanne; not all France, nor England, nor Holland,[1] but Germany, Russia, and America. Those who failed as Impressionists have found salvation in his example. Foremost amongst them is Van Gogh.

Some twenty years have passed since I last read extracts from the interesting letters of this man who suffered under the stress of religious and artistic mania. Do we quite realise the strange atmosphere in which some later-day art moves – the impatience to arrive, the exasperated and paradoxical theories in which Proto-Byzantinism has been possible? In Russia a blind painter, once an

Impressionist, gropes to-day with pins upon canvases; between the spaces he smears barbaric colour, thereby obtaining more curious and 'advanced' results than does Monsieur Matisse even. A strong German Impressionist – my informant is a museum official – proposes to work in the future in coloured wools because nothing 'new' can be done with pigment. Alas for the worsted pictures and flowers of our grandmothers, how gracious, new and instinctive was your gentle art!

Into the studio atmosphere which Zola described some twenty-five years ago in *L'Œuvre*, poor Van Gogh started 'to find himself' and to sink under the stress. With his religious frenzy we are not concerned: – that he should have cut off one of his ears in that kind of place in which Shakespeare has placed part of the fourth act of *Pericles*, to prove that physical pain does not exist, is interesting only to a doctor. Of his struggles and the mental 'strabism' of his art we can form some idea in his weird canvases, where the commonplaces of 'painting from nature' have become tortured by some horror of vision of which he may have been not wholly conscious. I have seen his fate compared to that of Nietzsche, in whose life we see the rapid and overwhelming accumulation of complex faculties crushing in the brain that could no longer bear their weight. The poor mad painter Van Gogh grasped at nothing; the soul did not bruise itself against the outer wall of adamant beyond which lies the unknowable. Compassion should make us silent upon the humble abnormality of his work and thoughts, had some critics not pointed to him as a master for our study and example.

What a subject for some Russian novelist in the account of this man, who sought his life, possibly in some chance moment of insight, when the aspect of the picture he was painting flashed upon him in the field where he shot himself. The shot failed, leaving him for two timeless hours to struggle some few paces to the roadside inn beyond – struggling without visible movement perhaps, stirring only inch by inch towards the place where he would creep to die. There he lingered on for two days more, his teeth clenched upon an unlit pipe, with face and soul locked in silence, till the pipe and its ashes dropped out upon the floor, and consciousness and agony had ceased.

This history of much recent art in France is like a blind battlefield. Impressionism, and the reaction within its ranks, has been embittered by the wish to owe nothing, save only to itself, or admit the debt which each artist pays willingly or unwillingly to his fore-runners in the past. How strange the thought that the language of art which has centuries of experience in its structure can be made anew, that the vocabulary and matter alike must change, lest we repeat what has already been said!

I would avoid all ludicrous pseudo-scientific exaggeration of those common vicissitudes in life which pass unnoticed in the many, but may be made significant in men who are before the public. I have no wish, like the followers of Lombroso, to open a grave to examine the dentation of Dante, to detect the fancied sign which might class him among potential criminals. Yet all critics have not noted or valued the obscure hints at folly, the signs of aphasia and the obliquity of vision in many later-day works of art. In other walks of life, these shortcomings have serious disadvantages – colour-blindness unfits a man for the navy. Some critics would seem to prize all accidents that can be written about; they praise individuality whatever its character, and novelty whatever its kind. There is an affinity between egotism, madness and anarchy which is but another form of madness. Many maniacs feel they have a vocation and set no store upon our common experience: this is the case with many of these painters. Have the English advocates of Post-Impressionism realised this? But of these it is difficult to speak who has not cast his lot upon a chance, and sacrificed the wisdom of centuries of thought to catch at opportunity. Their aim has nothing to do with art or its future, it is but a new phase of self-advertisement.

NOTE

1 Since this was written a collector has presented to the Rijks Museum, Amsterdam, a selection of Post-Impressionist pictures. He has since gone mad, and died in an asylum. [Ricketts's note]

38. T.B. Hyslop, 'Post-Illusionism and the Art of the Insane'

Nineteenth Century, February 1911, 270–81

Theodore Bulkeley Hyslop (d. 1933), Physician Superintendent to the Royal Hospitals of Bridewell and Bedlam, was invited to lecture to the Art Workers' Guild and the text was published in the *Nineteenth Century.* Though he never mentions the Post-Impressionist artists by name his talk was clearly a degrading attempt, on the model of Max Nordau (1849–1943) and the criminologist Cesare Lombroso (1836–1909), to suggest parallels between Post-Impressionist art and the art of the mentally disordered.

Of late we have both seen and heard so much of post-impressionism in art, and there appears to be so much doubt in the public mind as to the real meaning and significance of the works which have been exhibited and heralded as indicating the approach of a new era in art, that the time seems opportune to discuss the subject of post-illusionism as met with in degeneracy and in the insane.

It would be regarded as presumption and as beyond the legitimate province of the writer were he to attempt to criticise the artistic efforts of those who are not in asylums, so the following article will be confined to the consideration of what he has observed with regard to art and degeneration in asylum practice. Indeed, the only criticism with regard to post-impressionism now offered is quoted from an insane person who informed the writer that, in his opinion, only half of the post-impressionistic pictures recently exhibited were worthy of Bedlam, the remainder being, to his stable perception, but evidences of shamming degeneration or malingering.

THE IMPORTANCE OF BEING IN EARNEST

The insane artist is usually in dead earnest, and beyond what is prompted by his morbid rise in self-consciousness, intense egomania, and a desire to express or reflect the workings of his disordered mind, there is, as a general rule, no other or ulterior motive to tempt him to distort or misinterpret the evidence of his senses – *i.e.* he does not seek to deceive the critic or the public; and, although he may be an egomaniac, his artistic efforts are mostly for art's sake alone, and they merely reflect the character of his own imaginings.

In dealing with the work of an insane artist the positive manifestations of sensory or motor defects displayed therein do not demand our study as much as does the something, caused by disease, which prevents the artist from being able to recognise and correct such defects – *i.e.* our attention is apt to be arrested by faulty delineation, erroneous perspective, and perverted colouring, but these form only positive symptoms of decadence, and they do not give us in all cases the measure of the negative lesion which may be due to disease. This holds good not only for the insane artist but also for his critic; and, as we shall see presently, both the insane artist and the borderland critic have certain characteristics which are peculiar to them.

Degenerates often turn their unhealthy impulses towards art, and not only do they sometimes attain to an extraordinary degree of prominence but they may also be followed by enthusiastic admirers who herald them as creators of new eras in art. The insane depict in line and colour their interpretations of nature, and portray the reflections of their minds, as best they are able. Their efforts are usually not only genuine but there is also no wilful suppression of skill in technique, which, were it otherwise, would brand them as impostors. They do not themselves pose as prophets of new eras, and, so long as they are in asylums and recognised as insane, both they and their works are harmless, inasmuch as they do not make any impression on the unprotected borderland dwellers from whose ranks they otherwise might enrol a large following.

An art exhibition in an asylum excites as many cries of admiration as of pity, for here we find much to praise and profit by. Seldom is the artistic instinct or technique so far deteriorated as to leave no sense of beauty in line or colour, and, as a point of diagnosis, it is to be noted that, where no feature of beauty or workmanship exists in the work of one who is known to have formerly possessed both artistic instinct and skill in technique, the defective character of such work is due either to gross cerebral degeneration (such as we find in general paralysis of the insane or in organic dementia) or to imposture. As a matter of interest, the writer may state that he has never seen such an instance of wilful imposture in art by an insane artist.

Degeneracy in art sometimes takes a fairly definite course. A genius who is also a degenerate may influence the trend of art. His imitators, with their more limited capacities, form a subspecies, and they in their turn transmit in a continuously increasing degree the peculiarities and abnormalities which become ultimately merely evidences of gaps in development, malformations, or infirmities.

The artistic works of lunatics, however, do not always bear evidence of degeneration. The ideas of the paranoiac (or deluded person) may be grotesque and fanciful, but the artistic merits shown in his works may be great. Except in conditions of progressive paralytic dementia and of gross cerebral degeneration the evidences of deterioration may be merely manifestations of disordered thought and imagination. All merit is neither obscured nor lost. When, however, no artistic merit is observable to the fully qualified normal critic it usually means that there never has been any development of the artistic faculty, that the faculty has been lost through disease, or that there has been wilful imposture.

In some forms of progressive mental and physical degeneration (dementia and general paralysis of the insane) there is usually a retrogression or impairment of the highest evolved and latest acquirements. This impairment extends gradually back until the degenerative process affects even the most stable of the bodily and mental functions. In general paralysis the musician loses his power over his fingers, the linguist forgets the languages he has latest acquired, the elocutionist blurs his phrases, and the expert fails in the technique of his handicraft. In artists suffering from general paralysis there is a retrogression, both sensory and motor, of the artistic faculty. Sensation and perception of colour, form, and perspective become impaired. There is also loss of the tactile and of the so-called muscular senses so essential to the proper co-ordination of movement. Not only do they suffer from tremors, but also from failure to co-ordinate the various groups of muscular activities. Hence the executive mechanism becomes defective, faulty, and impotent. This gradual retrogression of the mental and physical functions results ultimately in a pathological return to the crude and rudimentary conditions of barbarism.

In sculpture, as portrayed by the paralytic in his early stages of degeneration, the work may be sensuously charming and excellently executed, and the perfection of its form may cover even what may be suggestively pornographic or even immoral. It may be attractive or repellent according to the mental bent of the critic. When, however, the work is prompted by ideas which are repugnant to good taste, and depicted in all its ugliness by a technique devoid of all artistic merit, and stripped of all evidences of those finger co-ordinations and adjustments acquired through education and practice, then the predilection in its favour of any critic is open to the charge of dishonesty or degeneracy.

The intellectually beautiful, consisting as it does of representations, concepts and judgments, with an accompanying tone of feeling elaborated in the subconscious, stands above the merely sensuously beautiful about which there can be but little scope for the higher processes of mentation. Insane æsthetics grow enthusiastic over their own creations, which, to the sane, are absurd or even repulsive. The insane sometimes take glory in the attention they excite, and there appears to be no limit to their eccentricities. So long as they are confined in asylums, however, they do not rank as cranks or charlatans, but as degenerates. They do not voluntarily shun the true and the natural as being incompatible with art. It is by reason of their disease that they ignore all contemporary ideals as to what is beautiful, significant, and worthy to be portrayed, and it is thus that free play is given to the workings of their defective minds, and whereby they evolve their absurd crudities, stupid distortions of natural objects, and obscure nebulous productions which, being merely reflections of their own diseased brains, bear no resemblance to anything known to the normal senses or intellect.

POST-ILLUSIONISM

The distorted representations of objects, or partial displacements of external facts, are known technically as 'illusions.' Their psycho-pathological significance is great, and they may arise in conse-quence of the fallacy of expectant attention (whereby the image of the expected becomes superimposed on that of the real) through toxic affection of the brain cells (as in alcoholic post-prandial illusionism) or as the result of faulty memory (paramnesia, distorted memory, whereby post-illusionism or false post-impressionism becomes manifest). Post-maniacal illusionism is almost invariably distorted, and the faulty representations bear little significance except as manifestations of disease.

One psychological (and æsthetical) fact to be noted is that, no matter how whimsical, absurd, perverted, or unreal in its nature or relations an illusion of the senses may be, it can never be constructed from data other than from those derived primarily from reality. The trouble does not lie with the varied aspects of nature, which feed the mind through the special senses, but with the diseased mind which fails to digest the sensory pabulum so derived. Nature itself frequently endeavours to treat such mental dyspeptics by its appeal for a simpler diet, and a taste for the perception of objects devoid of all condiments and the numerous unessential attributes of perception acquired by conventionality and civilisation. This craving for what is crude and elementary is nevertheless significant of a return to the primitive conditions of children, and sometimes betrays an atavistic trend towards barbar-ism. Certain of the insane exemplify this tendency in a marked degree. They lose not only their finer perception of linear dimensions, relative proportions and planes in perspective, shades of light and effects of atmosphere, but also the power of giving adequate expression to what is actually perceived. Thus the pathological process underlying reversion to a primitive type of simulation of barbaric art is frequently characteristic of brain degeneration. The works themselves reveal nature as reflected from distorted mirrors: the mirrors being but the psychical equivalents in consciousness of the morbid activities within the perceptive centres of the brain.

Many insane artists do not see nature as do the sane. The soul peeps from its dwelling-place devoid of all the conventionalities and harmonies of line and colour, and to the normal individual the result is disconcerting and incongruous. Were it not that the condition is pathological, and that disease prevents these unfortunates from recognising things as they really are, we should be tempted to lose our sense of toleration and say to them in parliamentary language 'enough of this tomfoolery.'

The artistic efforts of the insane, even when atavistic, almost invariably betray some indications either of something lost or of something to be gained – *i.e.* there is some trace of beauty or of technique left, like the mast, to show the wreck. Failure to find any such trace indicates either that the cerebral and mental devolution of the artist is well-nigh complete, or that there is a background of ignorance or deceit.

As cerebral degeneration progresses, the artistic representations become so negative in quality that for any person other than the artist himself they have no meaning and arouse no feelings other than those of pity. The works themselves have neither pictorial nor symbolic value, and their defections can be counterbalanced only by the hidden meanings in the minds of the insane artists themselves.

Sometimes the works are, in their defective drawing and awkward stiffness, reminiscent of the old masters; but, be it said to the credit of the insane, there is seldom any conscious or voluntary withholding of the skill they may have previously acquired. It is, as has been said of the old masters and some modern impressionists, the contrast between the first babbling of a thriving infant and the stammering of a mentally enfeebled grey-beard. This retrogression to first beginnings, and the affectation of simplicity is frequently seen in degenerates, and it has been described by Nordau as 'painted drivelling or echolalia of the brush.'

ANALYSTS AND SYNTHESISTS

When rightly prescribed, catharsis, purging, or purification of a system may be beneficial; but the love of wholesale depletion or destruction of the products of evolution, without due regard to their significance in the trend of life, society, and art, is merely evidence of wanton stupidity. When an artist reduces a composite whole to its component parts he becomes, not a synthesist, but an analyst. He leaves the reconstructive process to the imagination of the critic. He represents light, not in its composite form as perceived by the normal eye, but as dots, blobs, lines and squares of primary colours, leaving the task of synthesis to the imagination of others.

Women take their clothing to pieces with the object of reconstructing the various articles to suit the fashions of the moment. The insane, on the other hand, merely destroy: they do not reconstruct. So it is with some of the degenerate artists who divest themselves of all their acquirements, but are incapable, by reason of disease, of reconstructing a work of any artistic merit. It is easier to destroy than to construct, and the process of dissolution proceeds along the lines of least resistance.

The degenerate may be a genius, and he often is one; but seldom does he open up new paths which lead to true higher development. That hysterics and neurasthenics sometimes swear by him, and imitate his extravagances, goes for little. Glaring colours and extravagant forms have great attractions for hysterical persons. Charcot's researches into the visual derangements in degeneration and hysteria furnish us with an intelligible explanation of what Nordau terms 'impressionists,' 'stipplers,' 'Mosaists,' 'papilloteurs,' 'roaring colourists,' and dyers in grey and faded tints. Their efforts are genuine results of physical disease. Nystagmus (or quivering of the eyeball) is responsible for a want of firmness in outline, and affections of the retina for distorted zigzag lines and for defects in the perception of colour. There may be a predilection for neutral-tones or for glaring primaries: this predilection being due to the abnormal condition of the nerves and not to any observable aspect of nature.

The psychologist, however, is not in any way deceived by the glaring crudities of those artists who – disowning all factors other than sensations – present their works in the form of gross lines or blobs of primary colours. Acting on their knowledge of the complementary qualities of red, blue, and yellow, they present them, not as they perceive them in combination, but as primary reds, blues, and yellows. They utilise their knowledge gained through science to hoodwink themselves into the belief that by representing the ultimate and crude elements of colour it becomes easy for others to recombine them into a composite whole. This is known to be a fallacy psychologically, and such pseudo-art productions, instead of reflecting the realities of the external world, reflect but the pseudo-scientific mental conceptions of the artists. This post-illlusionistic resorting to a symbolic suggestion of what is merely known to consciousness is false, and the symbols are as frequently like their objects as the symbol H_2O is like water to the visual sense.

SYMBOLISM AND MYSTICISM IN ART

Symbolism is rife in the insane, who undoubtedly do perceive mysterious relations between colours and the sensations of the other senses. So-called secondary sensations, however, although occurring in great variety, are never theatrically displayed for the benefit of the public. Sane critics would liken such efforts to those of the decadent Gautier, or of Baudelaire who died of general paralysis of the insane. Symbolism in insane art is sometimes invested with a high significance by the artists themselves. Fortunately, however, both they and the public are protected from the vapid and sickly sentimentalism of the borderland critics: *i.e.*, those critics who, in order to arouse curiosity, make a noise with something new and sensational, and by pandering to the gaping uncritical attitude of the presumably sane endeavour to raise a market for the disposal of commodities of palpably fictitious value.

Many lunatics are mystics and imagine they perceive ususual relations amongst phenomena. They see signs of mysteries, and they regard ordinary external phenomena as but symbols of something beyond. Their earlier impressions become blurred and indistinct through disordered brain action. Faulty memory, and the superposition of distorted former imaginings, give to present objective facts a sense of mystery. Thus, a blue colour will arouse associations of many things of blue, such as the sea, the sky, a flower, &c., which become merged into the primary percept of blueness and invest it with other meanings or associations. It is, of course, well-nigh impossible to follow the suggestions aroused in the insane mind by a primary impression. The consciousness is befooled and wrecked by will-o'-the-wisps and inexplicable relations between things. Things are seen as through a mist and without recognisable form, and both the insane artist and his degenerate critic forge chaotic meaningless jargon to express what is seen or felt. The pseudo-depth of the mystic is all obscurity. Outlines of objects become obliterated, and everything which has no meaning becomes profound. The step from mysticism to ecstasy is short, and, with failure to suppress the wanderings from the real to the imaginary, there are produced for the onlookers such manifestations of imbecility as can find adequate expression only in pseudo-art, pseudo-music, so-called literature, or in the ravings of the insane.

The indifferently interpreted, blurred and nebulous, sensory impressions of early general paralysis are sometimes suggestive, not of a renaissance of mediæval feeling or of post-impressionism, but of a return to primitive barbarism. Inside asylums such a renaissance deludes neither the patients nor their attendants; nor does it provide an excuse for æsthetic snobs to found a fashion meriting little else than laughter, wrath, or contempt.

The works, although pitiable in themselves, are sources of self-congratulation to the artists, who boast freely as to their merits and hidden meanings. They estimate their value according to their own supersensuous imaginings rather than to any mastery of form or beauty of colour. The clumsier the technique, the deeper its meaning. Faulty drawing, deficient colour, and general artistic incapacity, stamp such works as pre-Adamite, eccentric or insane.

In maniacal states there is inability to fix the attention for long. The impressions of the external world as derived through the defectively operating senses become still more distorted by disordered consciousness. Hence the faulty representation of external realities and the exhibition of what are manifestly illusions.

When faulty memory is brought into play, the distortions become even more manifest, and the vagaries of the post-illusionists find therein their full expression. The conscious state of a person receiving impressions in the domain of one sense only has been termed 'impressionism.' The impressionist pretends to see before him merely masses of colour and light in varying qualities and degrees of intensity. In disease, purely optical perceptions may occur without any activity of the highest centres of ideation. This is also one of the first steps towards atavism. The concept is absent, and nothing remains but a simple sense stimulation. The undeveloped or mystically confused thought which exists in savages is fully exemplified in the childish or crazy atavistic anthropomorphism and symbolism so prevalent among degenerates. A predilection for coarseness in line or colour is symptomatic of degeneration, and obsessional explosions of obscenities, so characteristic of some forms of mental decay, show themselves as 'coprographia' – *i.e.* pertaining to lust, filth, or obscenity.

Most paranoiacs (deluded persons), who, as a rule, do not suffer from disorders of their physical or co-ordinative mechanism, present in their artistic works manifestations of genuine and fertile talent. In spite of the evident craziness of their ideas, their technique is usually too skilful to appeal to gaping simpletons as mysteries and revelations of genius. Their critics find in their works but little scope for the employment of words of empty sound and devoid of meaning. Asylums do not harbour such puppets, nor do their inmates in their intellectual darkness become the devotees of the snobs of fashion.

The ego-maniac has but little sympathy with, or capacity to adapt himself to, nature and humanity. His perverted instincts render him anti-social even in matters of art. Real lunatics do not form a league of minds, for the simple reason that they are concerned only with their own individual states and experiences. Some feel a passionate predilection for all that is hideous and evil, others are all for good.

EGO-MANIACS IN ART

The crude, barbarous splendour of the insane artist's productions is, as we have seen, often due to optical illusions. Egomaniacs sometimes become decadent, and surprise us by the increasing barbarity of their taste and technique. They banish from their horizon all that is natural and surround themselves by all that is artificial. Sometimes their perceptive powers are wholly inaccessible to the beauties in nature, or they suffer from a mania for contradiction of, or revolt from, the realities of things. The egomaniac regards himself as the super-man; whereas he is often merely a plagiarist or parasite of the lowest grade of atavism. He sometimes becomes a post-illusionist, and subordinates his highest nervous centres and consciousness to the perceptive centres and instinct. Sensations are perceived by him, but they go no further. The primary impressions are reflected in their distorted state. The beautiful things in nature have for him no existence. He himself is the creator of all that is wonderful and good, and the reflections from his turbid mind are, to his own way of thinking, examples of art for art's sake.

BORDERLAND IMITATORS, CRITICS, AND MALINGERERS, AND THEIR EFFECTS ON SOCIETY

Borderland dwellers, *Dégénérés Supérieurs*, or Mattoids, comprise the hosts of those who follow, what they are foolishly told to believe to be, new eras in art. The insane person differs from the borderland dweller in that his insanity prevents him from adapting himself to, or following, any new fashion in art. They have this in common, however. Their revolutionary effects on art may be not only pathetic, as evidences of ignorance and absurdity, but they may also be genuine. One point to be noticed is that borderland dwellers alone are inspired by the diseased ideas of the insane. As is the case with hypnotism, Christian Science, and many other crazes, neither the sane nor the insane are affected by them. The founder and his disciples may be sincere; but, sooner or later, the participants of the new doctrine form a rabble of incompetent imitators who lack initiative, and quacks who abuse their membership by reason of their greed for money or fame. These latter follow merely the dictates of their pockets and easily prey upon a too gullible public.

Sincere originators have even been followed by dishonest intriguers, who invent beauties where none exist. None of these movements herald really new eras, being merely attempts to destroy or suppress the advances and acquirements of the age and endeavours to hark back to the past when the æsthetic sense and skill in technique were but ill developed.

The insane, however, are emancipated from traditional discipline: they have, in fact, a contempt for traditional views of custom. Hence their departure from many of the ideals in art which for thousands of years have become gradually matured and more or less fully established. This departure being neither foolery nor knavery, but merely degeneracy, there gathers round it no concourse of gaping imbeciles greedily seeking for revelations.

In asylum practice, neither mysticism, symbolism, nor any other 'ism,' finds a foothold for advancement, and inasmuch as lunatics are free from sordid motives they are harmless in their ignorance and segregated in their snobbishness. They do not found so-called intellectual or æsthetic movements and by futile babbling and twaddle seek to propagate what may be, as a matter of fact, nothing else than idiocy or humbug.

To the borderland critic who is ignorant of disease and its symptoms the works of degenerates are sometimes more than mere sources of amusement; they may serve to provide inspiration for his own unbalanced judgment. They are seldom deliberate swindlers who play up as quacks for the ultimate gain of money. The truly insane critic is usually definite and significant in his language, and he seldom seeks to cover his ignorance by volubility in the use of obscure and purposeless words. Such being the case, there is no scope for the promotion of bubble-company swindles in asylums, and there is never any danger of leading the public by the nose.

The pseudo-artist is common in asylums and has aspirations which he is unable to justify; whereas pseudo-art is almost invariably the product of imposture – *i.e.* in asylums pseudo-artists are numerous, but pseudo-art is rare. In the former, their performance is quite unequal to their desire; whereas, in the latter, the works are usually products of deceit. The unbounded egoism of lunatics also prevents them from discovering in the works of others beauties in what are evidently the lowest and most repulsive things.

That the works of insane artists may be crude, absurd, or vile matters little so long as they exert no corrupting influence on society, and so long as society fully appreciates their pathological significance. Unfortunately, however, some creations which emanate from degenerates are revered by the borderland critic, blindly admired by the equally borderland public, and their real nature is not adequately dealt with by the correcting influence of the sane.

Moreover, the insane critic is honest in his criticisms, and views the works of his insane comrades of the brush with an honest and fearless eye and judges them from his own mental standpoint. Seldom or never does he conform to the artist's interpretations of nature, and, although he may recognise the artist as being an imbecile or dement, his courage seldom fails him in giving expression to his real convictions. This is characteristic of the insane, who know no fear, who have no conventionality or æsthetic fashion to conform to, and who have no axe to grid. Undoubtedly their intense egotism prevents them from perceiving their own shallowness and incompetence, and in asylums individualism pertains as in no other community. Seldom it is that the truly insane – those who have passed the borderland and have become certifiable lunatics – imitate each other in art. Rather do we find imitative tendencies in those who are technically and legally neither sane nor insane – *i.e.* in that enormous class which comprises the 'borderland.' This rabble of hysterics, neurasthenics, weaklings and degenerates have nothing of their own to say, but, by means of a superficial and easily acquired dexterity, they imitate and falsify the feeling of masters in all branches of art, and not only do they injure true art but they also tend to vitiate good taste among the majority of mankind.

Among this class are also to be found vast numbers of incompetent critics who, for reasons best known to themselves, welcome these bunglers of the brush and encourage – although inimical to society – abuses of true art which are in reality but instances of mean childishness and demoralisation.

To the physician who has devoted himself to the special study of nervous and mental maladies there is seldom any difficulty in recognising at a glance the manifestations of shamming degeneracy of the malingering. That there should be malingerers in art is, nevertheless, a question open to discussion. That malingering in art should occur in true degeneracy, apart from hysterical simulation, would appear improbable, and certain it is that in asylum practice there is but little evidence in favour of such a supposition. In malingering post-illusionism there is usually some evidence of higher mental activity, as shown by the artists' knowledge of the theory of colour vision, a knowledge of which they freely avail themselves in order to falsify the objective realities before them.

Stigmata of degeneration are not confined merely to artists and their works. Critics who fall into raptures and exhibit vehement emotions over works which are manifestly ridiculous and degrading are themselves either impostors or degenerates. Excessive emotionalism is a mental stigma of degeneration, and Max Nordau's criticisms apply very aptly to some critics whose own excitabilities appear to them to be marks of superiority. They believe themselves to be possessed by a peculiar spiritual insight lacking in other mortals, and they are fain to despise the vulgar herd for the dulness and narrowness of their minds.

> The unhappy creature [says Nordau] does not suspect that he is conceited about a disease and boasting of a derangement of the mind; and certain silly critics, when, through fear of being pronounced deficient in comprehension, they make desperate efforts to share the emotions of a degenerate in regard to some insipid or ridiculous production, or when they praise in exaggerated expressions the beauties which the degenerate asserts he finds therein, are unconsciously simulating one of the stigmata of semi-insanity.

THE REMEDY

The insane art critic never asks himself 'what sort of a bad joke is this?' – what does this artist want me to believe?

Morbid aberrations may serve as causal factors in the production of what is sensual, ugly, and loathsome in art, and without doubt the artists may have been quite genuine and sincere in their efforts; but, inasmuch as our asylums do not give shelter to all perpetrators of such mockeries or travesties of good taste and morality, it is difficult to suggest a remedy or means whereby they can be suppressed.

The insane art critic who scribbles incoherent nonsense for his fellow-sufferers is simply to be pitied and treated as an honest imbecile and not to be punished as a rogue. If he sees hidden meanings in mystically blurred and scarcely recognisable objects, the misfortune, and not the fault, is his, and for what to us may be abominable, ignoble, or laughable he may have some subtle sympathy or affection.

The borderland critics, however, must ever run the risk of being classed with rogues or degenerates. How best to treat them is another matter. From motives of humanity we are prompted to aid in the survival of those who are biologically unfit; but, with regard to the encouragement, or even toleration, of degenerate art, there may be, with justice, quite another opinion.

39. Huntly Carter,
'The Independents and the
New Institution in Paris'

New Age, 25 May 1911, 82–3

The first Post-Impressionist exhibition awoke a new and more critical interest amongst the British in what was happening in Paris. Huntly Carter (d. 1942) was a journalist, traveller, lecturer and writer on the arts who was later to publish *The New Spirit in Art and Drama* (London: Frank Palmer, 1912) (see no. 67). Like many of his contemporaries, he found the rapid changes in the art world bewildering yet exciting. He strongly admired the work of Fauve British ex-patriots (especially that of J.D. Fergusson) but he notices in this review that cubism (as yet unseen in England) had come into vogue in Paris.

Crossing to Paris I was given a suitable reminder of what the new movement in art and the new intuition really are. From New-haven to Dieppe the sea spread like a waveless plain saturated with vaporous air. Trailing rhythmically across this green plain were soft amethyst columns of vibrating light that dipped far above and below into the sea-dissolved air and the air-dissolved sky, seeking infinity islanded by the vast world of consciousness. From Dieppe to Paris there were corresponding symbols of the rhythm and continuity of life. In the passing landscapes stained a faint green, in hills shouldering the pink and amber of the westering sky, in blossoming orchards shining like pink snow under the sinking sun, in newly fledged fields and vital waters vaporous with the fragrant stream of life; in all these were signs of creative evolution. Art is the symbol of infinity; the new form of art is the perception and expression of continuity. The new intuition is the apprehension of Reality underlying forms of life, of things living and evolving.

At the exhibition of the Société des Artistes Indépendants in

Paris I came across a manifestation of art which purports to be living and evolving. The exhibition was held in an elongated and well-lighted tent running parallel with the Seine. There were seventy rooms or divisions containing 6,745 exhibits. At least 5,000 of the exhibits might walk into the Seine; they would never be missed. The first impression of this amazing exhibition was disastrous. For hours afterwards I felt as though I were walking in a fit. I was, however, fortunate, since critics with weak hearts are daily carried out dead. As I slowly recovered from the rush of the complete nude, and as the sensation of sea-sickness caused by the avalanche of paint decreased, I became aware of the true nature of the elements that affected me. I then saw they were the manifestation of a movement tremendously big, tremendously vital; not the mere outcome of trickery and charlatanism, but of intensity of vision carried to the highest pitch, almost to madness. I had indeed been caught in a tempest of diverse temperaments seeking with amazing courage to express feverishly, each in its own way, an astonishment at a new vision of life created by the massacre of tradition. Men and women were standing before the forms of nature and social life concentrating their attention, not on the object itself, but on the underlying idea, not on the so-called dead matter, but on matter living and evolving according to the eternal law of continuity; thus projecting, focussing and fixing personality in a moment of intense clear vision.

I noticed that the vision had paralysed the old forms of speech. To these artists words had wandered from their original meaning; and the current meaning meant nothing to them. So answering to the need, a new form of speech, new symbols were arising. From all sides came the voices of primitive, yet highly matured, men raised to frenzy before the happening of a miracle, the revelation of some hitherto unknown beauty of a star, of some hitherto unperceived majesty of the elements. On all sides artists were engaged in processes involving the production of inspired pieces of music, putting together new pictorial material, composing lyrics in colour, lyrics in line, lyrics in light to the new deity, rhythm. This search for new material, new language, new means of expression was not beyond criticism. Not all the elect painter-composers were masters, and not all the masters were perfect. I was conscious indeed that many of them were incoherent, many making a fearful hash of things. But it was, at least, a hash of the right sort. They were going wrong, but in the right direction. Out of so much that is wrong will emerge something supremely right.

Amongst the Indépendants who are exhibiting this year, and who are expressing the idea and the continuity of things, in continuity of line, distribution of colour and light, arrangement of mass, and space suggesting infinity, places are due to some English and American artists of distinction. In passing through the dozen or so rooms where the best work is happily centred, I paused to note how the tremendously vital line in the six vivid compositions by J.D. Fergusson – a pioneer and one of the really big men in the movement – goes swinging in a vast way across his canvases, enveloping nudes and still life alike in a broad harmony of moving outline, accentuating the finely orchestrated colour, and riveting attention on a statement of ideas, stamped with a quality that makes even some of the leaders sink into oblivion. I sought unsuccessfully for Peploe, whose recent work, I was told, had gone off to exhibit itself in Glasgow. Glasgow needs it. I came across Anne Estelle Rice putting together some musical material in paint, having due regard to movement of line and balance of colour. Her studies are fresh and invigorating in colour, and they are uttering the latest technical notions free from clichéisms and illiteracy. I wandered into Jessie Steward Dismore's *Jungle*, and saturated myself in its delicious harmonies of colour moving against a background of golden infinity. Miss Dismore should decorate London fairy plays. Lyonel Feininger's caricatures in colour and line nearly floored me. His figures are brilliantly drunk with movement. They tear down the gallery like men and women coming out of the Rat Mort at six a.m. Having recovered my breath, I turned to make the acquaintance of two astonishing Russians. Pierre Kantchalowsky's search for wonderful colours harmonies suggested abnormal demands on the painter's reserve of nervous force. Ilia Machkoff is a giant. He offered me a dish of fruit that dominates the entire exhibition. He had seized it with immense strength and understanding, and set it revolving in a perfect maze of pure colour.

Coming to the Frenchmen and others, I met the well-known back-number, Signac, and a batch of his followers, as well as the followers of other back-numbers, Sisley and Carrière. Flandrin, who is fairly flat, showed me a sample of his *Cavaliers au Bois*. Van Dongen was dressed in interesting colour. Chabaud was out of sorts. Friese, too, was hardly up to the mark, his best effort being a sketch of a forest and figures, in which the arrangement cleverly suggested the oneness of man and nature. I met Segonzac feverishly hammering out a black and white nude. There was a suggestion of immense power in his hammer strokes. Near by was Georges Rouault, exhibiting swirling decorative figures and landscapes caught in a mist of sweeping lines. Matisse was too busy booming his own greatness at tea-parties to contribute more than a washed-in idea of a very fresh interior, very simple and very big in treatment. It is a suggestion to be carried out in the theatre.

Apparently the star of Matisse is falling, and Picasso is the father of the new extremists. Perhaps the most remarkable feature of the exhibition was the latest development of Picassoism or Cubism. Here the attempt to express, not the object itself, but an analysis of the underlying idea, is carried to an extreme limit. The aim is to record the feeling or sensation created by an object, which is subtly expressed in a rhythm of shapes, cubes, prisms and so forth, just as Max Beerbohm expresses the idea of character in a rhythm of lines. Thus the dyspeptic is not represented, but the feeling created by the word dyspepsia. The impression of the latest works by Picasso and his followers is at first bewildering. Painted in a low key they have the appearances of masses of stone ruins piled one upon the other. Indeed it is as though an early Impressionist, Cezanne or Signac, had scientifically and laboriously built up a picture, bricklayer-fashion, with coloured bricks and one of the cubists had kicked it down in disgust and poured grey mortar over it. But as one looks at them the idea gradually disentangles itself. Thus from a tangle emerges a woman reading a book, and surrounded by the objects created by the feeling for the incident. In this way the expression of ideas or sensations created by the insight into an object is carried far beyond that of Matisse's attempt to do the same in colour. Picassoism is appealing to some of the strongest men in the new movement. Laurentin, Lèger, and Le Fauconnier have some admirable examples of this kind of picture-making. Their work is very distinguished, remarkable for harmonious and subtle colour, and for thoroughness of drawing. The anatomy of these pictures must be full of colour. Herbin, who is not represented, has also developed cubicitis.

The absence of Herbin reminds me that the best work of several new extremists is to be found at the dealers' shops where our own

Post-Impressionist exhibition came from. The late Post-Impressionist exhibition, which sent so many respectable persons mad, is the standing joke of Paris. It was organised from the dealers, where one may discover many of the bad and indifferent and ancient things that found their way to the Grafton Galleries. As a matter of fact each dealer's shop is a Post-Impressionist exhibition owing to the prevalent mania for running Les Groups, as they are called, of Impressionists. Each dealer or private speculator makes a corner in an extremist or group of extremists, writes up the work, and waits patiently for the absent-minded millionaires. Thus Stein has sold his Monets for £400,000 and is now busy buying Matisses. Vollard has got all the Cezannes and is so proud of it that he never opens his shop, which has the appearance of Dirty Dick's at Houndsditch. He has got so many Cezannes that occasionally, when a visitor calls, he uses one for a teatray. Durand-Ruel has no end of the eighties and nineties Impressionists. The Galerie Kahnweiler has made a corner in Georges Braque, and four others. Herbin is subsidised by Sagot, whose talk on the former's marvellous colour and extraordinary use of straight lines, cubes, prisms, and the placing of objects at an angle, goes on for ever. This system of subsidising extremists is both good and bad. It enables artists to produce their best work freely; and it also encourages some to exceed the limit of extravagance. But though the system is open to endless abuse, it affords the new tendencies a sanctuary and it enables artists to be themselves. We have nothing of the kind in England. Hence the reason why art is valet to the dealer.

As a supplementary note on the *Blue Bird*, at the Théâtre Rejane, I may mention the conscious attempt to apply the new Post-Impressionist principles of continuity, by (1) lines all composed to suggest rhythmical movement, (2) by colour distribution to suggest movement, (3) by space suggesting infinity, (4) by light and atmosphere suggesting mystery. The attempt to obtain rhythm and continuity is indeed the feature of the production. Rhythm is suggested by cloths and swaying columns fretted with innumerable circling and waving lines. This rhythm is, in one scene, caught up and repeated by the nicely related lines of the sweeping steps and the circular arches of the two rostrums down stage L. and R., as well as by the flowing draperies of innumerable children. Infinity is suggested by the frequent use of black fly-borders, black 'wings,' and black stage cloths. A peculiarity of some of the scenes is that they are set right up stage, the front part of the stage being blacked out, doubtless to suggest the infinite nature of these scenes. They have no beginning or end.

---

40. Roger Fry,
'The Salons and van Dougen'

Nation, 24 June 1911, 463–4

Fry consistently mis-spells the name of Kees van Dongen (1877–1968) in this article.

The two great salons are as disheartening as ever. One remains aghast at the immensity of the effort they represent compared with the paucity of the result. In the old Salon the whole of the immense lower gallery is filled, as usual, with sculpture, each piece of which gives proof of a special aptitude, highly trained. If professional skill sufficed to the making of a work of art, what a golden age this would be! But, alas, it is the least important factor, and the one thing needful, the possession of some definite feeling which calls for expression, strikes one as almost entirely wanting. What, one wonders, does the public get from this display of skill, which it can only dimly understand; what pleasure does it get from the contemplation of these thousands of successful, gratuitous, and brilliant declarations of the artist's complete indifference to reality; and how can it choose, except by the title and the nature of the subject, between one and another? One feels that such a sight calls for some cataclysmic remedy, but we scan the horizon in vain for the purifying hordes of vandals who will help us to destroy, or at least help us to stop producing, this yearly confession of our spiritual sterility and inane vacuity of heart. The new Salon offers a similar spectacle, a little chastened by culture, a little mollified by good taste, but still essentially the same. There is, it is true, some respect for beauty; the rhetorical appeals to outworn and unreal emotions are less blatant, and the general effect is not exasperating. Still it is far from exhilarating. There is scarcely anywhere a sense of the passionate need for expression, only the habit of making pictures that are more harmonious and agreeable than those of the old Salon. Here and there are placidly sentimental landscapes like M. Réné Ménard's or M. Dauchez's, which show a personal vision. Then there are M. Jacques Blaudie's interesting attempts to keep abreast of new ideas by treating scenes of barbaric or oriental splendor with all the familar elegance of his eighteenth-century manner. There is M. Albert Guillaume, elaborately witty, with an over-polished dexterity of finish which really expresses less than his drawings for the comic papers. Among the paintings, however, M. Maurice Denis's three compositions stand out from all the rest by their distinction and classic perfection. Every year M. Denis adds to the astonishing skill of his transpositions of tone and color, so that he can give, by the adjustment of half-a-dozen flat tones, an illusion of any particular natural effect; an illusion which is far more convincing than what the plein-artists attained by their minute analyses and complex constructions. In one of these sea-side pieces the sky is a flat mass of gold, the sea all pure delicate green with rose-colored surf, the shore violet, and the girls at play – ever so delightfully grouped and silhouetted upon the background – are all rendered in pure flat masses of local color, blue, rose, orange, and violet; and yet, for all the abandonment of light and shade, for all the suppression of intervals, the effect of figures moving in air saturated with the golden dust of a sunlit afternoon is irresistible. This is, indeed, a triumphant vindication of the method of translation from the actual tones of nature to the appropriate pictorial symbols. It is a feat that could never have been performed without the patient toil of the Impressionists, and although it contradicts their principles, it justifies them by the event. The triumph of this lies in the fact that by this method the feeling of a particular momentary effect of atmospheric color is transmitted, and yet the picture has the decorative unity that has hitherto seemed incompatible with such a performance. And yet M. Denis, too, seems to me to be sacrificing too much to the perfection of his method, using it, too, for slighter and less ambitious ends than he once did. He is getting to be almost too dainty, too prettily mondain, too easy. His fancy is more delicate, more playful than ever, but his work has lost the note of passionate intensity which did not recoil even at brusqueness or oddity in the endeavor to attain expressiveness. There is no sign here of the promise, that his Nativity gave some six years ago, of an artist who might recover the abrupt expressiveness of gesture of the medieval *imagiers*. All that dramatic force has been softened and blunted by the research for lyrical charm in these scenes of modern society.

Among the sculpture in the new Salon are one or two exhibits of interest; there is a large bronze figure by M. Bourdelle, of which it is difficult to speak fairly. It has every sign of astounding talent and intelligence; a real sense of what sculpture ought to be, of how to subordinate and how to unify the rhythm; of the quantities and disposition of relief. In short, it has all the stigmata of a serious work of art, and yet it leaves me sceptical and indifferent, with no means of explaining my instinctive reaction.

M. Rodin contributes a finely characteristic portrait bust and a figure in high relief upon an inclined plane of marble. It is a winged female figure, and may be – there is no help here in the catalogue – one of the many sculptured tributes to aviation. The *gaucherie* of the limbs suggests vividly, though with a sentimental exaggeration, the pathetic helplessness of a figure flung to earth like a discarded thing. As for the Société des Indépendants, it grows and flourishes year by year, unhampered by the restraining influence of a jury. Year by year more cubicles are added, and it progresses gaily on its way down the Seine towards Rouen. There are fifty odd rooms at present. There is unquestionably more rubbish here than in the other two Salons put together; there are no seats, no large empty spaces; nothing to help one through, but for all that it is not so tiring as its grander rivals. The best of it is that most of the rubbish is real downright chromolithographic rubbish that the justest critic can pass at four miles an hour without a pang of conscience. And what is but rubbish is often of fascinating interest. Perhaps there are no masterpieces. As M. Forain is reported to have said 'C'est fini maintenant; on ne veut plus de chefs-d'œuvre,' thereby paying a quite unintentional compliment to the efforts of the younger men, who are indeed not trying to make masterpieces, but rather at all costs to get something definitely said. What strikes one most about the new movement is the variety of individuality which it inspires. Impressionism was a way of looking at things, demanding a special technique, and its votaries tended to see the same things in the same way. Post-Impressionism implies merely the release from the tyranny of representation, has set free each individual to search for the expressive quality that corresponds to his personal feeling. True, this personal feeling must be intense enough and clear enough to demand expression; but it is surprising how many of the younger men have already arrived at a consciousness of their own personality, and how varied and interesting are the racial and individual characteristics they reveal. M. Fornerod is as clearly Swiss in the sober domesticity of his interiors as Archipenho and Mlle. Vassilieff are Russian in the morbid intensity of their feeling, and the tortured extravagence of their forms; while already among the Frenchmen a classic feeling for pure beauty has begun to infuse itself. One sees it in the work of Othon Friesz of Lhotè, and of Girieud. There are already many groups among the younger men, each exploring the possibilities of expression in a different direction, or combining the results already achieved in different ways. There are those who, like Herbin, are following Picasso in his search for an artistic philosopher's stone, endeavoring to get at the intellectual abstract of form, whereby they can recreate a world of pure significance; and there are those who, following Matisse, search for an intenser unity in the balance of directions and volumes, and the just disposition of intervals – some of M. Matisse's Russian followers are, by the by, becoming really incoherent – and those who, accepting the general texture of appearance, are content to treat it with a freedom of emphasis and elimination that would have seemed incredible to an older generation, with its prepossession in favor of a literal instead of a synthetic unity of construction. Of these last, one of the chief to-day is Van Dougen, the exhibition of whose work at Bernheim's Gallery enables one to get a clearer idea than hitherto of his personality. It must be admitted that it is anything but a sympathetic one, and we may forgive anyone for missing his remarkable talent out of disgust at his predilection for corruption and vice. And yet Van Dougen is not decadent or perverse, like Lautrec. He is, after all, of the race of Rembrandt, and, for all his brutality, a strange gleam of sympathy comes through the exasperating and outrageous display of wanton depravity. There is nothing of Degas's intellectual and ironic vision of the factitious life of our cities; Van Dougen is more instructive and more spontaneous. He claims no superiority, no indifference to the life he paints with such triumphant, terrible force; but he realizes the humanity of his models beneath the corrupt artificiality of their condition. And with what astonishing assurance and breadth of handling he places them on the canvas, what tremulous vitality is in their movements, and what solidity and force of modelling he can suggest by the mere brushing of a contour. It would be absurd to call Van Dougen a moralist, nor does he make any of the pretences which allowed Hogarth to treat of vice, but for all that his attitude is not base; he sees something magnificent, something sane and fundamentally human in the brazen insolence of his type. His is none the less a turbid and imperfect art, straining at a beauty that it cannot quite attain, leaving the problem with its raw edges of cruel discords, but stated with a candor and sincerity that one is forced to admire.

I must not omit to mention M. Matisse's picture of a corner of his studio at the Indépendants. It is a large canvas, but contains nothing but a bare wall, a window, a few screens and chairs, and on the floor a rug. The perspective is by no means correct, but the artist has managed by a subtle adjustment of these rectangular forms in a strikingly coherent and indestructible unity to arouse in the spectator a sense of the amplitude of his ideal space. And the color washed on with almost crude simplicity of handling is as satisfying as it is new and strange. The picture has that rarest of qualities, glamor, in that it gives to commonplace things a significance quite beyond what their ordinary associations imply.

Those who remember the outcry which Von Gogh's portrait of Dr. Gachet caused last year at the Post-Impressionist Exhibition will learn with some surprise that Dr. Swarzenshi has had the courage to buy it for the Stadel Institute at Frankfort, where it will take its place beside the masterpieces of Dutch and Italian art. We in England shall probably wait twenty years and then complain that there are no more Van Goghs to be had except at the price of a Rembrandt.

41. Michael T.H. Sadleir, 'Fauvism and a Fauve'

Rhythm, Summer 1911, 14–18

Sadler or Sadleir, as he later spelt his name (1888–1957), was a bibliographer and son of the educationalist and collector Michael Ernest Sadler (1861–1943). His father was converted to Post-Impressionism in the autumn of 1910 (see headnote for no. 45), and at about the same time Sadleir, fils, began to buy works by Kandinsky (see headnote for no. 53). His articles for *Rhythm* on fauvism serve as a reminder that it was this form of Post-Impressionism that was most acceptable to the British, particularly in the work of Anne Estelle Rice, S.J. Peploe (1871–1935) and J.D. Fergusson (1874–1961) who was art editor for the journal *Rhythm* and whose picture *Rhythm* appeared on its cover.

The limits of this paper make it folly to attempt to deal in any comprehensive manner with the vast field of theory and discussion opened up by the latest movement in painting. I shall, therefore, confine myself to two points. I shall try to sum up to what extent the movement is revolutionary, that is to say, against what theories and practices it is a protest; and then try to suggest a few of the new ideas it brings with it, taking as illustration and example the work of a particular artist.

But before beginning I would say a word about my title. The name – Post-Impressionism – with which the movement was baptized on its appearance in England strikes me as futile and misleading. It suggests at once connexion and no connexion with the preceding school; it implies mere chronological sequence or diluted similarity. As will be seen, the second of these implications is false, while the first is merely idle. What is needed is some meaningless label, which shall serve as *nom d'école* without pretending in any way to describe aims and theories. The new movement is far too complex in its aims, far too varied in its ideals, to allow of its being summed up in a single word. The nickname of

'Fauves,' given to the artists in Paris, seems in every way suitable. But it must be given no ulterior association; it must remain simply and solely a tag.

The revolutionary nature of Fauvism can be summed up as follows. It is a reaction on the one hand against the lifeless mechanism of Pointillism, on the other against the moribund flickerings of the æsthetic movement.

The coming of Pointillism rang the death knell of Impressionism. Monet's experiments in the representation of light had been reduced to a system. The division of tones from being a servant had become a master. Signac, Seurat, Cross, Luce, van Rysselberghe, by their mathematical arrangement of spots of pure colour did succeed in achieving an extraordinary brilliance of atmosphere, and in some cases dazzling flesh values, but they sacrificed everything to this one aim with the result that in many of their pictures, their peculiar technique once removed, nothing remains.

That Impressionism should have become so mechanical was proof enough of the need for fresh vision and inspiration. There were, however, other reasons why it ceased to satisfy. Monet and his followers sacrificed line to colour and light. Strength of form and beauty of curve were lost. Then, again, under an Impressionist régime there was no place for flat washes of colour, for the massing and balancing of tones. Colour was purposely divided so as to fuse in the eye from the proper distance. The actual value of each pure red or blue or yellow went to create another composite value, and there was no attempt beyond the creation of a suffused brilliance.

This then was the impasse reached by one section of the Impressionists. Another group, this time closely allied to literature, followed up the Baudelairism which had got hold of some of the poetry and novels of the time, and lost themselves in an orgy of Satanism. Skeletons, visions of the Black Mass, posturing nudities, strange pictorial conundrums, all the paraphernalia of horror and grotesqueness occupied their attention. Such work was barren. There can be no real stimulus in artificial sensations, and the sowing of new thrills to spur the jaded palate ended, as might have been foreseen, in this first and last tortured blossoming. The limitations of Rops, Beardsley, Odilon Redon, and their host of followers – for with all their genius they are as limited as any group of artists have ever been – created a demand for width, for blood, for fresh air. Their adherence to literature and the obscurity of the riddles they contrived, led them away from their true aim.

It was Baudelaire who urged that a man should regard his life as a work of art; sound advice but dangerous, for it lies at the back of the host of affectations and deliberate eccentricities which killed the æsthetic movement and obscured what was genuine in its pursuit of the beautiful by what was merely precious. Fauvism is a frank reaction from the precious. It stands for strength and decision, alike of line, colour and feeling. It remedies the formlessness of Impressionism but keeps the brilliance, it is art and not literature, it is erratically individual and not mechanical.

But do not believe those outraged conservatives who raise the cry of anarchy in art. There is a difference between Anarchy and Revolution, the difference between wanton destruction and constructive enthusiasm. This movement is not a mere upheaval, a welter of destructive folly.

And what are the lines of advance? Their name is legion. That the development is so varied is the best of signs. There is no trace of fettering system or cramping formulæ. Almost every artist has his ideas and is working after his own plan, but is at the same time ready to welcome any new method of search, any fresh line of advance towards self-expression.

But has this motley crowd of individual workers any common aim and belief besides that of self-expression? I think so. There is one fundamental desire with which all start – the desire for rhythm. Be it of line or colour, be it simple or intricate, in every true product of Fauvism it will be present. And this rhythm is of a piece with the use of strong flowing line, of strong massed colour, of continuity. The work must be strong, must be alive, and must be rhythmical. Then there is another goal for which the Fauves are striving – decentralization of design. This aim is an important element in the wonderful decorative value of modern painting, painting which fills a space, which seems prepared to spread over any size of surface with the graceful continuity of its lines.

But I think these ideas can best be explained by reference to some particular artist's work, and no better example could be found than the work of Anne Estelle Rice, some of which has recently been on view at the Baillie Gallery.

Miss Rice, like every other leader of the Fauvist movement, is too individual to allow of her being classed wholly with anyone else. Her outlook is vigorous and personal, her methods definite and unhesitating. The stimulus derived from a visit to the exhibition in Bruton Street was frankly amazing. As one came in, one was faced by the artist's portrait of herself, a large square picture simply alive with the sweeping balance of its line and the brilliant vigour of its colour. The vitality and eagerness of the portrait are the artist's own vitality and eagerness. It is more than a likeness; it is like an intimate conversation.

The same force was apparent in the whole exhibition. Miss Rice has most kindly made a special drawing, which is reproduced at the head of this article, to express as plainly as possible the rhythm for which she strives. There is some similarity between it and one of the pictures shown in London, but here she has gained an added effect by the drooping band of decoration behind the figures. I think the skill with which the curves are related is too plain to need comment. The drawing is indeed typical of Miss Rice's tireless work, with its bold decorative planning and swift decided line, springing ever outwards and upwards.

It is not long since a large painting of Miss Rice's was pilloried in the London and Paris press as the extravagance of a lunatic. I can only hope that these critics of the *Egyptian Dancers* did not know what to look for. If they did, if they came tuned to receive an impression of gliding, continuous motion and did not receive it, there is no more to be said; but if they hoped for the rounded grace and frozen attitudes of Burlington House, these flat triangular forms might well startle. I wonder whether they would condemn for similar reasons – as in consistency they ought – such pictures as *Schéhérazade* or *The White Sail*. The use of line is the same; the subtle correlations of outline of the figures in the first case, of the sails and barge-prows in the second, have the same vital stillness, the same rhythmic repose always on the edge of action and always ready for action – to borrow a phrase from Mr Holbrook Jackson – as have the limbs of the dancers and the crouching forms in the background of the large picture which caused such an outcry.

There is no need for further analysis of the exhibition. Whether it is sunlight or moonlight she is painting, figures or landscape, still-life or boats on water, there is the same sense of surging design, the same bravery of colour, the same sincerity of vision. This is no *blague*, no craving for originality. It is very strong, very sane and – I think – very beautiful.

42. Frank Rutter,
'Round the Galleries:
the Autumn Salon'

Sunday Times, 1 October 1911, 13

Like Carter (no. 39), Rutter in his review of the *Salon d'Automne* recognises that cubism is the coming thing in France.

To-day is the *vernissage*, or private view, of the ninth annual Salon d'Automne, which opens to the public to-morrow (Sunday) in the Grand Palais on the Champs Elysées. In England, where its existence is only now becoming generally known, the Autumn Salon is commonly regarded as the home of all that is wildest and most extravagant in contemporary art, a view which, though not absolutely correct, was to some extent supported by last year's exhibition. In 1910 the *fauves*, if not in an actual majority, were certainly predominant; this year, though the 'wild men' are not wholly absent, their works are far less conspicuous. It must not be thought from this that there is as yet any slackening of or reaction from the *fauviste* movement, for its attenuation at the Autumn Salon is due not to any abandonment of their ideals and practice on the part of the extremists, but to the fact that comparatively few of the extremists are exhibiting this year. The Autumn Salon has a jury, and as the composition of the jury alters from year to year, so also changes to some extent the character of the exhibition. All the *socictaires* serve in turn on the jury in alphabetical order, thirty at a time, and the *fauves* evidently are stronger in some sections than others.

A MIXED MEMBERSHIP

Although originally founded by a number of the most advanced members of the Société des Artistes Indépendants, the Société du Salon d'Automne is not limited in its membership to artists of any one group or style. For example you will find in the present Salon works not only by the better known members of the Indépendants but also by a number of exhibitors at the spring salons, the "old" as well as the "new" salon. That is what makes the Salon d'Automne especially interesting to the outsider; to some limited extent it is in its way a *résumé* of the three spring exhibitions, those of the Indépendants, of the Société Nationale, and of the Artistes Français. Naturally, the presence of certain exhibitors from these last two societies is not altogether pleasing to the advanced wing, whose members are already declaring that the Autumn Salon is 'not so good as usual' this year and that the show of the Indépendants this spring was the more interesting exhibition. And since the decreased interest of this Autumn Salon is due, in the opinion of many, to indiscretions on the part of the jury, there is here certainly material for arguing that it is only by adopting the open-door policy of a no-jury exhibition that a society can keep itself perennially youthful and ensure the presence in its midst of all that is most vital in the work of the younger generation. Without in any way committing oneself to an expression of opinion on the principle involved, it is certainly permissible to express deep regret that so strong and interesting a younger painter as Chabaud should this year have been a victim of the jury and have had all his works rejected. Chabaud, of whose painting I wrote at length last year, is not an extremist, and his pictures have a closer relation to the normal vision of man than many of the works admitted this year.

THE EXTREMISTS

To English readers it is almost impossible to convey by words any idea of what the paintings of the extremists in Paris are like. They are hardly to be found in the Salon d'Automne and to see them one must go to the Galerie Kahnweiler, in the Rue Vignon, where you will find the very latest works of Piccasso, Derain, and Vlamuick, or to M. Sagot's gallery in the Rue Laffitte, where you can see some of the earlier work of Piccasso as well as the latest work of Auguste Herbin, Chabaud, and many other fascinating young painters. And when you have seen the later Piccassos and the Herbins, with their geometrically simplified presentation of form, you will understand that in Paris painters like Machkoff and

Kantchalovsky – screamed about as extravagant extremists at the Albert Hall – appear quite moderate and not at all extreme. For obviously the work of Machkoff and Kantchalovsky, as also that of Chabaud and Vlamuick, does bear some relation to normal vision and is therefore the more easily intelligible; while the later work of Piccasso really has no relation to the normal at all, and it is a mental exercise demanding the sternest concentration to discover whether one of his paintings is a portrait or a landscape or a still life. Let me not be thought to libel other advanced painters by suggesting that any of them has so commonplace an aim as representation. Of each one of them it might be said, as it has been said of Vlamuick, that he paints what he feels and not what he sees; *il peint ce qu'il sent et non ce qu'il voit*. But whereas Vlamuick and Herbin do appear to feel a bridge as a bridge, for example, Piccasso now feels everything as an assemblage of conic sections. At any rate this is the nearest I can go to an interpretation of his feelings, and it is useless as well as incorrect to dismiss his painting as *cubisme*. Some of his imitators may be *cubistes*, but Piccasso is not; his peculiar analysis of form is no mere matter of cubes now, but of cylinders, cones, and any other geometrical forms he desires, and its elucidation, I feel, requires a more profound knowledge of mathematics than I possess. Consistently as a lover of liberty, holding that everyone should be allowed to have and practice the art that appeals to him, I canot be so narrow-minded as to object to an art for higher mathematicians, but mere words being in any case inadequate, it is useless to pursue the subject further, and there is really nothing more to be said till the later works of Piccasso are shown in England.

Notwithstanding the absence of the more eminent extremists, the English visitor is likely to find even the present Salon sufficiently startling. Of its contents in detail I shall speak next week, but in the meanwhile I can assure the expectant that though Piccasso is absent, traces of his influence may be found. For the Autumn Salon's jury, like all other juries, frequently admits the disciple while keeping the pioneer originator outside. English painting, as usual, is very poorly represented. Mr. Walter Sickert has not sent this year, and practically the only conspicuous British exhibit is the group of full-coloured decorative paintings by Mr. J.D. Fergusson. The retrospective sections are also less interesting than usual, a feature has been made of seventeenth and eighteenth century engravers, with special collections of the work of Henri de Groux, Iturino, and Camille Pissarro, Interesting as the etchings of the last are, it is disappointing that none of his paintings is shown, and the forthcoming exhibition of his pictures at the Stafford Gallery, London, will undoubtedly tell us far more of the mastery, wide range, and indefatigable experimentalism of this master than the relatively small collection of his etched work at the Salon d'Automne. Another feature of the exhibition is the series of twenty-seven rooms devoted to modern decorative art, but few of these were to-day really ready for review.

43. J[ames] B[olivar] M[anson], 'The Autumn Salon and the "Cubists"'

Outlook, 14 October 1911, 485

Manson (1879–1945) was the secretary of the Camden Town Group where he regularly exhibited his own pictures. He became Director of the Tate Gallery (1930–8). As a friend of Pissarro he had considerable sympathy for the early phases of Post-Impressionism, but in spite of a conversation with Picasso (which he records in this article) he remained unsympathetic to cubism.

It is not often that Gallic wit is the victim of its own acuteness; but the latest movement in French art is one of the rare instances of this. Among certain groups of French artists the idea has become prevalent that, in order to recover from the decadence into which, as they hold, French art has fallen, it is necessary to cultivate a spirit of primitiveness. The idea itself, a truly French one, was amusing enough; it was only when it became serious that it began to be ridiculous. The idea of very sophisticated Parisians in the twentieth century, after partaking of the latest drink in a café on the

Boulevards, solemnly retiring to their ateliers in order to think, feel and paint in a primitive manner is decidedly humorous. The results, to be seen at the Salon d'Automne, are either funny or nauseating according as one chooses to take them. Were they conceived in a joking spirit – pour épater le bourgeois – it could truly be said that the exhibition just opened at the Grand Palais in the Champs-Elysées was a complete success. But, strange as it may seem, they are meant to be taken au grand sérieux. In that case one is compelled to speculate on the possible outcome of it all. Can it lead to anything but madness?

In these days one may search and search in vain for the great men among French artists. Where is the Pissarro, the Renoir, the Puvis de Chavannes of the present day among the younger men? However, let it not be supposed that there are not fine things to be seen in the Autumn Salon. There is a room devoted to the etchings and lithographs of Camille Pissarro, and yet another to the sculpture and paintings of Henry de Groux, a painter of vivid and astonishing imagination, little known in England; but of these more anon. To the student in the quartier the present exhibition represents the triumph of the 'cubists' over the followers of Matisse: a triumph in a battle of banalities. In England our knowledge of 'cubism,' as it is called, is limited to the few specimens by Picasso shown in the Exhibition of Post-Impressionists. Here it provoked mirth, but in Paris cubism is hailed by the critics as an important discovery, epoch-making, in the art of painting. It is a difficult matter to discover wherein its virtues lie.

If it mean anything at all, it would seem to be the glorification of a means – a crude stage in the construction of a painting – into an end. The merest student knows that the superficial aspect of anything is divided into more or less definitely defined planes. These the cubists seek for with infinite patience and exaggerate, expressing them in the form of triangles, circles, squares, etc. To simplify matters they practically do away with colour, limiting themselves to black, white, and dirty yellow.

In order to discover what occult meaning these things might convey to their creators, I sought Mr. Picasso, the originator of cubism, in his studio in Montmartre. I found him surrounded by canvases of all sizes, all covered with these strange hieroglyphics, the results of his geometrical dreams. 'It is simply development of impressionism,' he said. 'It is an impressionism of form. One seeks for the most characteristic form of an object and intensifies it. Take an apple, for example. It is more or less round – that is its fundamental form. I make it completely round. Perhaps you would want to model it, to paint it in light and shade, but that would be to destroy the charming simplicity of its original form.' It is all quite simple. But, according to the theory, to be successful such an expression should give a very intense impression of nature; whereas, on the contrary, incomprehensible chaos is the result. It is the same thing with the colourists. An object appears to be red; it is painted still more red, and so on. The result may be imagined – or seen at the Autumn Salon. A certain decorative effect may be obtained and very often is, but at a sacrifice of almost every other quality.

It was interesting to note that Steinlen, the strong yet subtle draughtsman so well known for his drawings and paintings of the ouvriers of Paris, has become a follower of Gauguin. His picture *Deux Négresses et un Chat* has much of the quality of a Gauguin. He displays the same fine feeling of decorativeness and simplicity of expression. English art is scarcely represented by Mr. Harrington Mann's portrait *L'Echarpe Bleue* in the Sargent manner, or Mr. Lavery's two full-lengths which remind one of a spirit-less Boldini and call to mind Degas's mot about Besnard: 'C'est un homme qui veut danser avec des semelles de plomb.' Probably the most remarkable things in the whole exhibition are the pictures of Henry de Groux, and these are not noteworthy as paintings, for technically their merits are quite ordinary, but as very poignant and dramatic expressions of a tremendous and often quite terrifying imagination.

Artistically, he might be considered the child of Blake and of Gustave Moreau, although he has not the real mysticism of Blake nor the sense of design of either. Nevertheless there is in his work, which is almost always on a large scale, more than a reminiscence of both. His *Christ aux Outrages*, painted in 1890, is a very moving picture, rather a vivid realisation of human passions of every description than a painting of human beings. Maybe it is somewhat sensational, perhaps even a little suggestive of transpontine drama, but nevertheless it is a work of great power and remarkable imagination. His sculpture (his busts of Beethoven, Tolstoy,

Wagner, Baudelaire) is probably superior, technically, but far less interesting.

The work of Camille Pissarro is invariably profound and beautiful, but a further opportunity of writing about it will occur in connection with the exhibition of his pictures shortly to be held in London. The sane and sound things in this Autumn Salon form, it is true, a small minority, but the rest need not be taken very seriously, although it is incredible that any one with French wit could do anything, so bête as to attempt to cultivate a self-conscious naïveté.

Two events in November 1911 demonstrate very clearly just how attitudes in England to Post-Impressionism had changed since November 1910. The first was an exhibition of paintings by Cézanne and Gauguin at the Stafford Gallery; the second was the reproduction in the *New Age* of two cubist pictures, one by Picasso, *La Mandoline et le Pernod* (spring 1911: private collection, Prague) the other by Herbin. The first event was greeted with total equanimity – within a year Cézanne and Gauguin had been accepted as modern 'old masters'. The second, however, raised a storm of letters and comment in the press almost as great as that which had surrounded the first Post-Impressionist exhibition a year previously. The hostile critics were led once again by Ebenezer Wake Cook, but it is notable that the group of writers and critics who sprang to the defence of cubism was very different from that which had lent support to the exhibition at the Grafton Galleries in 1910, and that both Fry and Rutter were largely silent throughout the controversy.

44. J.B. M[anson], 'The Paintings of Cézanne and Gauguin'

Outlook, 2 December 1911, 785–6

For Manson see headnote to no. 43.

The exhibition of the works of Paul Cézanne (1839–1906) and Paul Gauguin (1848–1903) which opened on November 23 at the

Stafford Gallery in Duke Street offers a much-needed opportunity for the calmer study of the two most interesting personalities in modern art who shared a full measure of the opprobrium which was poured upon the exhibition of so-called Post-Impressionists at the Grafton Gallery a year ago by certain indiscriminate sections of the public. Now that the works of these two painters may be judged in judicious calm, it becomes clear how unsensational, how sane and logical they are. Of no other painters has so much nonsense, both favourable and antagonistic, been written. They have been praised without discrimination and damned without understanding.

The work of Gauguin is derived from Impressionism. For many years he was a pupil of Camille Pissarro, under whose instruction he obtained an invaluable knowledge of colours. Many of his pictures in this exhibition bear remarkable resemblance to the master's work, although they lack the finer qualities of subtlety and delicacy which so beautifully characterise the pictures of the great Impressionist. All Gauguin's later work, since he definitely abandoned the direct Impressionist methods, possesses certain qualities of literary interest. The subjects have a decided literary flavour, they tell a story or they are concerned with human emotions which more properly belong to the province of literature. His work might even be regarded as a protest against the entire lack of literary interest in the paintings of the Impressionist movement.

But this quality, although its existence is an interesting fact, adds no real value to his work from the point of view of art. The greatness of his work lies in other directions. His feeling for nature is pre-eminently decorative. He had particularly a genius for decoration, and his use of decoration afforded him an opportunity for strict selection and for emphatic statement of such aspects of life as strongly appealed to him, for the expression of which there would be but limited scope in the practice of naturalistic painting. He had an unusually strong and sensitive gift of drawing; a gift essential to the production of fine decoration. The drawing of the young girl's figure in the well-known *L'Esprit Veille* (No. 22) is strong, fluent, and simple, and shows a power of selecting significant facts. The wonderful solidity of the painting is a tour de force; the repletion and variety of the colour and the subtle analysis of the tones give a vivid impression of life, but the most complete and most satisfying quality of the picture is its admirable decorative composition. It is great as a whole and great in all its component parts.

Charges of deliberate sensationalism have often been brought against Cézanne and Gauguin and other workers in the same movement. An unprejudiced examination and more knowledge of their aims reveal the fact that they are instinct with the salubrity of genius and the simple sincerity of men who work without thought of publicity. The same parrot-cry of 'insanity' greeted the birth of Impressionism, the most vital movement in modern art, and one now fully accepted in intellectual circles. Gauguin's work is a logical development of Impressionism from the expression of the elements of Nature as revealed in a definite scene to their use in a forcible decorative convention. His colour had been subtly developed under Pissarro, and he had naturally a fine appreciation of form. Note the articulate and intelligent rendering of form in the heads and hands of the Breton women in *The Struggle of Jacob and the Angel* (No. 16). The arabesque of this picture is very original and impressive.

In the case of Cézanne this charge of sensationalism becomes exceedingly ridiculous when it is known that with his extraordinary diffidence he frequently left his pictures behind him when he removed from one house to another. He painted entirely for his own satisfaction, and being satisfied lost interest. His favourite picture was always his next. In *Christ at the Mount of Olives* (No. 20) Gauguin expresses his knowledge of synthetic use of colour. It is a suave harmony of warm greys from green to purple which are held together by the dynamic note of vermilion in the hair and beard of the Christ. His painting of his friend, the artist Schuffencker, and family (No. 11), represents the transition from the subtlety of Impressionism to the broader statement of decorative convention. It has pleasing qualities of naïveté, and also a certain human feeling which is lacking in his later work.

Cézanne, as regards actual use of pigment, was a finer painter than Gauguin. His outlook was much more circumscribed. He had not the same unrestrained vigour of Gauguin, nor the same rather barbaric love of colour, nor the same primitive power of seizing significant form. In landscape he had a fine feeling for structural composition, and a finely developed sense of colour which found satisfaction in the analysis of colour tones presented by any definite aspect of nature. His still-life pictures probably show the beautiful qualities of his paint to greater advantage than does any other branch of his work. The painting of the oranges in *Still Life* (No. 2) is a wonderful piece of work great in its sensibility to subtlety of tone and variations of colour. His acute faculty of analysis is well demonstrated in the fine painting of *The Bridge* (No. 7).

Cézanne has had a great many imitators who, as is usual with such people, seize upon characteristics in his painting which are usually accidental or the result of lack of finish. One of their pet tricks is to outline heavily the objects they are painting. Cézanne drew in his work carefully with a brush with fluid paint, and occasionally in working on the picture he lost the drawing through the freedom of his handling, and he then used a definite line simply to correct the drawing. Followers of any artist lacking the genius of their prototype search for characteristics they can understand, and make a cult of the weaknesses of the artists they profess to admire.

---

45. P.G. Konody, 'Cézanne and Gauguin'

Observer, 3 December 1911, 10

Paul G. Konody (1872–1933) was art critic for the *Observer* and the *Daily Mail*. He wrote many books on art and artists and in 1903 published a translation of Camille Mauclair's *The French Impressionists*.

In September 1911 Michael Sadleir bought from the dealer John Neville Cézanne's *La Maison Abandonnée* (1892–4: private collection, USA), and four pictures by Gauguin: *L'Esprit Veille* (or *Manaó Tupapau*, 1892: Albright-Knox Art Gallery, Buffalo), *La Lutte de Jacob avec L'Ange* (c. 1888: National Gallery of Scotland, Edinburgh), a self-portrait and a pastel of a Tahitian girl. These, together with other pictures by the same artists, went on show in Neville's Stafford Gallery in November and December 1911. In his review of the exhibition, Konody, no admirer of Post-Impressionist painting, was forced to admit the intrinsic power of these paintings.

Scarcely a year has gone by since Gauguin, Van Gogh, Cézanne, Matisse and their artistic kinship burst upon the peace of the land at that memorable Grafton Galleries exhibition which caused so much derision, amusement, indignation and exaggerated enthusiasm. Scarcely a year – but the battle is over and won, and Post-Impressionism has taken firm root among us. We have become so accustomed to it – and to the new jargon it has introduced into art criticism – that the exhibition of paintings by Paul Cézanne and Paul Gauguin which has just been opened at the Stafford Gallery, in Duke-street, St. James's, is in no way bewildering and does not impress one as at all abnormal. Another year or so, and the intelligent public, including the enthusiastic apostles of Post-Impressionism, will be able to judge between the successes and failures of this 'new art,' instead of accepting or rejecting everything indiscriminately. For the leaders of the movement – if the striving for individual expression can be described as a movement – were terribly unequal, and their posthumous reputation is bound to suffer from the fact that the growing appreciation of their art has caused their very studio sweepings to be dragged into the glaring light of public exhibitions. Of these, the collection at the Stafford Gallery, fortunately, contains but an insignificant number.

One of the most jeered at Gauguin pictures at the Grafton was *L'Espritveille*, which now appears so logically carried out a piece of decoration, beautiful in design as well as in colour, and intense in expression, that it well deserves to be called a masterpiece. Even more strange is the fascination of *The Struggle of Jacob and the Angel*. Everything here is contrary to preconceived notions, and it would be easy to poke fun at this prize-ring scene, where Jacob and the Angel wrestle on a scarlet lawn before an audience of white-capped Breton women and a priest, whose face is simplified to the appearance of a Dutch doll. In its very absurdity this picture affords striking proof of the extremely loose connection between a fine work of art, which is as expressive as it is sumptuously decorative, and the representation of obvious facts. Of course, we all know that grass is green and not scarlet, but Gauguin's picture demanded that expanse of red, and strange to say, the more one looks at it, the less it appears incongruous. Gauguin passed through many phases, which are here all well represented. Some of his Breton pictures are absolutely normal and thoroughly competent exercises in impressionism. His portrait of himself marks the transition to his later manner. It is a masterpiece of simplification and rhythmic design. As a still life painter he is vastly inferior to Cézanne, whose fruit-piece (No. 2) is a superb achievement – a Chardin with extraordinarily increased vitality, if that term can be applied to lifeless objects. But, as in so many other similar pieces, he utterly fails in suggesting the texture and folds of a crumpled table-cloth. His linen invariably resembles a woollen blanket. Quite admirable is his landscape, *The Bridge*, which is, indeed, the finest and most completely realised work from his brush that it has so far been my fortune to see.

46. Huntly Carter, 'The Plato-Picasso Idea'

New Age, 23 November 1911, 88

For Huntly Carter see headnote to no. 39.

The *New Age* represents the new age. Picassoism is not of the new age, but in the new age. Accordingly there is presented with this week's issue a reproduction of an advanced study by a painter who is one of the most advanced spirits in Paris to-day. The *New Age* is the first journal in this country to show an intelligent appreciation of the latest stage in M. Picasso's remarkable development, that is at present generally misunderstood and derided, just as the comparatively commonplace early work of the Pre-Raffaelites was jeered at and spat upon. The work of M. Picasso and his followers has been so associated by the ignorant ha'penny critics with cubism that it has become the constant habit of these persons to discuss everything produced by these painters in terms of geometry. This is how the famous Mr. Lewis Hind lets himself go in the *Daily Chronicle*: 'The Cubists, those drear, reviled folk, who are *geometricians* first and painters second, arouse interest with their figures and architecture, and still-lifes emerging from canvases that look like coloured, symbolical frontis-pieces to editions of Euclid.' Here Mr. Hind develops his cube-root in a manner of which only Mr. Hind is capable. In another of his outbursts he asks whether any lover of the old masters can avoid feeling 'displeasure before a geometrical, cubical landscape by Picasso?' He is apparently quite ignorant of the fact that the old masters at least saw light reflected at angles just as Mr. Hind's cubists do, but they were not intelligent enough to give their vision the Picasso wideness of expression. Picassoism is thus summarily brushed aside to the satisfaction of Mr. Lewis Hind, whose efforts during recent years to make board and lodging in the daily press out of the advanced movement in painting has probably done that movement more harm than he (Mr. Hind) will ever be able to repair.

As a clue to what Picassoism really is and to what little extent it is related to geometry, I may quote from a letter which Mr. Middleton Murray sent me while in Paris. It seems that Oxford, no less than Paris and New York, is greatly impressed with the profoundly intellectual character of the French painter's work, and during a discussion on the subject Mr. Murray was led to put forward the following Plato-Picasso idea: 'It will be remembered that Plato, in the sixth book of the Republic, turns all artists out of his ideal state on the ground that they merely copy objects in Nature, which are in their turn copies of the real reality – the Eternal Idea. Plato, who was a great artist and lover of art, did not turn artists out because he was a Philistine, but because he thought their form of art was superficial; "photographic" we should call it now. There was no inward mastery of the profound meaning of the object expressed, so that the expression was merely "a copy of a copy." The fact is, Plato was looking for a different form of art, and that form was Picasso's art of essentials.' Mr. Murray's contention is that Picassoism is the first intelligent advance upon Platoism, seeing that it is a practical application of Plato's theory. Thus the study submitted to the readers of this journal, and chosen for the purpose by M. Picasso from the Galerie Kahnweiler, demonstrates that painting has arrived at the point when, by extreme concentration, the artist attains an abstraction which to him is the soul of the subject, though this subject be composed only of ordinary objects – mandoline, wine-glass and table, as in the present instance. It indicates, too, that painting is at the point of its greatest development. It is on the threshold of the will, and not at a halting-place of men sick with inertia.

---

47. John Middleton Murry, 'The Art of Pablo Picasso'

New Age, 30 November 1911, 115

Murry (1889–1957), together with Katherine Mansfield (whom he married in 1913) and J.D. Fergusson, had started the magazine *Rhythm* in the summer of 1911. He was aware that Picasso had created a considerable stir in Paris with his cubist painting but had no critical vocabulary to account for the power of the pictures.

Mr. Huntly Carter has quoted some words of a letter of mine on the subject of Picasso's work; and as I read them again I am struck by a suspicion of intellectual arrogance and assumed finality from which I wish to clear myself.

At the outset, modernist, ultra-modernist, as I am in my artistic sympathies, I frankly disclaim any pretension to an understanding or even an appreciation of Picasso. I am awed by him. I do not treat him as other critics are inclined to do, as a madman. His work is not a blague. Of that I am assured; and anyone who has spoken to him will share my assurance. Picasso has to live by his work, and a man who depends for his bread and butter on his work in paint does not paint unsellable nonsense for a blague. That his later work is unsaleable confirms my conviction that Picasso is one of those spirits who have progressed beyond their age. As with Plato and Leonardo, there are some paths along which pedestrian souls cannot follow, and Picasso is impelled along one of these.

Picasso has done everything. He has painted delicate water-colours of an infinite subtlety and charm. He has made drawings with a magical line that leaves one amazed by its sheer and simple beauty – and yet he has reached a point where none have explained and none, as far as I know, have truly understood. Yet he declares 'J'irai jusqu'au but.' It is because I am convinced of the genius of the man, because I know what he has done in the past, that I stand aside, knowing too much to condemn, knowing too little to praise – for praise needs understanding if it be more than empty mouthing.

A great friend of mine, a leader of the Modernists in Paris, a woman gifted with an æsthetic sensibility far profounder than my own, said once as she was looking at a Picasso, 'I don't know what it is – I feel as though my brain had been sandpapered.' And some such feeling as this is what affects me in his pictures. I feel that Picasso is in some way greater than the greatest because he is trying to do something more; when Plato speaks in transparent and wonderful terms of the Idea of the Good; when Leonardo speaks of the serpentine line; when Hegel makes toys of the categories, I stand aside, unconvinced because I am not great enough to be convinced.

I recognise fully that a speculation such as mine on the relationship between the art of Picasso and the æsthetic of Plato is perhaps of no great value in itself; but to those who have read and wondered at the seeming contradiction in the greatest of all philosophers, to those who have a living interest in living art, the work of Picasso offers the suggestion of vistas through which we can never see. I am still convinced that for men who endeavour to think at all profoundly Plato will always be found to be of all philosophers and artists most valuable in the attempt to appreciate and understand the developments of modern art. I would suggest for the curious in such speculations who have some knowledge of the development of Egyptian art, through the most realistic realism the world will ever know, to an intellectual art which is so near to many of the finer modern developments, that his travels in Egypt may suggest the reason for his condemnation of the 'realistic' art of contemporary Greece. I feel that thence came his new attitude: he looked for a closer approach to essential realities in art, and the art he saw seemed to him to take him further away from the eternal verities. Hence my tentative suggestion that Plato was seeking for a Picasso. Not for one moment do I wish to suggest that these two artists are on the same plane. But in each of them there is so much that I understand and value that I feel convinced that it is but my weakness that prevents my following them to the heights they reach.

They who condemn Picasso condemn him because they cannot understand what he has done in the past, and are content to assume that all that is beyond their feeble comprehension is utterly bad. All that I can say for myself is that I understand too much to be guilty of that crime. In the meanwhile Picasso must needs wait for another Plato to understand; but the world will never have strength to follow.

---

48. G.K. Chesterton, 'The Unutterable'

Daily News, 9 December 1911, 6

Gilbert Keith Chesterton (1874–1936) began his career reviewing art books for the *Bookman* and the *Speaker*. In 1904 he published a book on G.F. Watts. The reproduction of Picasso's *La Mandoline et le Pernod* (1911: private collection, Prague) generated a massive and hostile response in the *New Age* and elsewhere. G.K. Chesterton's abuse is levelled primarily at what he saw as the poverty of contemporary criticism, but he has equal contempt for the work of Picasso itself.

Whenever you hear much of things being unutterable and indefinable and impalpable and unnamable and subtly indescribable, then elevate your aristocratic nose towards heaven and snuff up the smell of decay. It is perfectly true that there is something in all good things that is beyond all speech or figure of speech. But it is also true that there is in all good things a perpetual desire for expression and concrete embodiment; and though the attempt to embody it is always inadequate, the attempt is always made. If the idea does not seek to be the word, the chances are that it is an evil idea. If the word is not made flesh it is a bad word.

Thus Giotto or Fra Angelico would have at once admitted theologically that God was too good to be painted; but they would always try to paint him. And they felt (very rightly) that representing him as a rather quaint old man with a gold crown and a white beard, like a king of the elves, was less profane than resisting the sacred impulse to express him in some way. That is why the Christian world is full of gaudy pictures and twisted statues which seem, to many refined persons, more blasphemous than the secret volumes of an atheist. The trend of good is always towards Incarnation. But, on the other hand, those refined thinkers who worship the Devil, whether in the swamps of Jamaica or the salons of Paris, always insist upon the shapelessness, the wordlessness, the unutterable character of the abomination. They call him 'horror of emptiness,' as did the black witch in Stevenson's *Dynamiter*: they worship him as the unspeakable name; as the unbearable silence. They think of him as the void in the heart of the whirlwind; the cloud on the brain of the maniac; the toppling turrets of vertigo or the endless corridors of nightmare. It was the Christians who gave Satan a grotesque and energetic outline, with sharp horns and spiked tail. It was the saints who drew the Devil as comic and even lively. The Satanists never drew him at all.

And as it is with moral good and evil, so it is also with mental clarity and mental confusion. There is one very valid test by which we may separate genuine, if perverse and unbalanced, originality and revolt from mere impudent innovation and bluff. The man who really thinks he has an idea will always try to explain that idea.

The charlatan who has no idea will always confine himself to explaining that it is much too subtle to be explained. The first idea may really be very *outrée* or specialist; it may really be very difficult to express to ordinary people. But because the man is trying to express it, it is most probable that there is something in it after all. The honest man is he who is always trying to utter the unutterable, to describe the indescribable; but the quack lives not by plunging into mystery but by refusing to come out of it.

Perhaps this distinction is most comically plain in the case of the thing called Art, and the people called Art Critics. It is obvious that an attractive landscape or a living face can only half express the holy cunning that has made them what they are. It is equally obvious that a landscape painter expresses only half of the landscape; a portrait painter only half of the person; they are lucky if they express so much. And again it is yet more obvious that any literary description of the pictures can only express half of them, and that the less important half. Still, it does express something; the thread is not broken that connects God with Nature, or Nature with men, or men with critics. The *Mona Lisa* was in some respects (not all, I fancy) what God meant her to be. Leonardo's picture was, in some respects, like the lady. And Walter Pater's rich description was, in some respects, like the picture. Thus we come to the consoling reflection that even literature, in the last resort, can express something other than its own unhappy self.

Now the modern critic is a humbug, because he professes to be entirely inarticulate. Speech is his whole business; and he boasts of being speechless. Before Botticelli he is mute. But if there is any good in Botticelli (there is much good, and much evil too) it is emphatically the critic's business to explain it; to translate it from terms of painting into terms of diction. Of course, the rendering will be inadequate – but so is Botticelli. It is a fact he would be the first to admit. But anything which has been intelligently received can at least be intelligently suggested. Pater does suggest an intelligent cause for the cadaverous colour of Botticelli's *Venus Rising from the Sea*. Ruskin does suggest an intelligent motive for Turner destroying forests and falsifying landscapes. These two great critics were far too fastidious for my taste; they urged to excess the idea that a sense of art was a sort of secret to be patiently taught and slowly learnt. Still, they thought it could be taught: they thought it could be learnt. They constrained themselves, with considerable creative fatigue, to find the exact adjectives which might parallel in English prose what had been done in Italian painting. The same is true of Whistler and R.A.M. Stevenson and many others in the exposition of Velasquez. They had something to say about the pictures; they knew it was unworthy of the pictures; but they said it.

Now the eulogists of the latest artistic insanities (Cubism and Post-Impressionism and Mr. Picasso) are eulogists and nothing else. They are not critics; least of all creative critics. They do not attempt to translate beauty into language; they merely tell you that it is untranslatable – that is, unutterable, indefinable, indescribable, impalpable, ineffable, and all the rest of it. The cloud is their banner; they cry to chaos and old night. They circulate a piece of paper on which Mr. Picasso has had the misfortune to upset the ink and tried to dry it with his boots, and they seek to terrify democracy by the good old antidemocratic muddlements; that 'the public' does not understand these things; that 'the likes of us' cannot dare to question the dark decisions of our lords.

I venture to suggest that we resist all this rubbish by the very simple test mentioned above. If there were anything intelligent in such art, something of its at least could be made intelligible in literature. Man is made with one head, not with two or three. No criticism of Rembrandt is as good as Rembrandt; but it can be so written as to make a man go back, and look at his pictures. If there is a curious and fantastic art, it is the business of the art critics to create a curious and fantastic literary expression for it; inferior to it, doubtless, but still akin to it. If they cannot do this as they cannot; if there is nothing except eulogy – then they are quacks; or the high priests of the unutterable. If the art critics can say nothing about the artists except that they are good it is because the artists are bad. They can explain nothing because they have found nothing; and they have found nothing because there is nothing to be found.

49. Walter Sickert, 'The Old Ladies of Etching-Needle Street'

English Review, January 1912, 311–12

For Sickert see headnote to no. 29. In the course of a general survey of art in the latter part of 1911, Sickert developed an *ad hominem* attack on Matisse and Picasso.

The conspiracy of semi-unconscious 'spoof,' which is looked upon by some as an alarming symptom of the artistic health of the present day, is in reality a very small and unimportant manifestation. In the story of the 'Emperor's New Clothes,' it was the whole nation that affected not to see that his Majesty was naked. The modern cult of post-impressionism is localised mainly in the pockets of one or two dealers holding large remainders of incompetent work. They have conceived the genial idea that if the values of criticism could only be reversed – if efficiency could be considered a fault, and incompetence alone sublime – a roaring and easy trade could be driven. Sweating would certainly become easier with a post-impressionist *personnel* than with competent hands, since efficient artists are limited in number; whereas Piccassos and Matisses could be painted by all the coachmen that the rise of the motor traffic has thrown out of employment. It is, after all, an extremely small circle of very unoccupied ladies who find amusement and excitement in going one better than the other in ecstasy at the incomprehensible.

Students will still continue to struggle with the pedestrian difficulties of objective drawing, and to register with satisfaction the minute degrees of increase in power and in insight which is all that human effort may expect to accomplish. When they turn, in their daily paper, to accounts of what is being done by workers in other fields, they will find recorded, as a matter of course, such statements as the following:

'Heydweiller, of Berlin, in repeated experiments, found discrepancies ranging from the two thousand five-hundredth part of a grain. Landot therefore began again, and spent five years over his investigations.'

Read by the side of this paragraph the adventures experienced by the souls of Mr. Lewis Hind and Mr. Berenson, in their voyage through masterpieces of nonsense, seem to be lacking in interest, and even in actuality.

---

50. Unsigned review, 'The Goupil Gallery'

Pall Mall Gazette, 15 January 1912, 4

In a review of the work of Denis, Laprade, Desvallières and Sérusier, the critic of the *Pall Mall Gazette* pointed out how deeply Post-Impressionism had affected British art.

Of the four French painters who fill the two ground-floor rooms of the Goupil Gallery, only M. Maurice Denis is a familiar figure at London exhibitions, although isolated examples of M. Pierre Laprade's art have been shown at the epoch-making Grafton Gallery Post-Impressionist exhibition a little over a year ago. Barely twelve months – but what a revolution they have witnessed, not only in the art production of this country, but even more in the critical attitude of the public! How many of the scoffers have turned into admirers; how many of those who were most abusive and bitter in their denunciation have come to look with indulgence upon what only twelve months ago appeared to them intolerable! It is no good closing one's eyes to the fact that the principles of Post-Impressionism in its milder form have permeated British art and renewed its vitality. Scores of exhibitions have familiarised us with the new rhythm, reconciled us to the synthetic, in place of the imitative, rendering of the facts of nature. And thus it is scarcely surprising that the work of Denis and Laprade, of G. Dervallières and Paul Sérusier, which a couple of years ago would have caused an outburst of indignation, appears now perfectly normal and sane. Indeed, Dervallières' decorative impressionism might strike the extremists as tame were it not for his very personal vision of colour harmonies. Laprade proves himself a master of still life, who steers a clear course between, and is equally far removed from, the Dutch still life masters' minute and precious imitation of form, surfaces and textures, and from the childish affectations and wilful distortions of the fashionable cubists.

Maurice Denis is unquestionably the most interesting of this little group. He almost convinces one of the sincerity of his archaism, and he has a well-nigh infallible instinct for decorative design and colour. *Soir de Septembre* is a picture of haunting beauty. It is a beach scene, with bathing and tennis-playing women lighted by the evening sun. The synthetic simplification of form and colour is carried just to the point where any further step would lead to caricature or unintelligibility. The rose-coloured sky, the applegreen sea, the blue heath, the reflection of the sunset glow on the nude bathers, form a colour pattern of entrancing beauty. The profoundly touching intensity of feeling in his *Adoration des Mages* is only marred by the incongruous jumble of modern and ancient costume, which makes the three magi appear like mummers in fancy dress. Paul Sérusier follows closely in Gauguin's footsteps. Indeed, in his group of pictures may be traced the whole evolution of Gauguin's art, the *Souper des Enfants* corresponding to Gauguin's Pont-Aven period, the *Filles du Pélichtin* to his Tahitian triumphs, the hideously grimacing *Salomé* to his excesses and failures. Wholly admirable is the vibrant and yet vigorously painted lamp-light, *Nature Morte*.

51. J.B. M[anson], 'Four Modern French Painters'

Outlook, 24 February 1912, 281

For Manson see headnote to no. 43.

The bitter struggle of the French Impressionists for their ideals, which started some fifty years ago, seems to have cleared a permanent path through the jungle of popular prejudice, along which fresh reforms and new movements may proceed with comparative ease. To this is due the fact that the so-called Post-Impressionism is establishing itself in this country with but little difficulty. Now that the dust stirred up by the ci-devant artistic autocrats, with Sir W.B. Richmond at their head, is clearing off in the light of more impersonal opinion, one is able to observe some of the very real qualities of sincerity and originality which grace many of the works of the followers of the new reform. Any new movement which is contrary to the established order of things meets with an opposition whose virulence is in ratio to the newness of its doctrines and to the incapacity of those already in authority to regard it with intelligent detachment. Messrs. Marchant are offering another opportunity for the calm and judicious study of the work of four modern painters, two of whom are distinctly followers of the latest development and are, by virtue of their mature achievement, worthy of the most serious consideration; whilst of the other two, one practises a sort of bastard academism disguised by reckless looseness of handling, and the other hides his really academic spirit under certain mannerisms of treatment and subject.

The two most genuinely modern, and therefore the most interesting, of the painters at present showing their work under the auspices of the Goupil Gallery are MM. Maurice Denis and Paul Sérusier. The work of the former is the direct outcome of Impressionism modified by an intelligent admiration of the work of Puvis de Chavannes. His largest and, as it happens, most important pictures are examples of a highly synthetical use of

Impressionism. In *La Plage Ensoleillée* (No. 20) he has made use, by strict selection, of the results of close analytic observation of colour in the building up of a forcible expression of an effect of light, which is a much intensified rendering of the effect as seen in Nature. M. Denis, then, has not confined himself to using tones of colour in the exact degrees and values as they existed to his perception in Nature, but he has selected certain tones, excluded others, and intensified those that suited his purpose. In the result he has produced an intense and glowing effect of sunlight which may not be (probably is not) literally true to a particular effect of Nature under given conditions, but which is eminently true to Nature in the abstract. The picture is a poem of colour. In his treatment of form M. Denis does not follow the customary practice of Impressionism. In this he is quite academic, after the manner of Puvis. His composition too is influenced by the same master of decoration.

Of his companion picture, *Soir de Septembre* (No. 26), much the same thing may be observed. His daring selection of certain colours to the exclusion of others is most successful and effective. Observing that blue tones predominated in the shadowed sand of the foreground, he has selected such tones alone to serve his purpose. The effect in the picture seen as a whole is one of great breadth and truth. The fault to be found with the picture is that it embodies no big general design. Both these paintings are remarkable achievements and would repay prolonged study. Two of his smaller pictures, *Procession* (No. 22) and *Les Gondoles à Venise*, are simply essays in Impressionism – slight, but truthful and suggestive. *Maternité* (No. 28) is a charming and practically faultless picture. Its arabesque is excellent and its scheme of colour rich and haunting. The beautiful quality of the tones of colour – the crimson-purple, the yellow, and the green – show research and a fine feeling of harmony, in which the use of the pink in the child's frock is a master-touch.

In M. Paul Sérusier's work one is not often conscious of a definite intention. He is much under the influence of Cézanne and of Gauguin, and in one or two cases this operates to his disadvantage. In *Nature Morte – Oranges et Pommes* (No. 3), for example, he has made use of an awkward line running diagonally across the picture which is not integrally in the design. It seems to me that his purpose was to expose this corner of background for the display of a very good imitation of the brushwork of Cézanne. Both in this picture and in the *Nature Morte – Roses* (No. 1) he follows Cézanne's practice of separating masses of colour by a definite line, with good decorative effect. In the former picture his planes are not clearly defined – a fault which will not be found in any of Cézanne's work. The other picture – *Roses* (No. 1) – has many weak passages. The pot and the roses are not well studied; indeed they have the appearance of being introduced as an afterthought. The pot is poor in colour through lack of analysis, and therefore is weak in modelling.

Both pictures are pleasing on the whole and rich and harmonious in general conception of colour. M. Sérusier has a fine feeling for rhythm which often lacks restraint. His *Baigneuses* (No. 6), which shows the direct Gauguin influence, is rather spoilt by abundance of rhythmic intentions. They are not subdued entirely to the main theme. This is here a small fault, as it is only a question of degree. There is an effect of rhythm obtained by the repetition of the attitudes of the arms. From the left the position of the arms in the first figure is repeated in the third figure, whilst that of the second figure is repeated exactly in the figure in the background. The effect is very pleasing and telling. Then there is a definite effect of rhythm obtained by the placing of the trunks of the trees and also by the repetition of the curious blue tones of the drapery. It is a very successful picture; admirably decorative, it would make a fine design for tapestry. The subject of the *Filles du Pélichtin* (No. 12) seems to have been chosen for the opportunity it offered of rhythmic treatment of the arms raised in adoration. The picture has no general design, but in the smaller designs formed by the play of lines of the raised arms there is enough material for two or three pictures. *Salomé* is a subject that has attracted many painters, but I doubt if it has received such realistic treatment – I use realistic as meaning truth of character – from any other painter. The ugly brutality of Salome gloating over the head of the Baptist is more convincing than the attempts at bewitching beauty with which one is familiar in most renderings of the subject. The exaggerated glow of her jewels forms a fitting note of vulgarity. It is not a striking design. The mass of town in the background occupies too much the same amount of space as the corner of the terrace which forms the foreground. The artist has been mainly occupied with what are usually called the literary qualities of the picture.

There is a pseudo-romantic feeling about the work of M. Laprade due to the vague looseness of his use of pigment. His work displays abundant facility, but no study or analysis nor any intention other than that of astonishing by dexterity. The inclusion of his work in the exhibition was a mistake, for nothing is to be gained from the study of it. He paints after the usual methods of the schools, which he modifies by sheer recklessness and want of care.

For the most part M. George Desvallières' work has a certain literary interest. Occasionally one can detect a tendency to mysticism. His still-life paintings are as commonplace as those of M. Blanche, but more personal in colour and less accomplished. In *La Lecture* (No. 35) he is limited to a certain space for decorative purpose. His drawing and painting of the figure in this picture is mere school-work such as hundreds of students are doing in Paris to-day. It is not clear whether it was Messrs. Marchant's intention to show the pictures of these four artists as being characteristic examples of the 'New' movement or whether the work of MM. Desvallières and Laprade was included as a contrast to that of the others: in the latter case the object was sufficiently subtle and quite effective in the result.

---

52. D. S. MacColl,
'A Year of Post-Impressionism'

Nineteenth Century, February 1912, 285–302. Reprinted in *Confessions of a Keeper*, 1931, pp. 202–28.

MacColl (1859–1948) was ill when the first Post-Impressionist exhibition opened in 1910, but in this, his first contribution to the debate about Post-Impressionism, he offers one of the most closely reasoned attacks on the concept of Post-Impressionism as Fry presented it to the British public. He had studied at both London and Oxford universities and then under the painter Fred Brown in 1889. As art critic first for the *Spectator* (1890–6) then the *Saturday Review*, MacColl was one of the most staunch supporters of Impressionism. He regularly exhibited at the New English Art Club and the Goupil Gallery and in 1911 became Keeper of the Wallace Collection. In this article he points out some of the contradictions in the view of Post-Impressionism as, on one hand, 'classical in its tendencies', and, on the other, a version of 'romantic' expressionism.

When, a little over a year ago, 'Post-Impressionism' burst upon the town, I was in no condition to take a hand in the vast discussion that followed; but I did just stagger round the Grafton Gallery before I was despatched to a safe distance from work. When I left London the critics were disconcerted, but nervously determined, after so many mistakes, to be this time on the winning side. A few bravely, if wistfully, did declare themselves fossils; some were uneasily upon the fence; the rest were practising, a little asthamatically, the phrases of an unknown tongue. As it happened, one of the few critics on the Press with anything that can be called a mind, one of the fewer with a gift for persuasion and for writing, Mr. Roger Fry, had declared for the new aesthetic, or religion, and the impressionable could but wheel desperately after him on this sudden tack. Three months later I found the new religion established, the old gods being bundled without ceremony into the lumber-room, and the ardent weathercocks of the Press pointing steadily for the moment into the paulo-post-futurum. So easy a victory for a new creed is delightful, if it is deserved, but it tempts the obstinately critical mind to ask a few questions. I propose, after the fair run that the new faith has enjoyed, to look a little closely at its theories and its productions.

The Grafton Exhibition was not quite the beginning of things. Mr. Fry had played with the very reasonable speculation that the explorations in colour of the Impressionists might be employed by imaginative decorators not limited to a scramble for effect. It may be said, by the way, that this was precisely the programme already carried out by Puvis de Chavannes in wall-paintings like *L'Hiver*. But Mr. Fry, up till half-past eleven before the noon of the Exhibition struck, did not appear to have convinced himself that the expected method and the masters had been found; for he exhibited in Suffolk Street a ceiling that looked back to Guido, of all the Pre-Impressionists: there had been indications, however, that his vote was nearly cast. In the chaste pages of the *Burlington Magazine*, barely tainted with modern art, there appeared, with Mr. Fry's editorial blessing, a startling rhapsody on Cézanne. Its author affirmed a faith already orthodox in Germany, where the enthusiastic, if chaotic, Meier Graefe leads the song. The Germans, so enviably endowed for music, for science, and for business, are eager for all the arts. Denied almost entirely an instinct for the art of painting, they study it, they 'encourage it,' egg it on, adore, and even buy. Nor do they stop there. They have town-planned whole towns out of the back-pages of *The Studio* in styles that put to shame the cosiest corners of Mrs. Barnett's architects. They dine, they sleep, they commit every act of life in 'Art Nouveau.' And to their serious bosoms they have taken each extravagance of Montmartre and added an 'ismus' to its name. Wonderful Montmartre, that seethes and blazes for the duller world with the fire and fevers of youth and art! I remember, one summer morning in the early nineties, climbing the sacred hill. At the summit was a little shop that was a symbol of the place. There stood, with ancient *berets* on their heads, 'le père' and 'la mère Tanguy,' like two figures in the old Box-and-Cox barometers. They sold colours and canvases, if selling it could be called, since they were seldom paid. It was reported that they had long ceased to eat, so that there might be more colours for the young ferocious of that day, whose methods called for a huge quantity; and there, under their hands, was piled a heap of canvases returned with the colours thick encrusted, waiting in patient faith for the rare customer. There were flowers by 'Vincent' (Van Gogh), and landscapes by youths from Pont-Aven, who announced day by day that 'black was red' or 'violet was green.' Then we went from one house to another, of artist and collector. We had begun in another quarter with Comte Camondo. He had just bought the picture by Degas that so shocked Sir William Richmond and all the professionally and periodically *scandalisables* of London who write letters to the papers. Two people were drinking absinthe and coffee: had the scene been laid in London and called *Afternoon Tea* no one have been shocked; as it was, the picture was hooted out of the country and is now, by the Count's bequest, one of the treasures of the

Louvre. On the hill we found Degas fuming because he had been written about in the papers, 'like Whistler,' and said to paint 'comme un cochon.' Last we visited the rooms of an ancient Jewish collector, and, when we had gone through them all, we crossed the street with him and plunged into a 'dive' like Mammon's, a cellar in which he had 'laid down' hundreds of 'Impressionist' pictures to mature, and pictures twenty years later to be 'Post-Impressionist.' There they were, stacked on trucks, and he was hoarding them. Manet was then beginning to sell at Durand-Ruel's; Monet was dribbling through to America; the day of the others was to come later, when Vollard opened shop in the Rue Laffitte and held up to admiration scores of still-lifes by Cézanne, sparely constituted of an apple or two and a metallic napkin. Anquetin had just abandoned his 'synthetic' manner, that of strong outline and flat tint, and the real master of the Japanese convention, Toulouse-Lautrec, was terrifying the hoardings.

But I must return to the *Burlington*. In its numbers for February and March 1910 appeared the eulogy on Cézanne by M. Maurice Denis, with reproductions of the artist's work. The main line of M. Denis' argument was that Cézanne is a 'classic,' because in his painting the spectator is not preponderantly moved by the object itself, nor by the artist's personality, but by a balance of the two. This sounds a promising description of classicism, to which I will return later. But M. Denis goes on to affirm, of this 'classic' painter, that his painting is painting and nothing more, that it 'imitates objects'

> without any exactitude and without any accessory interest of sentiment or thought. When he imagines a sketch, he assembles colours and forms without any literary preoccupation: his aim is nearer to that of a carpet-weaver than of a Delacroix, transforming into coloured harmony, but with dramatic or lyric intention, a scene of the Bible or of Shakespeare.

Sérusier is quoted in support:

> One thing must be noted, that is the absence of subject . . . The purpose, even the concept of the object represented, disappears before the charm of his coloured forms.

After these explanations we seem to be already in difficulties with our 'classic' painter. The balance of object and subject we have just heard about means that the object is inexactly rendered, and that there is no subject at all. And M. Denis, a painter himself, in a pretty convention, shallow sentiment and villainous colour, of religious and legendary 'subjects,' adopts, for his eulogy of Cézanne, the theory of poetry attributed to Mallarmé, that its beauty consists mainly in sound, of painting that its beauty is limited to the 'carpet' aspect of it, and of imagination that it works properly in the vehicle of words. The name alone, 'imagination,' might have stopped him. A scene such as is recorded in the Bible or in the pages of Shakespeare is only 'imaged' when it is seen; that is to say, when it lends itself to the art of vision, which is painting: it can only be referred to and evoked, not rendered, by the symbols of words. A scene, therefore, in the Bible or Shakespeare is at least as much the natural subject of painting as of writing, and there is nothing 'literary' in painting it. The real distinction between literature and painting is that writing, being indefinitely continuous, can evoke a chain of successive actions, and is therefore the fit medium of narrative; but it cannot represent those actions or any one point of them; painting can actually render a fixed point. The stage, within certain limits, can reproduce the whole chain that narrative evokes and comments upon.

So much for the general confusion. If we take the two arts separately we shall find that their virtue is never a simple thing: it depends on a union of two elements, beauty and significance. This is easily tested, because in the case of poetry we can cut off significance and retain the mere beauty of sound. We have only to ask a good reader to recite a poem in a language known to him and unknown to ourselves. The result, if the language is sonorous, is gently pleasant for a very short time; soon, even for the most poetical, it becomes unbearably monotonous, so much is the virtue of poetry a combination of sound-beauty, fit and ingenious arrangement of words and ideas, weight of feeling and significance. The same is true of painting. The most decorative of our oil-paintings, if we see them at such an angle that the 'subject' is not grasped, are poor things beside a rich carpet or enamel, and the really good carpets themselves are a kind of picture, dependent for the sting of their beauty on the remote 'subject' that went to their design. If, then, Cézanne had ever succeeded in getting rid of subject, he would not thereby have become a 'classic' painter, or anything like it; he would have ceased to be a painter at all.

But that is not all of this queer eulogy. Cézanne, it appears, abolishes tone in favour of colour,

> substitutes contrasts of tint for contrasts of tone.... In all this conversation he never once mentioned the word 'values.' His system assuredly excludes relations of values in the sense accepted in the schools.

Unfortunately this is not Post-Impressionism at all, but the Impressionism of Turner and of Monet. It depends on the fact that no one, even if he wish to, can render the values truly of a sunlit landscape, because pigments do not cover so great a range. The upper notes must be sacrificed in any case, and the convention Turner and Monet adopted, to gain a general brilliance, was to omit the lower as well, to leave out not only the real sun, which no one could put in, but also the shadows, the tones, of the lower notes, rendering only their difference of colour or tint, and that in an exaggerated way. Monet's 'purple shadow' is as famous as Turner's vermilion. Our 'classic,' therefore, is on this ground a pure impressionist.

But there is a more mysterious business. By his modelling (or 'modulating,' for the first word is not permitted), Cézanne arrives at the 'volumes' of objects, and puts their contours in afterwards. These 'volumes' are an 'abstraction' from objects. That is intelligible enough, but something comes in at this point, some sort of bee in the bonnet of Cézanne or of his admirers, that was to play havoc later, and produce whole 'schools' and sects. 'All his faculty for abstraction,' we are told,

> permits him to distinguish only among notable forms 'the sphere, the cone, and the cylinder.' All forms are referred to those, which he is alone capable of thinking. The multiplicity of his colour schemes varies them infinitely. But still he never reaches the conception of the circle, the triangle, the parallelogram: those are abstractions which his eye and brain refuse to admit. Forms are for him volumes.

On a first reading this would appear to mean that by some lesion of his classic brain our painter could not conceive a parallelogram, and that of solid bodies he could only cope with three. But probably the author expresses himself badly or is ill translated. What he means is that Cézanne thinks in the solid, not in the parallelogram but in the cube – or am I wrong, and was the cube, afterwards to be so sacred, anathema at this period? If so, the less painter he! for the complete painter must think in both. He must imagine, extending back behind his canvas, a space containing solid bodies; and this space and these bodies he must render on the flat surface of his canvas. But he must also remember that these solid shapes, projected on the flat, will set up a certain pattern among themselves of forms in two dimensions, and that this pattern and its relation to the frame constitute the 'decorative' side of his art. Since the frame is normally either a parallelogram or a circle, he is a strange artist who cannot conceive of either; and we are more puzzled than ever by Mr. Fry's announcement, in his preface to this article, of an art 'in which the decorative elements preponderate at the expense of the representative.' The apostle of the new art is absorbed, it appears, in the 'representative' side (the rendering of depth), and knows nothing about the 'decorative' (the planning of the surface).

Mr. Fry himself speaks of Cézanne's 'compact unity' built up by 'a calculated emphasis on a rhythmic balance of directions.' But M. Denis describes one of these figure-landscape pieces in the making:

> The dimensions of the figures were often readjusted; sometimes they were life-size, sometimes contracted to half: the arms, the torsos, the legs were enlarged and diminished in unimaginable proportions.

Calculation was missing or erratic here: and with every variation in size the 'rhythmic balance of directions' must have altered.

But it is needless to pursue further these rather elementary confusions; let us take farewell of the article with a touching little phrase of the painter himself. He spoke, not of any of the great things enumerated, but of his 'petite sensation,' the little *sensation* that he was trying to preserve and render. I remember, in those same early nineties, a discussion among a group of American short-story writers, very earnest, constipated artists. One of them had been out for a walk, and his contribution was the statement that in coming home through the trees he had 'quite a little mood.' 'I did not write it down at the time,' he said, 'but perhaps I ought to have done so.' Cézanne was not a great classic: he was an artist, often clumsy, always in difficulties, very limited in his range, absurdly so in his most numerous productions, but with 'quite a little mood,' and the haunting idea of an art built upon the early Manet, at which he could only hint. He oscillated between Manet's earlier and finer manner, that of dark contours and broadly divided colour, and a painting based on the early Monet, all colour in a high key. In this manner he produced certain landscapes, tender and beautiful in colour, but the figure was too difficult for him, and from difficulties of all sorts he escaped into the still-lifes I have spoken of, flattened jugs, apples, and napkins like blue tin that would clank if they fell. What is fatal to the claim set up for him as a deliberate designer, creating eternal images out of the momentary lights of the Impressionists, is the fact that his technique remains that of the Impressionists, a sketcher's technique, adapted for snatching hurriedly at effects that will not wait. Hence his touch, hence those slops of form out of which he tries to throw a figure together. No one was ever further from logical 'classic' construction, if that is what we are looking for; none of the Impressionists was so uncertain in his shots at a shape. And when we come to fundamentals, to rhythm, whether it be the rhythm of the thing seen, or the rhythm of the picture imagined, or these two combined, as they are in great art, Cézanne is helpless. We have only to turn to the illustrations to appreciate this. Cut away the theories and the verbiage, and what is actually before us? A forcible head of the painter is the best of them; but even that has only one valid eye; the other portraits are blocks of wood. The vaunted landscapes with figures, the *Bathers* and *Satyrs*, are the work of a man who could not command the construction or the expressive gesture of a single figure, could not combine them together, or fit them reasonably into a landscape setting. What a blinding power has theory for the ingenious mind!

The Grafton Exhibition included many things. There was Manet as well as Cézanne. There was a group of the more tiresome 'Neo-Impressionists,' but including the inventor among them, Seurat, who introduced 'pointillisme.' The others turned the infinitesimal dots of primary colour that the theory required into large bricks of colour that could not possibly fuse at any distance. I suppose, by the way, it will be impossible to the end of time to persuade people that Monet never at any time used 'divisionism,' the splitting up of colour into its primary or even its rainbow constituents. Even so careful a writer as Mr. C.J. Holmes asserts this, against the evidence of all the pictures. I endeavoured years ago to explode the supposed scientific basis of the pointillist theory of painting, but all that came of it was a conviction among my critics that I was myself an Impressionist and advocate of pointillism. I perhaps deserved this for trying to give 'Impression-ism' a wider than its historical meaning.

But this by way of digression. Next after Cézanne among the painters new to London, and that London was grateful for seeing, came Gauguin, who was well represented. This painter, beginning as a rather dull Impressionist, in the wake of Pissarro, developed, for the handling of exotic scenes, a more nervous drawing and vivid colour, reverting to the Oriental decoration that was already implicit in Manet, Degas, and Whistler. There is nothing revo-lutionary in the drawing of the Tahiti figures; it is the drawing of Degas, stiffer, and less flexible, as might be expected from the painter who began work at thirty; and there is an illogical modelling of the figures in light and shade that does not extend to other parts of the picture. But the pose and grouping of the Tahiti pieces is finely felt, and his colour in these and certain still-lifes has original character and splendour. The fine period was short; it is a drop from *L'Esprit Veille* to fantastic rubbish like *Christ in the Garden of Olives*.

With the third name we come, I was going to say, to the real thing: but that would be unfair; to one of those spirits who break through the ordinary moulds, who survive, like the salamander, in a fiery element. Blake is the one English artist who did this and lived undestroyed at a perilous exaltation. Van Gogh had neither Blake's mental range nor his endurance, but in the short period of balance between the lethargic Dutch art of his beginning and the madness of his end he is very like the Blake of Thornton's *Pastorals*. The hallucination of a reality more intense than that of every day comes to some men by way of wine or drugs, to some by bodily fever, to others by the fever of the mind that production itself induces. Beginning like Gauguin flatly, Van Gogh worked up, like him, through Impressionism, and then, before madness overtook him, snatched at his startling landscape visions. Rain, a cornfield, a sunset, are discharged at us with heightened, hallucinatory intensity. The colour of flowers, too, thus excited him, and the portrait of himself, shown at the Grafton, the exasperated blondness of the tormented mattoid head against a flame of blue, was a masterpiece in its kind. Then he fell over on the other side, and the rest tells us merely the price he paid for a super-lucid interval.

But this was not all. We were asked to regard these three men as the initiators of an art which was carried a stage further by later artists, of whom two were the chief, Picasso and Matisse. Picasso appeared mainly in his early phase, as a Whistlerian, less certain even than Whistler in the construction of a painted figure, but with a delicate sense of colour; an etcher, too, of subtle line. But a portrait was shown in the sleeve of which (not yet in the face) some geometrical mania was at work. Of this more presently. Of Matisse there were only three small pieces: two insignificant landscapes and a silly doll, *La Femme aux yeux verts*, in which we were invited to find marvels of rhythm and harmony. I have seen landscapes by Matisse which had a certain barbaric strength of colour; I have not seen enough of his work to trace his history, and I am prepared to believe that he has given pledges elsewhere of good faith in these preposterous experiments; but I see no force in the argument that because drawing is very bad indeed, it must be very good because it is by a clever man, one who has been known, at other times and places, to draw pretty well. I pass over Herbin, Friesz, Vlaminck, and many more, all of them, like Baal's priests, cutting and maiming their forms in a desperate incantation of the fire that had touched Van Gogh. I return from the pictures to the theories.

The catalogue was prefaced by a brilliant piece of writing, unsigned, more closely knit than M. Denis' apologia, and a lecture was given during the exhibition by Mr. Fry, and printed in the May number of the *Fortnightly Review*. The writer of the preface tacitly showed M. Denis' theory about Cézanne to the door, and advanced a directly opposite account of those he christened 'Post-Impressionists.' M. Denis had claimed for Cézanne that he was 'classic,' meaning, as we may put it, that there is a fine balance in his painting between the desires of the painter and the rights of the object painted; that he renders the object justly but finely seen. If this is not a plausible description of Cézanne, it is a possible definition of classic painting. But now we were told that the methods of this school

> enable the individuality of the artist to find completer self-expression in his work than is possible to those who have committed themselves to representing objects more literally.

The school, in a word, render their emotions about objects rather than the objects themselves, and Mr. Fry makes it the definition of all drawing that it distorts the object. Personal feeling, then, is the note of the movement, and the 'post-Impressionists,' therefore, are not classic at all, but extreme Romantics. I was met by several ghosts of old controversies in this discussion. The 'rocking-horse' of the preface reminded me of the 'Noah's ark beasts' of the Glasgow School, 'better than Sidney Cooper,' and another old phrase, 'There is no such thing as correct drawing,' played its part. By that I meant that just as in literature writing can never be said to be finally 'correct,' nor even grammar, but only to approach perfection of expression, so with drawing. Imitation may be a large part of drawing, but the initial impulse is gesture, and 'correctness' of imitation by way of tracing is not only impossible, because contours must be amplified to suggest a third dimension, but the design of the picture, simplification for decorative breadth, sacrifice and emphasis for expressive force, also affect 'correct' copying. Again I am entirely with Mr. Fry in the stress he lays on the rhythmical basis of design. Perhaps I may be allowed to quote from an article that made people very angry twenty years ago.[1]

> Drawing is at bottom, like all the arts, a kind of gesture, a method of dancing upon paper. The dance may be mimetic; but the beauty and verve of the performance, not closeness of the imitation, impresses, and tame additions of truth will encumber and not convince. The dance must control the pantomime. Rivers and skies and faces are taken up by the painter as illustrations of a mood, and the lines of the image he creates are not meant to reproduce the thing, but to convey what he felt about the thing – the salutation, the caress he gave to it . . . He wishes to convince the imagination, not to delude the sense . . . In the lines of abstract ornament you will often get a more striking impression of conflict or repose than from the most document-supported picture of battle or of sleep; and it is this element, the music of space and form, that really plays to the imagination behind the images that represent person or thing. A division of the paper will do more to enthrone a figure or dignify a landscape than the dress of kings or the presence of palaces, and the drift or swing of a composition across the canvas be more eloquent of its motive than the particular attitude and occupation of its constituent persons.

Here, then, is common ground; but as they say in disputations at the Propaganda, *Distinguo*. When I came to work at the history of a period of drawing, I saw that there is a strong dividing-line between two schools, each of them great, by the degree in which they admit freedom of modification; and Mr. Fry's definition, and my old one, are not the account of all drawing, but of one school only. I have endeavoured elsewhere to bring this out,[2] but must risk a repetition here. 'Classic' drawing, conveniently so called because it is the drawing of Greek fifth-century sculptors, follows the model or 'nature' very closely, with a minimum of sacrifice and distortion for the sake of emotional emphasis, the expression of action, or the imposition of a rhythm conceived by the designer. It is realist among choice forms, aims at searching out the rhythm implicit in an object, and entrusts to a lucid statement of that rhythm the task of exciting in the spectator's mind the feeling already aroused in that of the artist. It sinks personality and renders the object.

Romantic drawing is not satisfied with this: it emphasises, caricatures, elongates, abbreviates, reshapes the form in accordance with a more violent emotion, a more tyrannic imposition of rhythm, a rhythm of the artist's excitement. The problem of both schools is at bottom the same – namely, to fit into the rhythm of the picture or of the sculptor's block the rhythms of the objects included; but classic art more humbly, more patiently and subtly waits upon the secrets of the object: it discovers a rhythm rather than invents. Romantic drawing shatters and reforms the object with its own passion and gesture, and introduces incidentally all manner of 'personal' elements of temper and touch. It follows, naturally enough, that classic art works, by preference, in presence of the object; romantic art tends to remake from memory. 'Nature puts them out,' these artists say; they borrow from her a shorthand of form, a scaffolding on which their system of expression may be hung.

These two tempers and systems exist side by side in varying force at all times.

Masaccio, Piero della Francesca, Dürer,[3] Holbein, Leonardo and Raphael, Titian till his later days, Velazquez are in the main classics. Giovanni Pisano and Michael Angelo are the great Romantics; Tintoretto, Rubens, and Il Greco are some of the followers. In the modern period Ingres and Stevens are classic; Goya, Géricault, Delacroix, Daumier, Millet, Rodin are Romantics. Almost all Englishmen of any account are Romantics; if we look for a classic among contemporary artists, English or French, it is hard to find any, except Mr. Havard Thomas, who is the extreme case (I am not now discussing relative merits but the completeness of the type). Mr. Walter Sickert is in theory a devout classic, all for the hairbreadth moulding of drawing after nature that is called Ingres: in practice he is a Cézanne, making uncertain shots at a real but elusive 'petite sensation' of his own.

But if the writer of the preface started out with the thesis that his artists were Romantics expressing less objects than themselves, he suddenly abandoned this, and threw out an entirely new and incompatible third theory, namely, that they painted not appearances, not even emotions about appearances, but the Thing in Itself. They paint, he says, 'the treeness of a tree,' and elsewhere, 'they draw a line round the concept of a thing.' Now, if there is one thing that painting certainly cannot do, it is this. You can *think* the concept of a tree, and you can talk about it, since words allude to ideas but do not represent, but you cannot imagine it, and you cannot draw it. The concept includes every kind and size of tree, the drawing must represent one. So, to take a simpler case, you can *think* the general idea of triangle, which includes equilateral, isosceles, and scalene triangles of all sizes; but you cannot draw it, because any triangle you draw must be one kind or another, one size or another. And the thing-in-itself, or 'substance,' being, by its nature, relieved of all particular appearances, cannot be drawn, because drawing is the art of visible appearance, not of invisible substance. It is true I can pick and choose among appearances those that, for my purpose, are most important; I can abstract from the total appearance of a tree; I can abstract its greenness or brownness, and draw it black; I can abstract its roundness, and draw it flat; I can abstract its leaves, and draw it bare; I can abstract its branches, and draw it a stump; but some part of the 'treeness of the tree' goes with each abstraction and resides entirely in no one of these particulars. A tree may, for the imagination, present forcibly one of its qualities at a time; it may be a green dome of shade on a hot day, a ladder of retreat for a man from the attentions of a mad bull, a peg on which an apple hangs, a screen for an assassin, a choir for birds; and its own business of spreading out its million pores to the air and propagating its kind, which comes nearest to being its 'treeness,' may be what occupies the artist least and bores him most. He deals with the accidents of its life, that serve the purposes of his own kind. But if, because of this, he scorns the tree's own idea of its main business, misunderstands and cramps the rhythm that mysteriously arises from the strains and expansions of its anchoring, its feeding, and its breathing, he loses, not perhaps the significance for his story that the ladder or the peg or the umbrella would sufficiently furnish, he loses beauty, the beauty implicated in the processes of life, and cannot replace it however he may cudgel his invention.

Mr. Fry, as one would expect, produced a more coherent theory than the other writers: he declared, as the object of the 'Post-Impressionists,' the 'discovery of the visual language of the imagination': a language analogous to music, and on this quest the abandonment of 'naturalism.' The 'distortion' already conceded to the Romantics is a part of this, but he goes on to demand the suppression of natural perspective and chiaroscuro. Naturalistic perspective, he contends, prevents a painter from giving the significance, say, of a pageant, because the policeman near at hand obscures by his comparative bulk the really important figures. That is so only if the painter takes his stand immediately behind the policeman. There is no reason why a more distant point of view should not be chosen, and Gentile Bellini, Carpaccio, Titian, and Veronese solved this difficulty without trouble. Chiaroscuro can also, within the limits of naturalism, be minimised and almost excluded by a lighting of the picture from the front, or reduced, for decorative breadth, to one step between light and dark, as by Manet, and for the matter of that Maurice Denis. But no reasonable man would deny to the artist, for special reasons of expression or decoration, a break with strict perspective, which, indeed, is seldom to be found in good pictures, or complete abstraction from shadow. What Mr. Fry seems to forget is that perspective and shadow are not mere science: in the hands of great artists they are instruments of expression, perspective a threatening power in the hands of a Mantegna, shadow an instrument of reverie and pathos in the hands of a Rembrandt; they also, like form, are matter for design. What we may well concede, and what I for one have often asserted, is that the full accumulation of natural effect, the total instrument of painting, is not only unfit for certain purposes, but is beyond the strength of all but very great artists. Many can play on the pipe who cannot control to purpose all the keyboards of the organ. If that is what Post-Impressionism means, the greater part of recent painting in this country has been Post-Impressionist. Strang, Lavery and Brangwyn, Nicholson and Pryde, Ricketts and Shannon, John and Holmes, all deliberately or less consciously throw overboard one or another element in the full range of representation so as to keep the ship floating. There are many varieties of such sacrifice. These sacrifices may be necessary, but the danger of any deliberate stereotype thus adopted is that an artist who once limits his traffic with nature not only cuts himself off from fresh sources, but is in danger of losing even that which he hath. If out of the whole alphabet of appearances he limits himself to ABC for the sake of A, B and C are apt to grow insolent and make an end of A. Mr. Fry, if I understand him aright, welcomes the possibility of 'genius' being no longer called for. He speaks of the possibilities of recovering an 'anonymous' art, as if that were the same thing; but in the anonymous mediaeval times it is easy to trace the points at which genius came in. He looks for the creation of a common language of imaginative expression which all might use, without any arduous training, without any wrestle with natural appearances, a language as direct as a child's in drawing. I agree that for such purposes laborious imitation is irrelevant, that the point of imitation reached in a thousand art schools is useless, because it will never be turned to imaginative use; but I hold with Black against Mr. Fry that a man must learn to copy nature if, to any high purpose, he would copy his imagination. The odd thing about this new language of the imagination is that once acquired it seems not to widen the imaginative range, but to limit it to an orange, an apple, a napkin, and a pot. These are subjects which of all others surely call for the full texture of vision to render them interesting, for the art of a Chardin or a Manet. Nature seems to revenge herself by allowing to the rebels not even 'nature *morte*'!

In the matter of Cézanne Mr. Fry holds, as does his able seconder Mr. Clutton Brock,[4] that we have 'classic' art. I have already dealt with this claim, but Mr. Fry has an obsession under this head, which calls for a word of examination. He appears to think that the residual element of reality, which renders painting 'classic,' is the expression of solidity, and that solidity is most fully expressed by the elimination of light and shade and the addition of a thick contour. We are reminded, at this point, of Mr. Berenson's famous 'tactile values.' The expression was ill chosen, because Mr. Berenson did not mean values of touch at all, but the sense of energy put forth and of resistance, which are quite different things; or else those appreciations of depth which (*pace* Berkeley) depend not on tactile but on visual machinery. The Florentine School of painting sprang from sculpture; hence its preoccupation with solidity; the Venetians made painting more distinctly a painter's art by their preoccupation with colour. This by way of parenthesis. Cézanne certainly 'blocks in' his forms with thick lines which give them a certain brutal force, but he does it indiscriminately with a flower-pot, which if solid is fragile, and with table-cloths, which are as little solid as objects may be. And Mr. Fry finds this magic of solidity in the most unlikely features. Cézanne sometimes draws the mouth of a circular vase or flower-pot seen in perspective not as an ellipse, but like a gutta-percha ellipse that has been squeezed till its sides are parallel; producing, Mr. Fry says, a greater effect of solidity. Why Cézanne did this it is idle to conjecture; in one piece he draws three flower-pots side by side, and their lower contours range from a straight line to a lop-sided curve. It is probable then that the flattened forms arise rather from thoughtless or clumsy shots at form than from an intention; but if intention there was, it must have been an intention to flatten the shape, not to expand it. The true shape of a circle in wide perspective has so straining an influence on the picture-field that designers are tempted to attenuate it; thus Puvis de Chavannes, in the foreground of one of his best-known mural paintings, draws a fountain-basin in the shape of Cézanne's flower-pot mouth. He, and perhaps Cézanne, was really flattening his form for decorative reasons. Mr. Fry finds the same 'classic' merits in the still-lifes with which Picasso has been rewarding the devout fervour of disciples. One of these I was privileged to see in Mr. Fry's company. In a 'design' that looked like fragments of stained glass pieced together could be made out the outline of a bashed flower-pot and a lemon, and other objects were explained to be a curtain and a piece of paper. In this case the mouth of the flower-pot reversed the formula of Cézanne; it had the shape of an irregular almond, with sharp ends; but was still affirmed to increase the solidity of the pot's reality: a sufficient *reductio ad absurdum*, one would think. The 'paper' was indeed solid, solid as iron; but then 'one must not look for imitation of nature.' Why then have paper at all? If my classic emotion before an orange may lead me to represent it, not as a sphere of orange-colour, but as a cube of green, need I look at oranges at all? And if the 'balance of directions,' as I was told, requires that the flower-pot should be mutilated, why take this 'direction' at the expense of a flower-pot? If all we want is a play of 'directions' leading nowhere, why do the flower-pot and lemon linger on the field, like indestructible properties saved, in the wreck of the universe, from the old still-lifes? The truth is that these painters have never betrayed the faintest capacity for the most rudimentary exercises in pattern design. They have, on the contrary, in this direction, an appalling taste; witness the mess by Herbin, recently served up for the readers of the *New Age*, which looked like a number of scraps from bad wax-cloth patterns stitched together. The admirers of these things are hypnotised, exactly as a hen may be, held over a chalk-mark.

Mr. Fry was perhaps at heart not quite satisfied with his artists, of whom his eulogy was a little disappointing: in default of existing examples, he took the heroic course of producing them. With a sporting spirit I cannot sufficiently applaud, the authorities of a respectable educational institution, the Borough Polytechnic, committed to him and those about him the painting of their walls. It was really a magnificent thing for a committee to do, and if committees elsewhere will show the same adventurous spirit we shall get on with the necessary experimental stages of a fresh period in mural decoration. I am going to be critical about these paintings, but there is not a doubt that in two of the artists employed Mr. Fry singled out new talents of which a great deal may be expected, Mr. Duncan Grant and Mr. Etchells. I have seen pictures elsewhere by Mr. Grant that give this conviction more certainly than the wall-paintings, and the flowerpiece recently at the Carfax Gallery was enough to prove him a fine colourist. Mr. Etchells' panel at Southwark was the most striking in its assertion of a bald, forcible rhythm, and this assertion of abstract rhythm was almost all that the new artists provided. That, however, is the fundamental, and we may look for a development into something richer. I mean that if a painter takes a Hampstead Heath Bank Holiday as his theme, he is making very little of it if he sets up figures that might be women anywhere reduced to the lowest common terms of humanity and action. Surely the dress, the fantastic hats, the Cockney character have something to supply that need not conflict with a deliberate structure in the design. There were absurdities of treatment besides. The ground and background were painted as if built up of tesserae; why should wall-painting imitate mosaic? Why, again, should Mr. Grant's figures look like diagrams of anatomy when their anatomy is obviously fantastic? And what suggestion of the rhythm of water do we gain from a treatment of its surface that looks like slabs of a marble floor tilted at angles to one another? Mr. Albert Rothenstein's design was more reasonable in its simplifications, less ruthless in its abstractions, and more subtle in colour; but the practice of small drawings told in a comparative failure of mass and general silhouette. Mr. Fry's own contribution had a genial idea, but calls for revision, since his little girl giving a bun to an elephant was neither standing on her legs, doing what she is supposed to do, nor attending to what she is supposed to be doing. But the total effect in a dismal room was gay, and has stirred a great deal of wholesome speculation.

A curious thing about those designs is that probably without any knowledge on the part of their painters they are much more nearly in the vein of Seurat, the 'Neo-Impressionist' leader, than of any of the 'Post-Impressionists.' Seurat, an artist of rather vulgar temperament, was a man of ideas. It was he who brought in the mosaic of dots, and it was he who produced a series of designs of dancing and other figures in severely repeated parallelism. Mr. Fry might revive him for another Grafton Exhibition.

Since then Mr. Fry has opened an exhibition of his own paintings on a less ambitious scale. As I look back on his production before it took this sudden turn, I remember a succession of phases. There was a good deal of *pastiche*, ingenious exercises, now in one form of the older art, now in another. But every now and then there would peep out something of his own, a 'petite sensation' trying to get itself expressed, shivering a little because the borrowed clothes were cast aside, but much more interesting than the borrowings. And in the recent exhibition I find the same contrast. A great deal of it is *pastiche* of a new set of models. These are what I should call toy-box pictures, theoretical reductions of sketches to block forms through which, as through the reading of a picture in chunks of wood or in large wool stitches, the sketch may be vaguely seen. Others, like the flood scenes at Guildford, are a development of something more truly apprehended, and set down with a greater confidence than of old. The toy-box system may be useful for an artist who wishes to strip his design down to its simplest terms in the process of work: it is certainly not a method by which the artist can say anything individual: pictures painted on this system are as like one another as the work of the 'Ripolin' painters, to use the image that a witty friend has suggested. The best painting in the exhibition was a portrait of Mr. McTaggart, the Cambridge philosopher, who once wrote on 'The Further Determination of the Absolute,' an explanation of what really constitutes the Thing-in-itself. Here, Mr. Fry must have thought, is a subject for the new art; but if the colour is negative and the background a needless reminiscence of Van Gogh's patterns, the head is drawn with no more of caricature than the character of the sitter reasonably suggested.

Here I might stop, if account were taken only of the merits of the pictures that have been put forward and of the theories that have been spun about them. But it has been noticeable that the attraction of these pictures for many people was not what they positively were, but rather what they negatively excluded. They were welcomed in the degree in which they renounced with violence the world as it is seen. They were accepted as a promise, queer and doubtful, of a painting that should render the world beneath appearances, the world *unseen*. There cropped up again and again in discussion the word 'symbol.' These distortions of reality were thought, in some unexplained way, to give us 'symbols' of a deeper reality than the painter ordinarily represents. Mr. Fry, indeed, started in chase of the will-of-the-wisp of a painting that should use symbols free from particularity like words, as Mallarmé sought for verse that should use words and their associations freely, almost like music. The difficulty is that painted symbols can supply no 'argument' like words: that they represent the nouns of speech, but not the verbs, nor any conjunctions except 'and'; while Mr. Fry and his school are taking away most of the adjectives. It is a muddle-headed condition of mind that sees 'symbols' in the still-lifes we have been dealing with. A colourless sphere or a circle may be used as the symbol of an orange; an orange can hardly be called the symbol of a circle.

But the hankering thus incongruously revealed for symbols in painting, for this paradoxical use of an art whose natural field is the superficial beauty of visible reality in all its infinite variety, this need is after all a need of the religious spirit, calling for help from the imagination to picture what is strictly unimaginable and therefore cannot be painted. The sterner religions, Hebrew and Mohammedan, have forbidden such a use of art, a traducing of the unseen by idols; but the weakness of humanity has demanded some equivalent, in terms of the despised visible beauty, for the life of the soul and the superhuman beings of its adoration. It was this hankering, so entirely unsatisfied, so actively repelled by the disgusting pictures ordinarily called religious, that was anew excited by the rumour of a return in painting to symbolic art.

When I drew the distinction referred to above between 'Classic' and 'Romantic' drawing, and defined the attitudes behind these, by names from the Greek myth of Olympians and Titans, I set over against these two, as the third dominating attitude of the imagination, the Mystic – I did not develop the consequences of the last for drawing so far as the other two – partly because of the obscurity of the inquiry, partly because illustration of the attitude in modern art is so scanty. But the question at least calls for definite posing even at the end of a short article like this: as there is a Classic and a Romantic drawing, is there also a Mystic drawing? Can we trace the laws that govern the artist who attempts to render the superhuman in some sort of visible terms? Symbolic, evidently, the drawing must be; that is to say the image given will be there not fully to represent anything, but to mediate with the Unseen, as Incarnation with God, to stand for something beyond itself. In what ways will the drawing suggest this?

The modern romantic temper tends to confound with mystic vision two words whose sound favours the confusion, 'mystery' and 'mist.' The first of these was originally the mystic's drama, the rite of initiation, but it has been worn down till it means little more than something misty; and mistiness is the romantic evasion for mystic vision. But this is the reverse of the character we find in the images of really religious times. Definiteness of outline, massive form, are their characteristics, as of forces imperishable and unchanging. And we may put this more generally by saying that as much as possible every element of contingency must be excluded, all those features that made Plato distrust the art of painting because they render the idea a shifting thing. For this reason perspective will be minimised, for this reason changing light and shadow, the mirage of atmosphere, the decomposition of reflected lights; in composition the studied confusion of the picturesque, in expression all transitory emotion will be banished for severe symmetry and solemn calm. The illusion of the passing world will be reduced to its lowest term of abstraction, and for this reason sculpture, in what is obviously not flesh, will be preferred to painting. Detail and accessory will be as rigorously dealt with; such incident and detail as is admitted will be admitted reluctantly only because it is forced upon the artist to enhance significance. And symbolic realities thus admitted will wear some mark of strangeness, as by the faint tradition of religion people still 'dress for church.' It shows how far this idea has been perverted that the modern does not put on a dress, like a surplice, that would sink his individuality; Mrs. Brown does not wear a veil, but affirms herself not Mrs. Jones by her competitive hat. That is not surprising, since for so many centuries religious art has been lost, has been ebbing with the receding wave that withdrew religion itself to the East from which it came. Just as in Greek art the 'classic' period is too realistic and human to be religious, so in Gothic figures like *Le beau Christ* of Amiens are already outside, and in painting we must go back to 'primitives' behind Titian for examples of what we are in search of. In early Greek and Gothic blocks, in mosaic on non-illusive golden grounds, in Egyptian granite, in oriental bronze, something of the divine and eternal was communicated. And the drawing of such images differs from the choice realism of classic art, the curiosity and personal emphasis of romantic; it sweeps over the minor points of representation that in portrait, in the drama, in genre and still-life are properly sought out and enforced. In the native lands of religion this synthetic drawing has extended itself beyond the religious subject, has checked the portrait-painter when he deals with the individual, and even the landscape painter, tied to symbols when he seeks the freedom of clouds or sea.

That only a religious revival could restore the conditions in which even the other great kinds of painting might grow again to their highest stature is, I think, an inevitable conclusion from history; and monumental art of any kind calls for 'sacrifices' of small imitation. But the 'sacrifices' of the 'Post-Impressionists' seem to me to be sacrifices in the wrong place, and not to be laid on the altar even of an absent god.

NOTES

1 'Painting and Imitation,' *Spectator*, June 18, 1892. [MacColl's note]
2 *Nineteenth Century Art*. (Maclehose, 1902.) [MacColl's note]
3 A check upon classic drawing is obedience to a normal form or canon

of proportion. Some years ago Mr. Sturge Moore published a pretentious book on Albert Dürer, in which he laid it down that Dürer's canon was used only to be departed from. He showed no acquaintance with Dürer's own writings on the subject, except a short passage translated by Sir Martin Conway, and that he had misunderstood. His view was accepted by all his critics, who evidently had not read Dürer's book. But a patient German, L. Justi, at the same time was publishing a treatise, showing that on the back of some of Dürer's drawings the construction from the canon was to be found and was followed. [MacColl's note]

4 *Burlington Magazine*, January 1911. [MacColl's note]

53. Michael T.H. Sadleir, 'After Gauguin'

Rhythm, Spring 1912, 21–9

Sadleir wrote an article on Kandinsky's *Über das Geistige in der Kunst* for *Art News* (9 March 1912) and in this piece he elaborates further on Kandinsky's colour theories. Kandinsky's painting was first seen in London at the Allied Artists' exhibitions in 1909 and 1910. In 1911 Sadleir bought some Kandinsky woodcuts from the same exhibition, obtained the artist's address, then in 1912, accompanied by his father, visited him at his home in Munich. Sadleir went on to translate Kandinsky's book which was published as *Concerning the Spiritual in Art* (1914).

The followers of Gauguin fall into two distinct groups. The one is closely related to him, being partly composed of actual pupils; the other has little apparent connexion, but owes nevertheless the essentials of its origin to the art of the master. It is the Gauguin of the Brittany period that has inspired the men of the first group – naturally enough, since after Pont Aven was no more teaching. Their pictures are full of the grave reverence of *Le Calvaire* and of *Le Christ au jardin des Oliviers*. Of their leaders, Serusier shows fine design and a true sense of the values of quiet tones. His less successful work has a tendency to dullness. Girieud is more definitely ecclesiastical, and has caught the spirit of early Italian religious painting in his views of Siena, his portraits of monks, his biblical scenes.

But it is the second and more difficult group of Gauguin's disciples that call for attention here. These – for want of a better name – I shall call the neo-primitives, because they have arrived in their search for expression at a technique reminiscent of primitive and savage art. It is idle to blame them as archaistic, unless one can prove the old contention that primitive art was mere ineptitude and was inspired by an emotion differing only in incoherence from that of the renaissance. The opposite opinion, that the ideals of primitive art were of a different and purer quality, that therefore different methods of expression were evolved, admits the possibility of similar ideals to-day with, consequently, similar expression. Gauguin himself affords a proof of this possibility, for only by denying the sincerity of his life's work is it possible to overlook the savage basis of his nature.

André Derain, of whose work a fine example was given in the third number of *Rhythm* (p. 28), has only recently 'found' himself. His earlier pictures showed little merit beyond fierce effort. Those who visited an exhibition of French art held at Brighton in the summer of 1910, will remember a painting of the Thames Embankment, coloured with a wild disregard for nature, which revealed a strong decorative sense struggling unsuccessfully for utterance. But Derain has found his expression in wood-cuts and decorative pottery. The illustrations to *L'Enchanteur Pourrissant* are less naturalistic even than Gauguin, but they have an essential and forceful realism. The human form is sometimes a series of angles, sometimes merely a movement, but in every case one aspect of reality is retained and its truth emphasized by the skilful simplification. Derain's work has the rugged sufficiency of early religious carving. It shows the primitive force of Gauguin's Tahitian pictures freed from the limitations of time and place.

Even more abstracted is the work of Wassily Kandinsky, a Polish artist who leads a small coterie of enthusiasts in Munich. He has just published a book of practical theory,[1] which, besides being an aid to the appreciation of the author's ideals as an artist, is a valuable commentary on the whole modern movement.

From this book two main contentions arise. The first is virtually a statement of Pantheism, that there exists a 'something' behind externals, common to nature and humanity alike. This is what Wordsworth believed, but while he approached the question subjectively, contenting himself with describing the experiences of his personal mood-communion with the nature round him, the new art is to act as intermediary for others, to harmonize the *inneres klang* of external nature with that of humanity, it being the artist's task to divine and elicit the common essentials underlying both.

The second contention follows naturally from the first. An art intent on expressing the *inner* soul of persons and things will inevitably stray from the *outer* conventions of form and colour; that is to say, it will be definitely unnaturalistic, anti-materialist.

In his development of the first of these beliefs Kandinsky maintains – and few will disagree with him – that just as the common soul of man and nature has always existed, so through the art of all ages runs a common spirit, the artist being a seer, able, in varying degree, to realize what lies underneath the life which inspires him. It follows then, to return to the question mentioned above, that the differences between the art of one time and another are merely differences of attitude, and not of fundamental quality, depending on the clearness or otherwise of the artist's conception of his duty as harmonizer of internal truth. Kandinsky argues that primitive art is a more direct expression of the soul of externals than is that of later periods, because it belongs to a time when life was simple, experience single-eyed, and when the fundamental *zusammenklang* was less obscured by the noise of the naturalistic appeal. For this reason the new art inclines at present to primitive technique, but such a tendency will be temporary only, because the feeling of the primitive artist for the inner reality, being more instinctive than educated, was soon overlaid, while the art of the future, once it has thrown off the chains of naturalism, can develop fresh methods of its own, unhampered by tradition.

Once more – just as the painting of all ages has a common element, so equally have the various arts between themselves. Music, poetry, painting, architecture are all able in their different way to reach the essential soul, and the coming era will see them brought together, mutually striving to the great attainment. As an example of how this is already being done, Kandinsky mentions Skrjabin's symphony, in which sight, smell and hearing are all enlisted to intensify the impression received.

This is a very important question because it introduces the psychological problem of the possibility of hearing colour, seeing sound, touching rhythms, and so forth. Kandinsky has an elaborate discussion of the psychological effects on the observer of various colours, likening them to musical instruments, and tabulating the movements, repellent and otherwise, of pure colours, both singly and in combination. It should be remembered that the Symbolistes realized the value of music and painting to poetry. Verlaine and his followers got wonderful effects by a weaving of assonances, while Verhaeren and Paul Fort have experimented in the working of cross-metres and trailing repetitions. The danger of this interesting theory – a danger to which I feel Kandinsky is sometimes perilously near – is of its becoming a system. Indeed in the hands of Réné Ghil it became one more than twenty years ago. In his extraordinary book *L'œuvre, en méthode à l'œuvre* (a revised edition of the *Traité du Verbe*) Ghil reduces the musical value of vowel sounds to a science. He writes poetry that can be played by an orchestra. Rimbaud's famous vowel-sonnet, which assigns to the vowels different colours, is not analogous. While there is no reason to believe that for him the vowels had not such colour-values, there is considerable reason to put these values down to previous association. Kandinsky admits such possible association in the psychological working of colours, and I cannot help feeling that such an admission tends to vitiate the whole theory. What seems more valuable is the idea that colour can convey a more immediate and subtle appeal to the inner soul than words, and this not only for the reason given by Kandinsky – that it is capable of greater variety of tone – but also from its greater freedom from this same association of ideas. The desire to escape the ordinary association of language, inspired Mallarmé's great theory that the impression given by certain arrangements of words can be quite different to their ordinary meaning taken as separate units. In *L'Aprèsmidi d'un Faune* are examples of this synthetic word-painting. On these lines poetry may come to strive, side by side with Kandinsky's painting, for the new ideal. For the anti-naturalism of the two methods is the same.

While encouraging the new art to abandon the accepted aspect of nature, not to fear in fact the charge of bad drawing and impossible colour, Kandinsky gives a wise and necessary warning. The process is attended by a double danger. Anti-naturalism may become pure pattern-making, and form and colour mere symbols. He should have added that the chief fault of such an art is that it leads nowhere. It is barren because it never touches reality, and reality is as essential as naturalism is deplorable. The other danger is the creation of an imaginary dream-world, which being also divorced from life is equally worthless to the future of art. The true way is the way of the inner *notwendigkeit*, and that can be found only by the true artist.

From a book full of suggestive thought, which touches every aspect of the modern effort, I have extracted a few main ideas, with a view to making clear the aim towards which a large part of the art of today is striving; for though the book is one man's, he has voiced the inarticulate ideals of a multitude.

NOTE

1 Kandinsky. *Über das geistige in der Kunst*. Piper-Verlag, Munich, 1912. The same firm will shortly publish the first volume of *Der blaue Reiter*, an occasional periodical devoted to modern art and edited by Kandinsky and Franz Marc. [Sadleir's note]

FUTURISM AT THE SACKVILLE GALLERY, MARCH 1912

Marinetti published the first Futurist manifesto in the Paris journal, *Le Figaro*, in February 1909. Extracts were printed in the August issue of the English journal the *Tramp* (pp. 487–8) and again in the August issue of *Art News*. Outside Italy the paintings of Boccioni, Carrà, Severini and Russolo were seen first in Paris in February 1912 before coming on to the Sackville Gallery in March. Boccioni showed among other paintings *Leave Taking: Those Who Go Away* (1911: private collection, New York), and *Those Who Remain Behind* (1911: Galleria Civica di Arte Moderna, Milan), and *Rising City* (1910–11: Museum of Modern Art, New York). Carrà showed *Leaving the Theatre* (1910–11: private collection, London), but the most highly appreciated painting was Severini's *The Pan Pan Dance at the Monico* (1910–11: destroyed in the Second World War). Severini put on a one-man show of his work in London early in 1913.

The British, already inured to shocks in art from the Continent, had heard of the disturbances created by Futurism in Paris and treated this exhibition with a mixture of curiosity and circumspection.

54. P.G. K[onody], 'The Italian Futurists: Nightmare Exhibition at the Sackville Gallery'

Pall Mall Gazette, 1 March 1912, 5

For Konody see headnote to no. 45.

Faint echoes of the ridicule that has been showered in Paris upon the Italian Futurists have filtered through to London, and have prepared the English public for the shocks that await them at the Sackville Gallery, in Sackville-street, whither the astounding exhibition of the Futurists has been transferred *en bloc* from M. Bernheimjeune's gallery in Paris. Fortunately, the exponents of Post-Impressionism and Cubism have trained our faculties to accept the new and the revolutionary without going into hysterics of indignation, and it is not likely that the visitors to the gallery will be roused to anything but mild amusement, or that Sir W.B. Richmond will throw vitriol at the offending canvases, although he may in future substitute a new joke 'Futilism' for the somewhat stale reiteration of his 'Post-prandial Impressionism.'

The Futurists, perhaps in the consciousness of their inability to make themselves intelligible by means of the painter's brush, have pursued their propaganda with literary weapons, and have adopted the wise precaution to brand those who laugh at them as imbeciles. The critic's only escape is to take them seriously and to avoid the obvious temptation to poke fun at their inextricable jigsaw puzzles. Now and then, in the three rooms filled by these paintings, the artist extends a helping hand, in so far as he introduces a hint of tangible form, whereon, by aid of imagination and printed explanation, the critic may build some theory. But as often the absence of such a clue leaves him helpless in front of a kind of gaily-coloured representation of a patchwork-quilt composed of triangles, squares, and polygons, like Severini's *The Rhythm of My Room*.

To analyse these Futurist pictures is simply impossible. The majority of them strike one as the pictorial rendering of confused nightmares, in which all objects are not only in motion but in dissolution under the impulse of violent forces. And it is, indeed, one of the guiding principles of Futurism to search for the *force-lines* which express the manner in which inanimate objects would disintegrate, had they the power to follow their inherent impulse. Another principle is the necessity of crowding into one canvas not only the painter's visual impressions, but the associations evoked by these impressions in the mind of the artist, the person depicted and the spectator. The result is an effect, in colour, similar to that in black and white, of a Kodak film on which three or four different views have been exposed – say a portrait, a landscape, a still life, and a street scene – in other words: chaos.

Symbolism of line and colour plays a leading part in the Futurist's pictorial conception. Thus, in Boccioni's trilogy, *Leave-taking, Those who go away*, and *Those who remain behind*, oblique lines represent the state of mind of those who are going away, whilst the colour indicates 'loneliness, anguish, and dazed confusion'; the perpendicular lines indicate the depressed state of mind of those who remain behind, and 'their infinite sadness dragging everything down towards the earth. The mathematically spiritualised silhouettes render the distressing melancholy' of their soul. The catalogue provides no end of similar explanations, but the pictures themselves present merely a tangle of things that look like curled wood-shavings, cut across by sharp lines of varying direction, the number 6943 painted with the mechanical precision of a fascia-board painter, bits of rails and engines, 'mangled telegraph posts,' and the like.

There is something like real grandeur and a suggestion of volcanic, irresistible force in the same artist's *The Rising City*, in which 'the immense horses' – composed of red and blue flames – symbolise the growth and the desperate labour of the great city thrusting her scaffoldings towards the sky. How an artist who develops such power could be guilty of such commonplace vulgarity as *Laughter* and *The Police Raid* is a greater puzzle than the most wilful incomprehensibility of Futurist incoherence. These two pictures, like Russolo's *Portrait of the Artist*, being an interpretation of the artist's own state of mind (!), suggest a very ordinary academic painter's mind disguised in eccentric fancy dress.

Another picture that might be hung, without giving offence to the normal person, at an ordinary exhibition, is Carra's *A Swim*. It is true the forms of the swimmers are chipped out in square blocks, not unlike Caran d'Ache's wooden toys – and this in spite of the Futurists' vigorously expressed contempt for such academic movements as Cubism – but here for once the picture fully explains itself and expresses with extraordinary intensity 'The sensuality and the coolness of a bathe in the Mediterranean.' Quite the most amusing picture of this strange assembly is Severini's huge geometrical pattern of angular segments, *The 'Pan-pan' Dance at the Monico*, depicting the sensation of the bustle and hubbub created by the Tsiganes, the champagne-sodden crowd, etc., at the famous night-tavern at Montmartre. The leisurely person may spend a pleasant half-hour in attempting to detach the different forms from this kaleidoscopic pattern.

At any rate, the Futurists, with all their recognition of tradition, are not only sincere, but are endowed with over-abundant imagination and are free from *the fumistieri* of such painters as some young members of our own Friday Club. They want to evolve an art that breaks every link with the past. In addition to condemning tradition, the study of perspective, the nude in art, the picture without subject, the Futurists maintain that the use of the model is an absurdity and a mental cowardice; that a portrait should not resemble the sitter; that space does not exist; that solid objects are not opaque, as has been clearly proved by the Röntgen rays; that it is idle and foolish to paint the left shoulder and ear of a person of whom you have painted the right shoulder and ear; that you have to paint the invisible – that which happens behind you and at your left and your right, instead of arbitrarily confining your vision to the little square of life before your eyes.

We have seen many new movements in recent years. We have had our eyes trained to see nature in a new way. We have had spectral analysis introduced into art. We have seen many a violent break in the continuity of tradition, and many an arbitrary return to an earlier and more archaic tradition; we have even seen the Cubists turn the human body into geometrical figures. But the Futurists go beyond all this. They not only break the continuity of artistic evolution, but they declare war upon all art of the past and the present. They deny everything, every axiom of science, every rule of common sense. They believe in a new art of unbridled and undisciplined imagination. It is unnecessary to refute their theories in detail, for while they admit the one truth that art is a language for communicating emotions, they are, one and all, oblivious of the equally important truth that the communication of their emotions can only be accounted art if it is delivered in intelligible language.

55. C.H. Collins Baker, 'Futurist Academics'

Saturday Review, 9 March 1912, 300–1

Charles Henry Collins Baker (1880–1959) was art critic for the *Outlook* and the *Saturday Review* and in 1911 was secretary to the Director of the National Gallery. He later became a Keeper at the National Gallery and Secretary of the New English Art Club (1921–5).

In our plausible sad world pretensions have ceased to be enough; they are so inexpensive, so ingenious, and usually so inexact. Thus we instantly suspect exhibitions whose catalogues are prefaced with introductions and concluded with manifestoes. For experience teaches that painters who are so tremendously impressed by their rebelliousness that they think the public and the silly critics must be educated and initiated by lengthy explanations, metaphorically speaking do not wash. Their very self-consciousness, as I have before endeavoured to explain, inevitably crowds out sincerity.

The Futurists, starting off to the heroic strains of 'Arma virumque cano', up to a point are plausible enough. This sort of thing – 'in the rendering of Nature the first essential is sincerity and purity' – is quite sound and orthodox. Or again – 'To lend an allegorical significance to an ordinary nude figure, deriving the meaning from objects held by the model, is evidence of a traditional and academic mentality,... and must necessarily displease us'. Taken by itself, and not applied to the pictures, this reads well; I suppose it is as old as Art itself, and have no doubt that young palæolithic artist said just the same about academies. But so far as history records no anti-academic band, no nor any academy, was ever so fettered by conceptual canons as are these academic Futurists. I can best explain this by setting forth the 'laws of interior mathematics' that govern Futurist painting as well as the unfortunate spectator. – 'With the desire to intensify the æsthetic emotions by blending the painted canvas with the soul of the spectator we have declared that the latter must be placed in the centre of the picture.... The crowd bustling (in a riot), and the noisy onslaughts of cavalry are translated in sheaves of lines ... following the general law of violence of the picture. These *force-lines* must encircle the spectator.'... 'All objects, in accordance with physical transcendentalism, tend to the infinite by their *force-lines*, the continuity of which is measured by out intution. It is these *force-lines* we must draw to lead back the work of art to true painting'. 'In the pictural description of the states of mind of a leave-taking, perpendicular lines, undulating lines ... may well express languidness. On the other hand, horizontal lines ... brutally cutting into half lost profiles or crumbling and rebounding fragments of landscape will give the tumultuous feelings of the persons leaving.' Finally, we have this: 'One may remark in our pictures spots, lines, zones of colour which do not correspond to any reality, but which, in accordance with a law of our interior mathematics, musically prepare the emotion of the spectator'.

Thus I gather these pictures are 'futurist and the result of absolutely futurist conceptions'. The whole thing is so patently and pathetically misguided that one simply does not know where to start, or indeed whether one ought to start at all. I suppose the principal cause of the non-existence (it is not even failure) of these paintings, as art, is that radically and absolutely their aim is irreconcilable with art. Leaving aside these precious force-lines and the rigid laws and canons that 'do not correspond to any reality' but yet govern this academy, its other vaunted aim dooms it to still-birth: 'The simultaneousness of states of mind in the work of art: that is the intoxicating aim of our art'. What thin connexion is there between Art and the fleeting casual impressions we receive? 'How often', we are asked, 'have we not seen upon the cheek of the person with whom we were talking the horse which passes at the end of the street?' I don't remember noting this especial pheno-menon; perhaps my attention was too concentrated upon my interlocutor's cheek. But if I had, if everybody does, what has it to do with Art? To paint it were ultra-faithful imitation, and does not the Futurist creed say (vide p. 33), 'All forms of imitation must be despised'? Art as a manifestation of developed mentality is not chaotic, kaleidoscopic, watery; it is concentrated and selective. Inability to keep the mind still and the attention fixed is surely the symptom of low development. A stack of bricks or a packet of mixed seeds bears the relation to architecture or ordered flower-beds that drifting over-lapping impressions bear to art. Only optical researchers, students of consciousness, and various sorts of scientists may find the diagrams displayed in Sackville Street of interest.

That is assuming they are genuine records, taken as it were from the sensitive film of real impressions. My private opinion is that in most cases they are but deliberated 'fakes'. They do not represent almost sub-conscious sensations at all; they are carefully culled and concocted 'impressions'. For example, No. 28, 'The sensations of the artist's journey from his native house to Paris: the proportions and values being rendered in accordance with the emotion and mentality of the painter'. Taking this literally we see that the wheels of a cab struck Signor Severini in exactly the same way as did a horse; that a cocher's hat reproduced a sensation identical with that given by a motor bus. Of course, they did nothing of the sort, but Signor Severini thought it would be amusing if they had, or possibly after ingenious reflection he actually succeeded in remembering these trivialities. The question what public interest has the exposure of a mentality that records conspicuously such silly details I will not attempt to answer.

It is more complex to decide questions of purity in rendering Nature. I understand the same trouble arises in literary circles. Things may be made repulsive by sympathy or by antipathy; things like Mrs. Warren's profession. It is the attitude that counts, not the repulsion. Would it be unfair to class Signor Carra's *Impressions of a Courtesan* with the play. Or, taking another view, can we accept this as a genuine record of impressions, or should we rather class it as a raisonné sort of blood-and-thunder morality picture? It is perhaps delicate ground, though exhibition publicity challenges criticism. In any case the repulsive and material grossness of this impression (very different from the spiritual grossness of Beardsley) sickens one. The same painter's impressions of *A Swim, the Sensuality* (sic) *and Coolness of a Bathe in the Mediterranean* seem to me too much akin, in feeling, to the fleshly illustrations that adorn the covers of *Sketchy Bits* summer annuals to suggest clean, physical and sensuous bliss. Much in the same way Signor Boccioni's *Laughter* is gross and blaring, thus representing laughter in its coarsest manifestation. Of course, such mirth exists; but why is the selection of it especially Futurist?

Mr. Ihlee, at the Carfax Gallery, has more chance of appealing to posterity than these self-labelled Futurists. But he is content to occupy himself with the present instead of inventing future standards. Goya, Millet, Daumier, and of course Mr. John have set him thinking, and I dare say Lightfoot, whose recent loss to art is indeed deplorable, helped him to acquire a fuller perception of bulk. His earlier drawings are exercises in Daumier-like silhouettes and patterns, without much individual content. *Miner's Crouch* of a little later, though flat and bulkless, shows an unusual perception of vital character. The more developed drawings not only show a considerable quality of line and significant design, but also smoulder with hidden thoughts, motives, passions. In an indefinable way we seem to get the clues, almost apprehend the mysterious obsessions of these people. No. 20 is especially successful in its dramatic restraint; No. 23 shows, I think, Mr. Ihlee's obligation to the Slade School for his quality of line and sense of form. Of other good drawings I would select for mention *The Necklace* (No. 18), No. 23 and No. 4, *It Rains*.

At the present moment the Women's International Art Club, in the Grafton Galleries, makes one thoughtful. Why do nearly all women artists sacrifice their birthright in order to be masculine? They succeed but in throwing back effeminate and superficial reflections of the men they imitate – Messers. Brangwyn, Steer, Nicholson, Matisse, etc. Why do they not give themselves up to their sex's strange power of intuition, non-masculine perception and instinct. All the best known women painters have been weakly masculine, and so futile. When may we be entranced by a feminine art, as distinct from masculine as Oriental from European? Mlle. B.

de Jong, at the Grafton Gallery, is a notable exception. Her *Vieille Espagnole* expresses the profound impersonal penetration that only a few men attain after long years; the colour and design, too, are fine.

56. Roger Fry, 'Art: The Futurists'

Nation, 9 March 1912, 945–6

Fry is remarkably tolerant of the Futurist endeavour, though he decides that the pictures are more psychological or scientific experiment than... work[s] of art'.

At the Sackville Gallery is to be seen a small collection of work by this group of Italian painters. The catalogue contains a manifesto of their aims and beliefs. This, at least in the English translation, is by no means closely reasoned or clearly expressed; it might have been better to allow us the assistance of the original Italian.

It is interesting to find painters who regard their art as a necessary expression of a complete attitude to life. Whatever one thinks about the content of their strangely Nihilistic creed, one must admit that they hold it with a kind of religious fervor, and that they endeavor to find an expression for it in their art. Fortunately, too, their dogmas appear to allow of great variety of treatment or method, so that, as yet, no stereotyped formula has been evolved, and each of those artists pursues his researches along individual lines. None the less, admitting as one may the sincerity and courage of these artists and their serious endeavor to make of art a genuine expression of spiritual experience, I cannot accept without qualification their rash boast of complete and absolute originality, even supposing that such a thing were in itself desirable. Rather what strikes one is the prevalence in their work of a somewhat tired convention, one that never had much value and which lost with the freshness of novelty almost all its charm, the convention of Chéret, Besnard, and Boldini. It is quite true that the Futurist arranges his forms upon peculiar and original principles, breaking them up into fragments as though they were seen through the refracting prisms of a lighthouse, but the forms retain, even in this fragmentary condition, their well-worn familiarity.

Apparently what is common to the group is the belief in psychological painting. The idea of this is to paint not any particular external scene, but, turning the observation within, to paint the images which float across the *camera obscura* of the brain. And these images are to be made prominent in proportion to their significance, while their relations one to another have the spacelessness, the mere contiguity of mental visions. Thus, in rendering the state of mind of a journey, the artist jumbles together a number of more or less complete images of the home and friends he is leaving, of the country seen from the carriage window, and of anticipations of his journey's end.

These pictures are certainly more entertaining and interesting than one would expect to result from such an idea, and one or two of the painters, notably Boccioni (in his later works) and Severini, do manage to give a vivid pictorial echo of the vague complex of mental visions. If once they give up preconceived ideas of what sort of totality a picture ought to represent, most people would, I think, admit the verisimilitude of several of these pictures – would own that they do correspond in a curiously exact way to certain conditions of consciousness. Unfortunately, the result is much more of a psychological or scientific curiosity than a work of art, and for this reason that the states of mind which these artists investigate are not really at all interesting states of mind, but just those states of quite ordinary practical life when the images that beset us have no particular value or significance for the imagination.

The idea of painting from the mental image is no new one, though it is one that artists might well practise more than they do. Blake roundly declared that to draw from anything but a mental image was vain folly; but he drew from mental images only when, stimulated by some emotional exaltation, they attained to coherence and continuity of texture. Probably a great many of Rembrandt's sketches are the result of distinct mental imagery, but it was a mental imagery stimulated by reading the poetical prose of the Bible. The fact is that mental visions, though they tend always to be more distinctly colored by the visionaries' own personality than external visions, are almost as various in their quality, and are, as often as not, merely accidental and meaningless. Doubtless the Futurists aim at giving them meaning by their relations to one another, and in this they aim at a direct symbolism of form and color. Here, I think, they have got hold of a good idea, but one which it will be very hard to carry out; as yet their work seems for the most part too merely ingenious, too scientific and theoretic, too little inspired by concrete emotion. It is the work of bold and ingenious theorists expressing themselves in painted images rather than of men to whom paint is the natural, inevitable mode of self-revelation. One artist of the group, Severini, stands out, however, as an exception. He has a genuine and personal feeling for colors and pattern, and the quality of his paint is that of an unmistakable artist. His *Pan Pan* is a brilliant piece of design, and really does, to some extent, justify the curious methods adopted, in that it conveys at once a general idea of the scene and of the mental exasperation which it provokes. For all its apparently chaotic confusion, it is not without the order of a genuine feeling for design. Here, as elsewhere, the worst fault is a tendency to lapse into an old and commonplace convention in individual forms.

His *Yellow Dancers* is another charming design. The statement in the catalogue that it exemplifies the destruction of form and color by brilliant light, shows the curious scientific obsession of these people. Such a fact is æsthetically quite irrelevant, and the picture is good enough to appeal on its own merits. The same is true of his *Black Cats*, a novel and curious color harmony, which gains nothing from the purely autobiographical note in the catalogue. Whether Signor Severini arrived at his design by reading Edgar Allan Poe or not is immaterial; the spectator is only concerned with the result which, in this case, is certainly justified.

No amount of successful exposition of theory will make bad painting of any value, and, on the other hand, a good picture is none the worse because the artist thinks he painted it to prove a theory, only in that case the theory has served its turn before the picture was painted, and no one need be troubled with it again.

Apart from individual failures and successes, one result of these efforts stands out as having some possibilities for the future of pictorial design, namely, the effort to prove that it is not necessary that the images of a picture should have any fixed spatial relation to one another except that dictated by the needs of pure design. That, in fact, their relation to one another may be directly expressive of their imaginative importance.

In thus endeavoring to relate things not according to their actual spatial conditions but according to their imaginative purpose in the design, the Futurists are, no doubt all unconsciously, taking up once again the pictorial language of early art, for it was thus that Cimabue arranged his diminutive angels around the vast figure of the Virgin in the Rucellai *Madonna*.

What the Futurists have yet to learn, if their dogmas still retain the power of growth, is that great design depends upon emotion, and that, too, of a positive kind, which is nearer to love than hate. As yet the positive elements in their creed, their love of speed and of mechanism, have failed to produce that lyrical intensity of mood which alone might enable the spectator to share their feelings.

---

57. Frank Rutter,
'Round the Galleries:
The Futurist Painters'

Sunday Times, 10 March 1912, 19

For Rutter see headnote to no. 1.

'All London is talking of the Italian Futurist Painters at the Sackville Gallery,' so my morning paper says, and I suspect most of London is talking arrant nonsense. Of course, the gallery is thronged daily, for London is full of people who will gladly pay a shilling to see a 'sensation,' though they cannot recognise a work of art when they can see it for nothing. Hence the almost daily destruction of beautiful architecture, about which Miss Gertude Kingston writes so feelingly in the current *Ninteenth Century*. But in this unhappy, soulless city, where art is of no account, a ribald, well-dressed mob will always be found ready to have their laugh at paintings they neither understand nor wish to understand. And few are found to mourn that while our censored dramatists are minting the grim realities of modern life into works of the highest art, our painters, with the exception of a mere handful, are hopelessly astray in no-man's-land; producing with alarming facility pictures that are futile, purposeless, and insignificant. The truth is that we are in the midst of a pictorial reaction against nature, the nature worshipped, studied, and so wonderfully mastered by the earlier impressionist and neo-impressionist masters. It is possible that some of these, in making truth their god, were inattentive to the claims of design, also an important element in pictorial art; but, to my thinking, a younger generation is in far graver danger in making a god of design without reference to truth to nature.

THE DANGER OF DESIGN

The danger lurking in this one-sided pursuit of design is that young painters are apt to look for it in preceding paintings rather than in nature. That is the way to plagiarism, to fantastic theorising, mental arithmetic in paint, and all manner of foolishness. Nothing pleased me more last week than to hear that the National Gallery and other museums were closed to the public in view of the Suffragist disturbances. Here is a blessing in disguise, for what could be more helpful to the young painter of to-day than to close all the picture galleries and make him go and look at nature with his own eyes? In the early twenties of last century, when there was some talk of starting a National Gallery here, that great artist and reverent lover of nature, John Constable, viewed the project with the utmost concern. He feared that its establishment might seduce students into painting from pictures instead of painting from nature, and how terribly are his fears justified to-day. Not only here but all over Europe young men and women are painting from pictures instead of from nature. That is my first bone of contention with the Italian Futurists. I do not say that they never look at nature – one of them evidently has done so with good effect – but generally speaking they are painting from the extreme wing of the Salon d'Automne and the Galerie Kahnweiler. All their little novelties have already been paraded in Paris by men as able and abler than themselves. Further, while they 'repudiate impressionism' they make use, often ill-advised, of divisionist technique, and all their more intelligible works are more or less impressions of actualities. That many of their exhibits are inexpressively incoherent stands self-confessed by a catalogue which contains some thirty pages of text to explain thirty-four paintings. This orgy in letterpress immediately excites distrust in those who know that there are only two sorts of paintings worth looking at, these which show without comment, and those which themselves comment whilst they show.

AN HONOURABLE EXCEPTION

Few of either sort are shown by the 'Futurists,' and yet there is one really successful painting at the Sackville Gallery. It is painted by Russolo, entitled *Train at Full Speed* (24), and a descriptive note in the catalogue states that it is a 'synthesis of the ridge of light produced by an express train going sixty miles an hour.' Without boasting I may say that I should have apprehended all this, excepting the exact speed of the train – which if Italian I still question – without any help from the catalogue, which might also have added that the train is seen by night a fact more important than the mileage. Seriously, however, this work by Russolo is quite an admirable painting, conveying a truer expression of the rush of a train through darkness than I have ever seen in pictorial art. Its almost solitary success proves that there is always room for a painting based upon original observation of nature. But for the life of me I cannot see what good is to be got from an endeavour to convey abstract ideas by means of arbitrary arrangements of lines and colours. When Signor Boccioni tells us that in his painting *Those Who Remain Behind* (3) – 'the perpendicular lines indicate their depressed condition and their infinite sadness dragging everything down towards the earth' – I refuse to accept so dogmatic an utterance. A diagonal line sloping down from the left to the right may be accepted as signifying depression by a graphologist, and even he may be misdirected by the haste or carelessness of the writer; but I defy a graphologist or anyone else to establish any natural association of ideas between depression and perpendicularity. When Boccioni goes on to say that his "mathematically spiritualised silhouettes render the distressing melancholy of the soul of those that are left behind," I refuse to be cast down. To a rational observer they suggest something far more concrete and more hopeful, divers in orthodox garb groping through seaweed for the Tobermory Galleon. Signor Boccioni is full of cleverness; his *The Street Enters the House* (4) is quite an amusing caricature in colour, but we see how thin is the veneer of his originality when we contrast his emulation of Picasso in *Leavetaking* (1) with his quite academic painting of the garden hat worn by *A Modern Idol* (8), a hat which might have been painted at Newlyn or St. Ives quite as well as Milan. I will admit that his *Police Raid* (20) is full of 'go'; that the Futurists, speaking generally, are successful in rendering movement. When this is the object, as here or in Carra's *Leaving the Theatre* (18), it may be argued that the painters have obtained some measure of success. But though it is impossible to see Severini's big canvas *The Pan-Pan Dance at the Monico* (27) in the tiny rooms at 28, Sackville-street, his picture conveys to me rather the spasmodic, artificial oscillation of a cinematograph than the to-and-fro bustle of life itself. And this is true of so many of the other exhibits. It is a mistake for any painter to sacrifice all other virtues in the pursuit of one quality alone, but it would be better for the 'Futurists' to concentrate all their powers on the rendering of observable movement in nature than to play at visualising abstract metaphysics which their master, M. Bergson, can express far more lucidly and intelligently, than any of his conscious or unconscious pictorial disciples.

58. Unsigned review,
'The Aims of Futurism'

The Times, 21 March 1912, 2

At the Bechstein Hall on Tuesday Signor Marinetti, the founder of the Futurist movement, gave a lecture on Futurism in literature and art. He read his lectures in French with such an impassioned torrent of words that some of his audience begged for mercy, and of his sincerity there can be no question, but his doctrines are a morbid form of destructive revolution. There is no beauty according to the Futurist except in violence and strife; every museum and all the great works of the past should be utterly swept away. We have no longer time, said the lecturer, to weep over tombs, and he proceeded to outline his ideal world of the future and showed a place so stripped of all tenderness and beauty that an American was overheard to say that it would be like New York at its worst. He explained that he could say little of the pictures now on show at the Sackville Gallery, the painters of which claim that they have found the sense of speed and paint states of mind, not objects, without having them before them, and went on to read three Futurist poems. Two were in Italian and dealt with a suicide and his watch, and with a lunatic asylum; the third, a French one of his own composition, glorified an automobile in work of destruction. He ended with a passionate defence of war. Whatever element of truth may underlie doctrines deprecating an excessive veneration for the past the anarchical extravagances of the Futurists must deprive the movement of the sympathy of all reasonable men.

59. J.B. M[anson], 'The Italian Futurists and the X-Rays in Art'

Outlook, 23 March 1912, 439–40

As an evidence of the vitality of the Southern spirit, the outbreak of Italian 'Futurist' painting is interesting but not dangerous. There will be no need to inoculate the students in our various schools of art against infection, for the movement stands self-condemned by the ineptitude and incoherence of its results. It is not plausible for a moment. Nor is it likely to arouse anything but merriment in England, for, in spite of recent outbreaks of window-smashing on the part of certain infatuated people, the British nation is fairly hard-headed and certainly too prosaic to start chasing any fantastic will-o'-the-wisp of the Latin temperament. Nor is the nation to be supposed to be entirely imbecile, although it sits unmoved and goes on with its knitting while the trade of the country is ruined by dangerous strikes which a futile Government discusses as though they were a subject of merely academic interest. To an Italian Futurist we may well appear therefore to be fair game.

The pictures which this group of fervid Italian revolutionaries display at the Sackville Gallery as examples of the art of the future, are in fact less amusing than the initial Manifesto of Futurism printed in the catalogue. This starts in grandiose Walt Whitman style – 'We shall sing the love of danger, the habit of energy and boldness'; but it soon becomes feverish – 'We shall extol aggressive movement, feverish insomnia, the double-quick step, the somersault, the box on the ear, the fisticuff.' Theirs is obviously an art for transpontine music-halls rather than for a West End picture-gallery. But in the depths, beneath the froth and noxious gases which rise to the surface of this Italian hotch-potch, is a certain deposit of sanity and significance, but the art of painting (or any plastic art) is not a fit or even possible medium for its expression.

The plea of this group of painters for freedom from tradition, its emphatic statement of the need of thinking for oneself, the unworthiness of tying one's mind to a convention chiefly because the latter has a market value, are necessary and valuable, particularly in this country, where art has been crushed for many years beneath the weight of old fogeydom. But it is not by violence of this kind that evolution is stimulated, any more than a comatose Government is to be stirred into life by the breaking of other people's windows. The Futurists are determined nevertheless that our mental windows are to be shattered.

We are to be taught to walk in the proper artistic Futurist way. But one does not mount in an aeroplane in order to teach people to walk. There is about this band of cut-throat pioneers, a certain splendidly inconsistent and illogical spirit, but perhaps after all they are consistent, seeing that they disclaim any use for reason or logic. The future is to be free of all such hampering dogmas. It is a relief to know that the Post-Impressionists, synthetists, and cubists are now all superseded, for 'they obstinately continue to paint objects, motionless, frozen, and all the static aspects of Nature. We on the contrary seek for a style of motion.' The trouble is that it is impossible to express the degree of actual motion they wish to express on a canvas and remain coherent or comprehensible. They require a cinematograph machine which moves at express speed for the proper expression of their ideas. In their anxiety to surpass Impressionism the Futurists have fallen into a degenerate Impressionism. They pile one impression on top of another until the whole is one mad jumble without rhyme or reason. Art, after all, is a language; but a language which cannot be understood or which requires an elaborate key to its meaning, defeats its own ends and has little or no value.

In one respect the movement has a new interest, and that lies in its use of Röntgen rays as an element in art. Not only do some of these enterprising painters show us objects themselves, but objects behind them, objects all around them, inside them and through them, and to all this is sometimes added the painter's memory of what he saw yesterday, thought not as yet his anticipation of what he may see to-morrow. The picture *Laughter* by Boccioni, is explained: 'The scene is round the table of a restaurant where all are gay. The personages are studied from all sides, and both the objects in front and those at the back are to be seen, all these being present in the painter's memory, so that the principle of the Röntgen rays is applied to the picture.'

The painting of a *Train at Full Speed* by Russolo has been hailed with delight as being a comprehensible one. It is a 'synthesis of the ridge of light produced by an express train going at sixty miles an hour.' In this case the painter has made fixed lights on the landscape move equally with the train instead of in opposition to it, which seems to demonstrate the superficiality of his observation. There is not a single picture in the exhibition which is a good painting, as painting. Probably the Futurists are superior to such ridiculous considerations; possibly a decent, workman-like use of pigment requires too much study and practice. In any case the exhibition is assured of a succès de scandale.

60. Walter Sickert, 'The Futurist "Devil-among-the-Tailors"'

English Review, April 1912, 147–52

After his attack on Picasso and Matisse it comes as a surprise that Sickert should welcome the energy, iconoclasm and in some cases the 'sound workmanship' of the Futurists.

For Sickert see headnote to no. 29.

Let it be granted that to be a member of the Royal Academy does not in itself confer talent. Let it further be granted that to be a member of the New English Art Club does not in itself confer talent. Nor, further, can the impartial seeker after truth deny that talent is not conferred by affiliation to Mr. Fry, or by membership of the Kentish-town, or even the Somers-town group. Lest the writer be accussed of Xenophilism, let him add these further postulates. To be a member of the French Impressionists did not in itself confer talent. To call oneself a Post-Impressionist does not in itself confer talent, and, lastly, to be an avowed and militant Futurist is in itself no guarantee of pictorial ability.

It is curious how inexactly the Press will occasionally mirror the public opinion of the town. The *Morning Post* was lucky in the possession of one of the most amusing, readable and learned art-critics in the person of Mr. Robert Ross. Right or wrong in his opinions, his articles carried the organ of the gentry into numberless intellectual homes to whom fashionable intelligence is as naught. In the exercise of his duty it seems that Mr. Ross prepared an article on the Futurist Exhibition. I am told that this could not be published on the plea that the Futurist Exhibition was in itself an immorality, and must not be chronicled! I was, therefore, rather surprised to find the tiny galleries in Sackville Street packed with an orderly crowd, consisting mostly of the mothers of England, who were circulating slowly, and verifying, with reverence, the statements in the descriptive catalogue by the pictures on the walls. Here and there, a grandfather of distinguished appearance was pointing out to a delicious grand-daughter the obvious beauties of the pictures, and the sound reasoning of the accompanying tracts.

One thing the Futurist movement certainly is not, and that is, immoral. Austere, bracing, patriotic, nationalist, positive, anti-archaistic, anti-sentimental, anti-feminist, what Prudhon calls anti-pornocratic, the movement is one from which we in England have a good deal to learn. This is not to say that we are to accept the manifestoes in their literal entirety. We must remember that language in Italy is a far more florid and coloured thing than with us. Sir Claude Phillips is perhaps the only writer of *coloratura* we have left. Have we not been told that when it was necessary to have the Coronation 'written up' it had to be done by an Italian writer in Italian, and then translated into English?

The idea at the root of the Futurist movement is health itself. It aims at creating a 'contagion of courage.' It would teach us that a healthy intellectual life, a healthy political life, is based on active concern with the present and the future, and not on hypnotism by the past. In order that a salutary truth may penetrate the shell of inertia and habit in which humanity will ever lap itself, the most monstrous exaggerations may do good service. Fulminations that are easy to ridicule and confute, may, by the spirit they arouse and the atmosphere they create, serve as effective engines of beneficent force.

The Futurist movement is essentially a movement of Italian patriotism, and centres round the work, the person, and the ideas of F.T. Marinetti, the author of *Mafarka il Futurista, Distruzione, La Ville Charnelle*, and *La Conquête des Etoiles*. The prelude of the movement may be found in the columns of the Paris *Figaro*, and its continuation in Marinetti's five years' editorship of the international review *Poesia*. Among the Futurist poets are G.P. Lucini, Paolo Buzzi, Cavacchioli, Palazzeschi, Govoni, De Maria, Armando Mazza, Folgore, Libero Altomare, and Mario Betuda. And now the beauty-sleep of Sir Philip Burne-Jones has been disturbed for the third time by the sudden explosion in Sackville Street, next door to Whistler's tailor, of this strange artistic camorra, like the firework that is described in Messrs. Brock's catalogue as *The-Devil among-the-Tailors*.

The Futurist movement confesses to a literary origin; and the alliance of the pen with the brush has its dangers for both. No amount of explanatory doctrine and militant defence will make a bad draughtsman into a good one. Painting that requires literary explanation stands self-condemned. Here we have the condemnation of Matisse and Picasso, and even of most of Cézanne's canvases.

And on the same principle do some of the Futurists justify themselves in some of their pictures. I should myself, if I had a collection, like to possess the *Travelling Impressions* of Severini, or Boccioni's *The Street enters the House*. His large canvas, *The Rising City*, is workmanlike and lucid, and owes these qualities to an intelligent study of the past. Besnard and Constantin Meunier have had a part in it. His *Police Raid*, again is a vivid essay in lurid drama.

Simultaneous representations of successive impressions were not invented by the Futurists. You may find, by many a primitive painter, the birth, marriage and death of a person represented on the same canvas. I remember seeing in Brussels, twenty years ago, a well-known picture in the Wiertz Gallery that recounts the impressions received by the brain of a man in the act of being guillotined. These impressions are recorded as belonging to the first, second and third seconds after the severing of the head from the body. Whether the three impressions are united on one canvas, or not, I cannot remember. In Turner's later pictures simultaneous visions are painted of ice-cold passages juxtaposed to hot ones in a manner that is frankly impossible and unnatural. But they remain works of great beauty and interest, in spite of their fantastic nature. The Futurists may claim the same freedom as we allow the old masters. Indeed, the word 'allow' is ridiculous. The human race, in its ingenuity and curiosity, will rush through any porch where it seems to perceive light, whether we call the road forbidden wickedness or not. There are caves in art that lead somewhere, and others that are bottoms of sacks.

Cul-de-sac, I think, is the whole research represented by Boccioni's *Leave-taking*, *Those that Go Away*, *Those who Remain Behind*, because the results cannot be lucid, and require endless explanations. Painting, like speaking, is a form of expression, and a speaker who is incomprehensible cannot be said to be a speaker at all.

I was once standing with a Frenchman on the sea-front at Dieppe when we witnessed the meeting and parting of two early Edwardian grass-widows that we both knew. We also knew that it was the first time that they had met that season, and that they were fast friends. They passed each other hurriedly. The bells at the hotels had rung for luncheon. They nodded cordially, and their dialogue was compact, but, to me, quite lucid. 'Going strong?' said one. 'Full of beans,' was the reply, and the incident closed. My friend asked me to translate. 'Madame X,' I said, 'a dit "ça va fortement?" Et Madame Y a répondu "remplie de haricots."' 'Mais qui? Comment? Quoi? Remplie de haricots?'

I gave my friend the necessary explanation, which the reader can imagine. That Mrs. X, having suppressed, as we wisely do in England, what Casanova calls '*les compliments d'usage*,' had meant to convey that she was glad and surprised to see that Mrs. Y had arrived for the season. That she hoped Mrs. Y and family were well. That she hoped to see something of her. That she supposed she had been playing at the Casino. That she hoped, either that Mrs. Y had won, or at any rate, that she had not lost much. Mrs. Y, I further explained, by 'full of beans' meant to convey that she and her family were quite well. That they were delighted to be in Dieppe again. That they were delighted to see Mrs. X again. That they were staying at the *Bains*. That she was in a hurry as the luncheon bell had just rung, and she had to change. That she had been, as Mrs. Y supposed, playing *petits chevaux*. That, though she had lost a few francs, she was playing on an infallible system, which she hadn't time to explain, by which the nine must inevitably come up sooner or later. That she supposed she would see Mrs. X at the Gymkhana on the links that afternoon.

Now I maintain that, as literature, both Mrs. X and Mrs. Y's utterances were successful, since they were perfectly lucid. Both knew that the experiments of Professor Dewar are as child's play compared to the pressure that can be brought to bear upon language. That language can quite well be solidified, from a state of vaporous verbiage, into the solid instrument that is required to make their crowded and interesting lives as articulate as is needful. Each held the clue to the other's code, and I, lucky man, to that of both. If any of us three had needed a pamphlet to explain the conversation, we should have to consider the symbolism of these ladies to be a failure.

Both Severini and Boccioni, doctrine apart, are competent workmen. Severini's large canvas must be judged on another occasion, when it can be seen in a room from a distance at least three times its whole width from it. His 'Travelling Impressions' is an amusing and ingenious compilation. Cruikshank did something of the same sort in his big temperance picture that was the delight of my schooldays when it hung in the Kensington Museum. All his work is competent. Carrà is simply a bad painter, who happens to be a Futurist, inefficient and pretentious. He can be affiliated in sentiment, but not in efficiency, to Gustave Moreau's obsolete bombast, and Odilon Redon's portentous emptinesses. Russolo, on the other hand, has something to say.

There are one or two maxims embedded in Signor Marinetti's twentieth-century commination service that are worth retaining. It was a seventeenth-century bishop who laid down the too-much forgotten truth that both wisdom and interest recommend 'soft laws and strict excecution.' Are not the following versicles moderation itself?

'To lend an allegorical significance to an ordinary nude figure, deriving the meaning of the picture from the objects held by the model, or from those which are arranged about him, is to our mind the evidence of a traditional and academic mentality.'

'We protest against the nude in painting, as nauseous and as tedious as adultery in literature.'

It is always interesting to note any approach in thought of widely distant thinkers. Charles Keene once said to me, that one of the great faults in modern teaching was the incessant and dispiriting study of the nude. In an experience of teaching extending over some 20 years, I have become more and more convinced of the justice of this opinion. The nude is recommended, partly because it is, in theory, stable, while drapery is obviously unstable. Excessive restriction to drawing from the nude does not sufficiently cultivate rapidity, perhaps the chief requisite in study from the life. A sleeve or a skirt must be drawn in thirty to, at most, fifty minutes, whereas a leg or an arm, may, in theory, be continued from rest to rest.

My grandfather, who was a painter, and one of the earliest original lithographers, used often to end his letters to my father, who was a painter and draughtsman on wood, with the advice,

'Paint well and quickly,' somewhat, I gather, to his son's irritation. But my grandfather was right. If you are to draw from nature at all it must be quickly, and exclusive study from the nude dulls a student's observation, not only for the reasons I have given, but also because the nude corresponds to nothing in our modern habits of life. With all his powers of draughtsmanship, when Ingres essayed, in *le bain Turc*, a composition consisting entirely of nude figures, the effect is trivial and a little absurd. So far from giving an impression of life, the picture suggests an effect like a dish of spaghetti or maccheroni. Ingenuity has here strained excessive cerebration to the point of snapping, and learning becomes a form of folly.

It would be interesting to secure medical statistics on the average number of young women students a year whose health has been broken down by excessive hours of drawing from the nude, in the exhausted atmosphere of rooms heated to the temperature necessary for the model. 'The figure' spells emancipation and smacks of professionalism to the young. Sixty per cent of these young martyrs to carbonic oxide would be learning more from still life or the antique; and the remaining forty per cent would in these days be on a surer road to become artists if they learnt to make studies on the scale of drawings by Longhi, Watteau and Hogarth, from figures in costume.

61. Anthony M. Ludovici, from 'The Italian Futurists and their Traditionalism'

Oxford and Cambridge Magazine,
July 1912, 94–6 and 109–13

Ludovici, author of a number of books on Nietzsche including *Nietzsche and Art* (London: Constable, 1911), attributed what he saw to be the chaos in modern art to the rise of democracy. The salvation of art, he claimed, would come through the assertion of aristocratic forms, the expression of exceptional individuals. Consequently, though he welcomed the iconoclasm of the Futurist manifestos, he felt that the works of art themselves were deeply conventional and the epitome of 'democratic' art. Ludovici soon after became the art critic of the *New Age*.

There was a time when the Chinese used to reply to our punitive bombardments by lighting bonfires of sulphur, the fumes from which were intended to drive us from our position and to preserve them from further hurt. Slow as they are to abandon their old customs, however, they ultimately realised that this Oriental artifice was of no avail against the superior methods of Europeans, and they lost their faith in the efficacy of their stifling smoke.

Now, glancing through the replies that English art-critics made to the gallant onslaught of the Futurists, one cannot help being strangely reminded of the primitive methods used by the Chinese as late as the middle of last century. These young Italians came forward with a statement of their position and of their claims, which was as lucid and as piercing as the sunlight of the land from which they hail. Whether one approved or disapproved of their standpoint, one thing in any case was undeniable, namely, that they were perfectly clear as to what they themselves believed in and wanted. And how were they met? Either with articles like those in the *Daily Telegraph* of March 22, and in the *Spectator* of

March 16, *i.e.* with elaborate and non-committal nonsense; or else with accusations of madness – in fact with smoke, or more or less aromatic gas.

Sir Claude Phillips, in an article full of what some might call playful intellectual banter, implied that there was nothing to be learnt from the Futurists, and that he, at any rate, did not think it worth his while to take them seriously. He said: 'If we dilate on the exhibition at greater length, it is for the amusement of the public and ourselves [*sic*!], and not because it calls for any such extended notice.'

This was a pure evasion; for, if the Futurists deserved a column notice in the *Daily Telegraph*, surely they were considered serious enough to be treated seriously. And, if the duty of an art-critic is not to 'place' the particular phase of art under his consideration, then what on earth is it? The public do not want to be 'amused' by a serious journal if something apparently grotesque startles them in the art-world; they wish to be helped in 'placing' or classifying that grotesque manifestation.

The *Spectator* said: 'Anyway, we are *not sure*[1] that the mental pressure that is so desperately seeking through Post-Impressionism and Futurism to make modern art more expressive of our own strange times is all for the bad. Only, most of us feel that it is hard that it should fall to us to make the mental effort to find out what it is all about. And Futurism is especially difficult.'

We are perfectly used to the art-critics on first-rate journals being '*not sure*' about anything at all; but we must say that the ingenuous frankness of this one who sighed 'that it is hard that it should fall to us to make the mental effort to find out what it is all about,' and who then proceeded to say *nothing* concerning what it was all about, was a little refreshing. But, why then did he write a column of platitudes about it?

On the occasion of my first visit to the Futurists, I felt that it was of paramount importance that every one, opponents and sympathisers alike, should be as clear in their response to this new challenge from the world of art, as these five painters and their leader Marinetti actually were in bringing it forward. Subsequent visits did but strengthen this feeling in me, and now I am perfectly convinced that if art-critics are going to show any power at all in dealing with this new phenomenon – the Futurist movement – they must definitely settle the precise meaning of every term they use in discussing it. In the face of innovations thrust upon us with a shriek of such positive defiance, it is simply cowardly to be non-committal, it is merely idle and spinsterly to reply with accusations of madness, and it is actually dangerous to shrug one's shoulders and try to appear as if nothing had happened. If, however, it is admitted that something *has* happened, I maintain that in speaking about it one should be perfectly lucid and free from compromise – as lucid and as free from compromise as the Futurists themselves. One should be perfectly clear in regard to the meaning, the purpose and the terminology of art.

Any journalist who is used to writing column upon column concerning things he does not understand can ramble on like an untethered donkey, without necessarily getting anywhere; but we have had generations of this sort of thing, and the appallingly uncultured – or better still, muddled – state of the public mind on art matters is the obvious result of it. For it is in regard to the first principles of art that the reading public are so vague and doubtful to-day; and it is precisely concerning these things that the critics persistently fail to instruct them....

Now it is plain that, for scores of years, the art of Europe, and especially of England, has consisted chiefly of *Vulgar Realism* as defined above, and that it has been the expression of the man with the *slavish vision* or *grasp* of reality. An examination of the average work shown at the Royal Academy in any year during the last half of the nineteenth century, and a knowledge of other independent in art, prove conclusively that, if democracy and democratic principles have tended to come ever more and more to the fore in the sphere of politics during these years, this development has not remained unreflected in the art of the period. On the contrary, it might be said that all along art has prophesied the inevitable goal to which all social and political changes were tending. Indeed, if, as I maintain, art is invariably a symptom of a particular kind of life, it would be somewhat strange if the art of a nation or of a continent were not always the unconscious symbol of that nation's or continent's soul. And to turn to art, as to a prophetic oracle, with questions concerning the destiny of a people, would not prove such idle work as some might suppose, provided of course that the interpreter were there, who, like the Pythia at Delphi, understood the meaning of the signs.

Now let me approach this art of the Futurist without prejudice and, armed simply with the criteria which I have detailed above, ask, in the first place: Whose realism it was and whether it was a realism that mattered or was important? What inspired these men to the rare moments which were found enshrined in their pictures?

Taking the non-symbolic canvases first, I found that street scenes, railway scenes, courtesans, dancers, the mob, night-revellers – in fact all the things that any towns-man can see any day or night – were the portions of these painters' reality which they selected, and which they chose to perpetuate and to enshrine. In the face of such realism, I was confronted by three alternatives: These men were either inspired to express these portions of their reality by their actual love of these things themselves, by a higher idea connecting them and their art with these things, or by a love of the colour effects, decorative effects, and effects of form which they offered. As it is impossible to assume, however, without straining one's credulity to snapping point, that they did actually love such portions of their reality as a woman and absinthe, as the station at Milan, or as the impressions of another party in a tramcar and out of it (No. 14), and as, for the moment, I saw no higher idea connecting them and their art with the content of their pictures, I took it that it was the old love of the impressionists and neo-impressionists that inspired them to express their reality – a love of matters of form and of technique. But even if this were not so, and granting that they actually did love the things they presented, then they would have differed in no way from the ordinary *vulgar realist* of the past; for, in the first place, their reality was not a *rare* reality; it was too catholic to reveal a definite attitude towards life; it embraced too much to show any clean-cut passion, nor was it expressed in a realism which *gave* anything over and above the mere content of the picture itself. It was not a gift in any way, it neither placed, exalted nor defined.

The Futurists might point out that they did give something more, and that it was this 'something more' which, while differentiating their realism from the vuglar realists of the past, also bestowed a gift upon mankind and constituted the higher idea which connected their work and themselves with the content of their pictures. They might say there was a deeper love in their work than could concern itself with cabs jolting and railway stations, as ends in themselves. What was this love then, or higher idea? Was it a love that mattered? Was it a clear, definite, exalted love, inspired by a clear, definite idea which was greater than either themselves or their art? Was it a love so great, so masterful that it rescued its object from the thraldom of accident? Lastly, was it a new love – a new deep love that would inspire millions, who are now desperately adrift for want of some coordinating idea or desire, to live for something greater than that mere turn of the kaleidoscope which is their so-called 'life of enjoyment?' It was none of these things. As I have adumbrated above, if it was anything at all it was a vague love of the modern age of chaos;[2] of petty individualism;[3] of the subjection of man, not by a single grandeur, but by a complex and multiplied littleness[4] not by the obelisk, but by chimney-pots. It was a love of that which says 'Nay,' to man – the machine;[5] it was a love of that which says 'Nay' to society – the anarchist;[6] and withal it was a poor, weak love which wrapped itself in fiery words in order to seem fierce and ardent, but which had not even the power to rise above petty technical controversies and studio jargon; *which had not even the power to break with the past save in manifestoes – in words!*

For, suppose that the Futurists told me to turn my eyes from their words, and again to appeal to their work, what was it I saw? I did indeed observe something new in their reality, and if it is to be ascribed to any sort of love I do not hesitate to declare that I know what that love must be. For the realism depicted in *The Dancers at the Monico* (No. 27), *The Train at full Speed* (No. 24), *The Funeral of the Anarchist Galli* (No. 11) and *The Police Raid* (No. 10), spoke eloquently of no love that distinguishes these painters from any vulgar realists of the past, save the love of movement *as movement*. What, however, is the precise value of such a love? Why exalt movement *per se?* Is movement, *as movement*, necessarily inspiring? But our perplexity on this point is soon dispelled if we think of the age in which we live. This age idolises movement. The word 'Progress' is on everybody's lips. Change and movement are in everybody's mind. Mere change and mere movement are called 'progress' without any further ado.

If, however, you ask 'progress whither?' every one remains inarticulate. If you inquire 'change to what?' 'movement in what direction?' no one can vouchsafe an answer.

It is admitted that nothing is for two instants the same, it is granted that change is the only thing that is constant. But, hitherto, it has been the pride and noble endeavour of mankind to be master of change, to *direct* movement to the best of his ability, and to give it a goal. Progress, hitherto, has implied the idea of a definite direction, a half-realised bourne. Mere unguided movement, or so-called modern progress, means absolutely nothing, and to love it is to love nothing, unless it be progress for progress' sake! Or is it, perhaps, 'Gadarene Progress,' as a friend of mine, Dr. Wrench, has it? In its love of movement *per se*, then, Futurism, far from being original, merely reflects the most absurd phenomenon of the age. What extra virtue did it add to these pictures that they represented movement, any more than rest, any more than the "invisibility of high speed?" A picture of a peaceful sunlit valley in which nothing stirs, and in which there is no sign of a breath of wind, but which represents the space traversed by a cannon ball just fired, would be the image of the fastest movement as yet attained by human effort; but to represent such a scene, or *one's recollection of it*, and to call it *The Flight of the Cannon Ball*, seems to me to be making no violent breach with the past either in excessive futility or in extravagant originality.

NOTES

1 The italics are mine.
2 Catalogue of works at the Sackville Gallery (March 1912), p. 15.
3 Ibid., p. 15.
4 Ibid., p. 4.
5 Ibid., p. 3.
6 Ibid., p. 4.
[All notes are Ludovici's.]

62. Roger Fry, 'The International Society at the Grafton Gallery'

Nation, 20 April 1912, 87–8

Post-Impressionist techniques had blossomed in British art from the beginning of 1912, and this exhibition at the International Society anticipates the second Post-Impressionist exhibition in which, unlike the first Post-Impressionist exhibition, British and French painting appeared side by side.

'Eppur si muove.' Even the International Society, whose leading authorities were so refreshingly vehement in their denunciation of Post-Impressionism two years ago, have seen fit to include Van Gogh, Gauguin, and Flandrin in the present exhibition. Is this an unwilling concession to the change in public taste, or evidence of conversion? The fact that Flandrin's *Ballet*, one of the few really good pictures in the show, is hung in the very worst position, suggests the former; but we need not complain of the motives – any motive is good enough which leads to the exhibition of good painting; and though the Van Gogh is neither a very typical nor a very good example, several of the Gauguins are admirable, and set a standard by which, to tell the truth, the English work suffers.

But our gratitude is greatest for the pleasure afforded us of seeing two superb Manets, much finer examples than the very 'important' works seen on the same walls at the Post-Impressionist show. They illustrate the kinship of Manet to Cézanne, in a remarkable way. In the superb painting of *Flowers* (No. 8) there is already the essence of Cézanne's peculiar *facture*, the beginning of his peculiar crystalline interpretation of form. Nevertheless, the difference is fundamental, and the step taken by Cézanne remains one of the greatest in the history of art, for whereas in Manet the treatment is of the nature of a surface quality, beautiful in itself, and marvellously direct in its rendering of the play of color, in Cézanne the same treatment becomes austerely architectural, the basis of a pure structural design. The crystalline *facture* of the surface implies the plastic unity of the forms more than the accidents of light and color. Still the Manet remains a superb and masterly painting, and shows what a noble and impressive sense of style Manet had before his art was side-tracked by the scientific obsessions of the later Impressionists. In the other *Still Life* (No. 6) of a dead rabbit, there is the same breadth and ease of manner, the same beauty of handling, and a shimmering splendor of color; but perhaps even more than in the flowers we feel the absence of plastic quality, we realise that Manet's vision skims the surface of things, and prevents him from identifying himself with their objective existence.

A tiny little study by Renoir (No. 13) reminds us once more how great and sincere an artist he is. It represents a woman in a negligée white gown reading a book on a sofa. Nothing could be more unambitious, but so intimately is the movement of the figure realised, and with such a lyrical intensity of feeling, that this trifling and unpretentious *genre* piece more than holds its own beside the large J.F. Millet *Famille de Paysans* (No. 11) that hangs near by. At first sight one is carried away by the deliberate suggestion of monumental dignity; here at least, it would seem, is common life seen in terms of epic grandeur and sublime austerity, only as one looks more carefully, one comes to one term of phrase after another which is turgid and otiose, and in the end one feels that this has nothing to do with life, that it is a willed and premeditated formula of ostentatiously noble sentiment. We see that Millet has plastered his Michelangelesque conceptions on to his French peasant with more tact, no doubt; but much in the same way as Fred Walker was wont to Praxitelise the English laborer. The other Millet is scarcely even deceptive; it has the tiresome, inflated rhetoric of our own great Victorian academicians.

Jules Flandrin's *Figures de Ballet* (No. 155) shines out from its dark corner as one of the bright spots in this large collection of more or less capable, and almost entirely insignificant, mediocrity. Flandrin is never a sympathetic artist; he is hard, often gloomy and depressing, but he imposes nevertheless by the perfect lucidity of his construction, by the fact that he possesses, even more than most of his contemporaries, that high clear passion for form which Poussin implanted in the French tradition. This picture is for Flandrin curiously pleasing, and even attractive, in color; but it is by its plasticity, by the precision and comprehensive grasp of mass and movement, that it surprises and captivates the imagination.

Bonnard has never been well seen in England, and the International Society has never put his work forward as it might. It would be an appropriate and sagacious thing to do. He is the most distinguished of those artists whose work lies just on the Impressionist side of Post-Impressionism. His work would therefore fall more naturally into line with the general aims of the International Society than with those of the modern Exhibitions organised by the Grafton Gallery Company. And indeed he deserves to be known and studied in England, for his work is exquisitely fine and alert. One may indeed guess at his rare gifts from the little street scene here, with its delicate and witty observation.

Simon Bussy sends two pictures, a curious and recondite still life, and a remarkable portrait head. This is adroit and scholarly, with a subtle schematic treatment both of the form and color. But M. Bussy knows how to combine the demands of an almost abstract science of form with actual likeness, so that the look of verisimilitude, the realisation of life and character, are intense.

It is difficult to compare such work as this, work which implies the severest repression of all display of technical skill, with the mass of English work where technical dexterity seems to be the only aim. Mr. Nicholson still retains this, but his portrait (No. 7) shows that when once this becomes the main object of an artist, he loses the power of impressing even by his adroitness. Mr. Strang makes heroic efforts to keep pace with the times, but his efforts are purely stylistic; they imply no fundamental change in an essentially obvious and photographic vision. It is not enough to put an El Greco sky and a Cézanne landscape behind figures which betray in any lineament the teachings of the camera. His portrait of Mr. Festing Jones shows at least the advantages for purposes of record of his gift of literal observation, but this would perhaps be equally apparent if the coloring and handling were not summarised so wilfully. Mr. Walter Greaves shows that it was just as futile to mask literal dulness of feeling and observation beneath a Whistlerian formula.

63. Walter Sickert, 'The International Society'

English Review, May 1912, 316–22

For Sickert see headnote to no. 29.

It would be difficult to estimate how much students of modern painting owe to the indefatigable enterprise of Mr. Francis Howard in the collection and arrangement of old and modern paintings that the International Society of Sculptors, Painters, and Gravers has submitted to the public over a period of fourteen years. If ever there was a case of *l'état c'est moi*, Mr. Howard would have the right, if he chose, to say so. These exhibitions have had, I suppose, during the time covered by them, more effect in educating a generation of English students of art to modern realities than all the other societies put together. Reputations have been made, modified, and unmade; and now Mr. Howard's close association with a firm of picture dealers of such world-wide reputation as Messrs. Knödler's can only extend the field on which he will be able to draw for the exhibitions at the Grafton. The members of the society, with a courage and persistency that is above praise, expose themselves annually to the most searching comparisons with the mighty dead, and their individual positions, so far from suffering, have been rather strengthened thereby. Here and there, in the earlier days, one more timid or more modest, estimating himself an earthenware vessel, has resigned, fearing the impact in full stream with the brazen pottery of the classics. He has been shown to be in error.

Probably something like a double or treble audience has been created. A section of the public has been allowed to find a natural delight in the prettier and more amateurish productions to which they were accustomed. Social and commercial aspirations have received a measure of satisfaction at the hands of this section. This portion of the audience – divided, probably, fairly equally between the stalls and the pit, as it were – have been able to enjoy the exhibitions for what they liked, and to take the serious ossature of the shows on trust as 'very clever,' and not sufficiently overwhelming to be embarrassing. Did not Goethe, in his preface to *Faust*, recommend some such judicious mixture to capture at once the initiated and the profane.

Like a wise impresario, Mr. Howard's motto has been 'Works when possible. When not possible, at any rate, names.' The first president, with a playful cynicism which was characteristic, allowed pictures by Degas to figure on the walls, and the name of Degas on posters in Knightsbridge, not only without the great man's consent, but in defiance of his expressed wishes. It was the inauguration of a 'new style' in artistic etiquette, and if an individual here and there has been ruffled, the public has certainly been the gainer.

When I think of the difference in the opportunities of study for modern students now, and twenty or thirty years ago, I am inclined to say that there will be no excuse if the rising generation do not produce a small batch, at least, of masterly workmen.

Mr. Howard does well to lead us, at the height of the nonsense-boom, at once before the monumental Millet, *Une Famille de Paysans*. The sublime man and his stolid spouse face the spectator with all the gravity and symmetry of two caryatids, while the child essays, a baby Samson, the strength of the pillars of his house. Such works in their mighty simplicity are the ultimate words of art. There is only one way to draw and one way to paint, and here it is. There is no new art. There are no new methods. There is no new theory. The old one is great. It prevails, it has prevailed, and it will prevail. There can no more be a new art, a new painting, a new drawing than there can be new arithmetic, new dynamics, or a new morality.

When it seems that a new man or a new school have invented a new thing, it will only be found that the gifted among them have secured a firmer hold than usual of some old thing. Degas can only draw with the drawing of Mantegna and Holbein, if he applies that drawing to other subjects with other curiosities, and a point of cynicism that is other. Manet is only a blonde Ribera of weaker grasp. What is it that Pissarro, Sisley, and Monet have had to remind us? They have had to take us back the primitives; they have had to repurge painting of the hollow brown shadows. They have had to point to the Venetians, and recite again this law: if you use colour at all it is written in the immutable and mysterious laws of harmony that the colour in the shadow must be the sister of the colour in the light. You are only a painter on condition that you discover and maintain this particular relation with justice and decision.

What is the sum of the teaching of the impressionists in the matter of execution. Reacting from the tendency of much admirable work to become too set, they suggested forcibly that it were better to keep the execution of studies from nature in a state of greater tentativeness, or looseness, to sacrifice much of decision and authority in order to remain a passive conduit for the fluidity of the shifting impressions received from nature. The body of their achievement is not only of a very high order and of great intrinsic value, but they have left the art invigorated and refreshed. They have left it renewed and populated where there was a danger of sterility.

The error of the critical quidnunc is to suppose that the older things are *superseded*. They are not *superseded*. They have been *added to*. That is all.

I have met many a young neo-fervent who, I am sure, accounts Millet 'a back number.' 'Salt may have been very well,' he says, 'for our fathers; but these new salt substitutes save the waiters a lot of trouble.' I doubt if such a one would be at all impressed if I told him that Degas adores and reverences every line from the hand of Millet. We all know the story of the old Frenchman at his club, who overheard a youth speak of a certain lady as 'Marie.' 'How odd it is,' drawled the old man, 'when you are in the presence of Madame la comtesse you always call her "Madame la comtesse," and when you are not in her presence you call her "Marie." And I, who am her cousin, call her "Marie" when I am with her, and when I am not in her presence I call her "Madame la comtesse." *Comme c'est étrange!*' You take it from me, young man, the Futurists may be great fun, and neo-swank excellent journalism, but it will pay you to keep in with the old masters.

This Millet is particularly interesting, because it is finished enough to tell its story, and unfinished enough to serve, in its different passages, as an authoritative and magistral lesson on the phrase 'begun, continued, and ended.' It is like a great dissection, laying bare at one glance, for masters and students alike, the mechanism of a creation of genius. We see how Millet *brought about* the accomplishment of a work, '*Comme c'est amené,*' to use a favourite expression of Degas. You can follow the design from the state of linear expression revealed in the hands, through a stage of further modelling in the woman's torso, and on to the complete and exquisite finish of the man's head, triumphant in its placid achievement. Here is a chance for a patriotic millionaire. The place of this picture is in the National Gallery.

Carrière's is a reputation that critics are very shy of touching. The moment a man makes an obvious bid to cater for the most lachrymose sentimentality the timid critic feels himself, as it were, put upon his honour. To cavil is almost to be put into the position of the man in melodrama 'who lifts his hand against a woman.' To do so, even in the way of kindness, is to court a howl. Something analogous happened with our own Mr. Watts. He was interviewed and made to say that technique was not his aim (skilful though he was to a high degree); that his pictures were not paintings, but sermons; that he was warring against crimes and oppressions, and so on. If in the eighties any one had so much as squeaked in presence of a Watts he would have been apostrophised on all sides, somewhat thus: 'But, disgusting personage, you are, then, in favour of rapine and oppression! Learn that, on *this* side of the Channel,' & c. You see the pattern from here.

And so Carrière also has been immune, and to be immune is to be praised. The world of his canvases is a world of tears, of the moaning, partings, and eternal regrets of shadowy heads, in which nothing is material but the eyes – eyes like boot-buttons in their insistent and solid detachment from impalpable surroundings. To criticise Carrière in France would, I am sure, seem almost like a slight on maternity. But maternity, to most of us, is, fortunately, a sunny recollection, in essence both palpable and delightful. The plain man, unaided by æsthetic bear-leaders, does, I am certain, repudiate in his heart these howling and ghostly impressions of what has always seemed to him the happiest relation.

Technically, everything has been said about Carrière when one has said '*C'est de la peinture creuse.*' It is *hollow* painting. One law may be found running through good art in all the ages. From the Greeks and the Indians (look at the reliefs on the staircase of the British Museum) to Keene and Pissarro plastic excellence consists in the fact that the work is *plein*, as French sculptors and painters say, that is *full*. The forms are *convex* and not *concave*. Though in good carving material substance is cut away, the resulting forms are full, and have the character of substances built up and added to, and not of things from which subtraction has been made. I have not the verbal ability to make this clear on paper, but any one with a natural sense of the plastic in drawing or sculpture knows that this law exists.

Sound drawing and painting proceeds by cumulation, by addition; decadence sets in when subtraction plays a large part. An interesting example in the history of art has been the undoubted decadence of water-colour painting from the date when it was found that colour could be erased or scratched out, as well as put on. In the fine period of water-colour painting a thin bough was drawn by an outline. In the decadent period it begins to be formed by scratching out a light line from a darker ground. I remember a portrait of Verlaine by Carrière in which a white tuft of moustache is got by *wiping out*, and leaving a visible wipe to do the work that the real painter does by deliberately adding a touch of a certain tone knowingly concocted to a *nuance* decided on by himself.

The work of Carrière raises in a very interesting manner another question of immense importance, and that is the degree to which it is possible to eliminate colour from painting, and yet stop short of monochrome. We know that monochrome is a legitimate and satisfactory convention, whether it is drawn or painted mono-chrome. In fact, a monochrome is only a drawing done with the brush instead of with the point. Now there is a type of painter, over-sensitive, and dimly conscious of not possessing a sense of colour, who jibs at colour, just as certain neurasthenics, who are wanting in a talent for life, jib at life. Both embrace a kind of negative quietism. It is the fashion to call both refined, whereas they are in reality sick. Carrière was of these. His work would have taken a higher rank had he frankly foresworn even the tiny dose of colour to which he still clung. If you look at the picture entitled *The Mother's Kiss*, the hollow evocation in decadent form would have remained at least harmonious and consistent but for the intrusion of the odious pink tone which invades the province of colour only to utter a blasphemy on the immutable laws of harmony.

It is interesting to be able to turn round in this same Octagon Gallery to Vincent Van Gogh's portrait of a Zouave. Here we have the sane well-balanced grip of the born painter, the real unmistak-

able scratch of the lion's paw, what they call in French '*la griffe*.'
The accident that this sane talent lodged in a brain that broke down
is one of the ironic mysteries of fate, and in no way touches any
critical contention. Every stroke in this picture expresses fact and
form. The thing is said aloud in ringing tones, like the speech of an
old farmer, or a line in a comedy of Goldoni. It has all the stimulus
and snap of the crack of a whip or the shot of a rifle. These are the
real men, and even I can dimly apprehend it, though I belong to a
generation that has been debilitated by art-shades and terrorized by
Mr. McColl. To me the treatment of the white background
appears, I confess, what is called in the nursery 'rather rude.' I
believe that Gore, Gilman, and Ginner contemplate it without
flinching. And after all red is red, and be damned!

What a life-story in Gauguin's portrait of his painter-friend in
the unending list slippers, and family, in the blank depressing
studio, all north-light, at $12 a year! How humbly and honestly he
bows, and rubs his hands, and recommends himself to your
goodwill, your connoisseurship, or your charity, he does not much
mind which, so long as the little family are fed and the rent and the
colourman paid. No amphibious *arriviste* here, half-gentleman,
half-artist, with a West-end tailor, *liaisons*, and three clubs! One
cannot say that there is not progress in taste when these good things
are yearly more understood and appreciated. How much road our
eyes have made is evident in the manner we have left the virtuosity
of Stevens behind us. How *démodé* is this *peinture pour cocottes* with
its stale frocks, its bit of bric-a-brac, and the immobility of its fixed
figures in the eternal vacuum of the pretentious studio.

We turn in the younger generation to Vuillard and Maurice
Denis, who represent the two opposing tendencies between which
art has ever balanced. Vuillard seems to have convened to stay his
hand at the moment when he feels the freshness of his first
impression fade. His sense of colour is exquisite. It is almost as if he
said 'I am so utterly concerned that you shall attend alone to the
refinements of colour that I can show you, that I shall purposely
retain in my form a certain rudimentary quality, an almost
purposed negligence that you may, if you will, put down to
childish incapacity.' This in his easel pictures from nature; though,
in his elaborated decorations, his capacity for complicated
construction is unmistakably proclaimed.

In Maurice Denis I see one of several types of what one might call 'The complete Modern.' It places him to create a somewhat more concentrated world than the one we live in, but he observes, in creating it, the rules of the exactest realism. He seems to me to attain to the highest type of plastic drawing. Not only does he realize form with perfection, but it is form as it is lived, which is a very different thing from form as it is bottled in the studio or the museum. His gracious damsels move their marble limbs in the sunlight in obedience to the varied impulses of shyness, haste, or playfulness with an eloquence that is astonishing. To be able to say of an art that it is a lively classic, or that it is an art of naturalness is, I suppose, the highest possible praise.

And now, having commented on some of the guests of the International, I have left myself no room to express the admiration I feel for some of the hosts. But whose fault is that? Mr. Francis Howard's.

---

64. Unsigned review (probably by Robert Ross), 'M. Picasso's Drawings'

The Times, 27 April 1912, 13

The Stafford Gallery showed the drawings of Picasso and Joseph Simpson side by side in April and May 1912.

Simpson (1879–1939) was a leading caricaturist who published prolifically and worked for most of the prominent British and American periodicals.

M. Picasso, the leader of the most advanced school of French painting, has been called an incompetent charlatan. The exhibition of drawings by him, now to be seen at the Stafford Gallery in

Duke-street, proves that he is not that. There is only one Cubist work by him shown, and that, we confess, is quite unintelligible to us, and seems to have no abstract beauty of design or colour. It is like meaningless verse that does not even sound well. But there are many drawings shown which, though slight and fragmentary, are vivid and also rhythmically beautiful. We believe M. Picasso to be an accomplished and sincere artist possessed by an insatiable love of experiment and discovery and haunted by an incessant fear of the commonplace. Photography, with its boundless powers of inexpressive limitation, has filled many artists with this fear. They see the photograph in all pictures that attempt any close likeness to reality; and their own art is directed as much by the negative aim of avoiding the photographic as by the positive aim of expression. No living artist, perhaps, is so much troubled by this extreme fastidiousness as M. Picasso. He will not be literal, he will not be facile, he will not be a virtuoso. He is so determinedly artistic that he seems to see real objects as if they were already works of art before he begins to draw them. Thus his *Vieux miséreux* (16) seems to be drawn from a statue, and a very fine one; and other drawings seem to be taken from bas-reliefs. Some of the drawings shown are mere scrawls telling us little or nothing except what the artist has refused to do; but in others, as in the pen-drawing of a woman seated (20), there is an extraordinary sensitiveness of line combined with a rhythmical beauty the more remarkable because it is not at all fluent. The donkey's head (6), also a pen-drawing, is extraordinarily delicate and precise in character; the animal is treated in it with the imaginative seriousness of the great Chinese artists. We regret that M. Picasso does not manage to accomplish more, and we fear that he is now becoming a mere scientific experimenter; but we acknowledge his profound artistic seriousness, a seriousness which is shown by his attitude towards everything which he draws, by his desire to discover and express its essential character rather than to use it as rough material for his art.

It is just this kind of seriousness which Mr. Joseph Simpson lacks in common with so many clever English artists. The drawings of his, shown in the same gallery with M. Picasso's, are far more obviously attractive and at first sight look more skillful. He would probably be called a Post-Impressionist, but there is usually a good firm basis of commonplace to his most dashing displays of line. The conception of many of his heads is such as we have seen on a hundred posters, only it is treated in a new way. Needless to say this treatment, so summary and yet so telling, means very considerable skill. But Mr. Simpson has started being a virtuoso too early. The drawing of a dog (28) and of *An Englishwoman* (25) show how well he can do when he is absorbed in what he draws rather than in his manner of drawing it. His talent deserves to be rescued; but only he himself can rescue it.

65. Unsigned review, 'M. Pablo Picasso and Mr Joseph Simpson at the Stafford Gallery'

Athenaeum, 27 April 1912, 478

For Joseph Simpson see headnote to no. 64.

This exhibition will not lack visitors, because M. Picasso is perhaps the foreign artist most talked of among us and the least known. He has not always been fortunate in his advocates, who have frequently utilized their professed admiration of his work as a lofty position from which to pour derision on contemporary art in general – all of which, we are assured, is by comparison 'vieux jeu.' As in England there exists a large "press gang" who may be bullied into embarking on any adventure by the threat of being considered old-fashioned, London hears much of Picasso, and, seeing virtually nothing, is by so much the more impressed. While for these reasons we consider his already enormous reputation in England to be worthy of no respect whatever, it would be a mistake to assume that his work is necessarily unimportant. Indeed, by an unfortunate accident few of the better artists of the last quarter of a century have been able to 'arrive' without being advertised like patent-medicine vendors, so that from both points of view it is incumbent on the home-keeping Englishman to judge for himself of new arrivals.

The Stafford Gallery exhibition does not offer much opportunity for judging M. Picasso as the fundamental revolutionary he is usually painted. 'The real Picasso' is conspicuously missing, and, except in the not very impressive *Nature morte à la Bête [? Tête] de Mort* (25), we have no chance to determine whether his odd geometrical experiments are based on profound science or, as might seem to be the case in this instance, half-accidental whim. On the other hand, there is evidence in *Les deux Gymnastes* (2) of easy and expressive draughtsmanship of the old academic stamp, and this little drawing is certainly far superior to the large nude study by which he was introduced to us at the Grafton Gallery. *Tête égyptienne* (3) is another slight, but carefully drawn study, endowed with a 'weird' aspect by a cheap trick of exaggeration analogous to that by which M. Fernand Khnopff used to draw a head with scrupulous care and literalness, and then add an inch to the depth of the lower jaw, to the unspeakable delight of devout mystics; while in *Cheval avec jeune Homme en Bleu* (5) the horse is quite comic, from the way in which, by an exaggeration of Van Dyck's formula, its forequarters and the pose of the head suggest exactly the action of shrinking self-consicious modesty of the *Venus de' Medici*. The drawing of the figure, on the other hand, is firm and elastic, with a considerable grip on reality; and the same may be said of Nos. 14 and 16, in which a reasonable basis of scholarship is concealed beneath the unquestioning eye for facts which we usually find to-day only in a novice.

A slight lack of this *naïveté* mars our pleasure in the able drawings of Mr. Simpson, whose clever poster designs are generally and rightly esteemed. There is a suspicion of it perhaps in No. 25, *An Englishwoman*; and the challenging expressiveness of No. 11, *The New Hat*, shows an absorption in the human interest of the subject which dominates its cleverness. In others, such as Nos. 8 and 30, the designer's triumph of fluency of line is a little that of the virtuoso, No. 7 has a suggestion of painter's quality of a similar order, while No. 9, *The Hotel Window*, is admirably to the point as a study for the setting of a figure subject.

66. P.G. Konody,
'The Stafford Gallery'

Observer, 26 April 1912, 6

For Konody see headnote to no. 45.

The drawings shown by M. Pablo Picasso at the Stafford Gallery in Duke-street belong to the region vaguely described as 'the limit.' Our eyes have by now become attuned to all manner of *fauve* eccentricity. We may dislike Cubism and Futurism, but there is after all something to be said in defence of these new movements, whose chief fault is that they are too scientific. But what can be pleaded in excuse of M. Picasso? He shows, it is true, one brilliant gouache study of a young man with a horse, the action and play of the muscles as well observed as the play of light on the animal's coat; a rather commonplace pastel portrait, *Femme Assise*, and an early water-colour, *L' Apéritif*, which is neither better nor worse than the average illustrations on the front and back pages of *Le Rire*.

But the rest is merely an impertinent display of fumbling incompetence, which reaches its apogee in the water-colour amusingly catalogued as *Nature morte à la bête* (sic!) *de mort*. *Bête* is indeed the only term that can aptly describe this kaleidoscopic jumble, in which it is just possible to distinguish a skull that has obviously been under a steam roller. 'Expression, not impression,' has been claimed to summarise the aims of the modern group of which M. Picasso is a leading member. His childish pen-scribbles, *Tête de vieille femme, Femme nue debout, Femme accoudée*, and similar scraps that should never have been rescued from the wastepaper basket, have not even the expressiveness of a talented infant's drawings. To exhibit them as works of art is simply *fumisterie*, or, to use the popular slang term, 'spoof'.

It is an injustice to Mr. Joseph Simpson that his truly remarkable drawings should be shown in such company. After the futility of M. Picasso's scraps, Mr. Simpson appears a real master of the crayon, who knows how, with authoritative sureness, to reduce the complexities of form to their most significant elements. His nudes, heavily outlined with a broad, soft crayon that has the velvety quality of lithographic chalk, are drawn with such suppleness and clearness of accent that, in spite of a strict economy of means that dispenses altogether with light and shade and modelling, the sense of the roundness of the forms is invariably conveyed by the functional use of contour.

Mr. Simpson does not always confine himself to outline. In his portrait studies he relies entirely upon the massing of light and shade and makes his crayon yield effects of tone similar to the painter's preliminary laying-in of the shadows, half-shadows and lights with his brush. The effect of these drawings is essentially pictorial and does not in the least depend upon a substructure of line. Throughout, these drawings reveal closeness of observation, sympathy with throbbing life, and a search for the absolute essentials that make for the highest degree of expressiveness. It cannot be claimed for Mr. Simpson that he is an innovator or an initiator. He follows Mr. Fergusson as he once used to follow Mr. Nicholson and Mr. Peploe. He recalls Mr. Pryde in *A Spanish Dancer* and Mr. Sargent's *Avelegonfie* in *A Russian*. But, whatever the source of his inspiration, his own personality, invariably forces itself through; and, unlike the ordinary plagiarist who cannot rise beyond an enfeebled echo of his original, Mr. Simpson not infrequently surpasses the work of his prototypes.

67. Huntly Carter,
'The New Spirit in Painting'

The New Spirit in Drama and Art, 1912, pp. 209–11 and 215–21

Carter, who had already written reviews of the French salons
(see no. 39) and had written appreciatively of Picasso's cubism
(see no. 42), concluded that the uniting force behind all
successful modern art was its concern for rhythm.

The Autumn Salon and the relation between the advanced
movements in painting and in the theatre

When I started across Europe in quest of the golden sensation of
unity, continuity, and completeness in the Art Theatre, I was
prepared for disappointment. Before the coming of the Russian
Ballets I had looked for it in vain. I had, in fact, been led to the
conclusion that the prevalent view in the theatre is: Art is an adjunct
to the drama; it is a copy or fake of the emotional interest. The
view had robbed me of the big, complete sensation which every
play production should produce, for it had offered me dramatic
fare in detached masses, being unable to bring them together into
that organic relation which the sensation demanded. As I anticipated, the view was prevalent all over Europe. I was therefore
obliged, in order to realise the desired experience, to return to Paris
for it. Here I knew I should find it, not in the theatre, but in the
exhibition gallery. In the spring of the year I met the Post-
Expressionists, whose works once more proclaimed the fact that to
one body of artists, at least, art is not an accessory to life; it is life
itself, carried to the greatest heights of personal experience.

So I came to Paris in order to make a brief study of the methods
in use for attaining unity, continuity, and completeness, and beyond
this to trace the adoption of the new conception of decoration, and
to determine how it was likely to contribute to the artistic
movement in the theatre. I had previously noticed there was a vital
connection between the advanced movement in painting and the
movement in the theatre which, once established, would bring
about a union of the two, set them mutually acting and reacting

upon one another, and tend to remove all difficulties to the proposal to lift the theatre into the region of art. My subsequent observations in Paris led me to the conclusion that rhythm is the connecting link between plastic forms of art and the 'scene,' and the continuous and consistent search for this is hourly bringing them closer together.

The present point of approach between the artistic scene and the advanced movement in painting is to be found in the new conception of decoration. The object of the advanced movement in painting is to give the widest possible expression to the fundamental note or rhythm of each subject treated. The New Men, especially in Paris, are working on the assumption that the expressiveness of the immediately preceding centuries was far too narrow; and the Old Men did not give their subjects that psychological width of expression which the subjects themselves demanded, and they made the mistake of giving each object in a picture a structural unity of its own, and neglected to bind the whole composition together by one big unified design. They assume, too, that the Old Men made the fundamental error of elevating beauty where truth alone should be. With the New Men truth is the ruling passion. They maintain that truth is its own reward. If they seize the truth of a fundamental conception, their logical method of treatment will enable them to develop it truthfully and to give it the widest possible expression of truth. If they seize an untruth, their treatment of it will likewise detect it, and the result will be untruth. Further, they maintain the subject requires nothing beyond this truthful statement, none of the extraordinary adjuncts that the Old Men were accustomed to embody on canvas. Decoration, as we are accustomed to call it, must be nothing more than the reasonable development of its character. In fact, their decoration to them is not decoration, but the truthful expression of character. Thus the New Men are inclined to regard decoration from the point of view of the actor. The latter never speaks of decoration. He regards and refers to everything worn by a human being in terms of character. To him a dress, a piece of jewellery, the arrangement of the hair, are outward expressions of inner psychological states. In his opinion, anything worn that does not express character is superfluous. . . .

It was at the Salon d'Automne, amid the Rhythmists, I found the desired connection. The exuberant eagerness and vitality of their region, consisting of two rooms remotely situated, was a complete contrast to the rooms I was compelled to pass through in order to reach it. Though marked by extremes, it was clearly the starting-point of a new movement in painting, perhaps the most re-markable in modern times. It revealed not only that artists are beginning to recognise the unity of art and life, but that some of them have discovered life is based on rhythmic vitality, and underlying all things is the perfect rhythm that continues and unites them. Consciously or unconsciously, many are seeking for the perfect rhythm, and in so doing are attaining a liberty or wideness of expression unattained through several centuries of painting. By the time I had reached these conclusions, the Expansionists, as I may now call them, had sorted themselves into groups answering to the difference in expression of the general aim. These I will name, for convenience, Radiationists, Crystallisationists, Vibrationists, Rhapsodists.

I was compelled to place the Radiationists first. They grasped me so powerfully with their knowledge of unity and continuity carried to such a state of perfection that escape was impossible. Thus John D. Fergusson's *Rhythm* first swept me out of myself, away from the battling-ground of paint and canvas into immensity of the infinite. The splendid movement and vitality of this canvas was irresistible. It proclaimed the power with which this painter sets his seal upon his form of art, and singled him out as easily foremost among the strong men of Paris. It revealed his astonishing gift of seizing the fundamental rhythm of a character or scene, of concentrating on it, and of developing it in form and colour till the whole canvas rings with the magic of motion. Here the rhythm of the nude figure placed in the centre of the composition is felt, and the curves of line and colour flow out from it and on without end, creating a sense of an illimitable sea. Thus they pass from the powerfully drawn central motive to the arched tree of life, to the harmonious apples of discord, thence swelling out into the draperies, and so radiating out of the canvas in fulness and richness of a wide range of colours, of a balance of shapes, and of a related order of movement, producing a tremendous effect of power. It is a triumph of the expression of the universal in the particular. The same elements of unity and continuity proclaimed themselves in this painter's Still-Lifes, wherein his preference for lozenge-shape curves is strongly manifested, and his skill in giving a number of varied objects structural unity is clearly demonstrated. The ability to concentrate on the fundamental rhythm of a subject, and to develop it throughout as the main theme, is seen further in a small portrait study, broadly painted and very fine in movement and colour, though perhaps not so successful as another study of the same subject which I remember seeing elsewhere and cannot forget. In the latter, the swing of the green through the black hat, the spontaneity and splendid feeling of the whole thing, fascinated me as the subject must have fascinated the artist. From the portrait the eye wandered to an interesting study of flowers giving off radiations by means of subtly expressed colour. It afforded an instance of the new conception of a background created by the importance of the subject.

Once on the waves of rhythm, I was swept from canvas to canvas. Next the rhythmical music of Estelle Rice's *Nicoline*, penetrating and subtle, charmed me with its air of the infinite. Like a symphony, beautiful in movement and colour, the subject expressed the radiations of a brilliantly coloured mind, and the treatment revealed how such a mind may be given to the artist for decoration in the latest sense without fear that the truth of its character will be disgraced. It proved, indeed, that Miss Rice is the one strong woman painter in Paris who can subordinate decoration to truth and can cover a canvas with the essential facts of character brilliantly stated in line and colour. In *Nicoline* the circling waves of very subtle blues, pinks, and greens expand into the background, reflecting the woman's mind like coloured shadows thrown on illuminated discs, and thus fill not only the canvas but the mind and the world for the time being of the observer. Surely this is the purpose of a good picture – not merely to illuminate the soul of the subject-matter, but to lift the spectator out of himself, to link him with the universal, and so to blot out for fleeting moments the unattractiveness of life. At any rate, it is the effect of Miss Rice's pictures. She knows how to set one journeying through an exhilarating universe even on a note of beautiful flowers.

In the subsequent consideration of the work of these two painters, of the vision of the subjects chosen, and the manner of handling them, I found it offered three or four contributions to the subject of the composition of the 'scene.' First, it revealed the characteristics of Will-impressionism and Will-expressionism, shown in intense concentration and expansion. Thereby it suggested that the movement in the theatre should be based, like the advanced movement in painting, upon the supremacy of the Will in art. It showed how the Will may subject a theme to itself and thereafter give it the widest expansion. And it afforded, moreover, an example of a logical mind reducing an accidental jumble of objects to order, and binding them together in one big rhythmic design. A second contribution was the suggestion that the background should be created by the importance of the subject. Every important theme contains its own background, which the truthful touch of the poet's or painter's hand will realise. But unimportant themes have no background worth troubling about, and any attempt to supply one is a stupid waste of time. This consideration leads to the third contribution, to be found in the suggestion that the treatment of character in art and drama involves regard to unity and continuity in its development. Apparently Mr Fergusson, Miss Rice, and other rhythmists are of the opinion that the choice of characteristics of a subject should not be limited to intrinsic attractions, but extended to extrinsic attractions that will add the essential wideness of expression and completeness. It appears that the Old Masters confined themselves to an expression of the physical, moral, and intellectual attractions contained in an individual, without bothering to relate him to the environment. But the New Men claim they are justified in adding extrinsic attractions to which the individual is bound by the law of association. Hence the latter relate the central object to all surrounding objects, which become an essential part of its character, whereas the Old Men gave every single object in a composition an importance and structural unity of its own. A house is one thing, a tree another, a sky another, and each is so treated that we may view it in separation. The difference between the old and the new, vision and interpretation, is that of seeing and doing in detail and mass. The Old Men were concerned with detail-impressionism and detail-expressionism; the New Men with mass-impressionism and mass-expressionism. As with painting, so with the 'scene.' At present the latter is mostly seen and composed in detail. It is a jumble of odds and ends that contribute nothing to the main theme. It should be conceived and executed in mass, and everything selected and adapted to the main end in view. This treatment will be adopted as soon as its emotional bearing is understood.

Many useful suggestions were also forthcoming from the new Vibrationists, represented by M. Picabia. I was not greatly impressed with this painter's big study. It had nothing at the edges to carry off its four or five figures. There they stood vibrating on the yellowish-white sands, against the wide stretch of blue, flat sea. But its brightness was supreme; and it is in the expression of such brightness that M. Picabia exceeds all the scientific Impressionists. I could imagine the latter persons coming to examine *Sur la Plage* with astonishment, and being obliged to peer at it through smoked goggles. I much preferred the same painter's *Jardin*. It very successfully illustrated the new expression of vibrating light by the apparent interchange of masses. *Jardin* is full of the clash of direction and of colour. The juxtaposed masses, deep purples, violaceous shadows, fawn half-lights, fleeting clouds of greens and yellows, come and go, cross and recross, with the force of piston-rods. It is as though one took the flat sails of a fishing-fleet, gave each set a different colour – red, green, blue, yellow, violet, – and made them move against each other on a heavy ground swell. The clash and movement would be sufficient to rivet the most untutored mind. The canvas throws its will upon the spectator as few canvases do. The action of de Segonzac's *Boxers* was another successful example of the will to power. There was no mistake about the strength and completeness of this canvas. When I looked at it I was conscious of a succession of stinging blows. They rained upon me from all parts of the canvas. Not only from the combatants in the foreground, but from the mind of the audience symbolised in the background by a tremendous direction of swishing line.

68. O. Raymond Drey, 'The Autumn Salon'

Rhythm, December 1912, 327–31

This year the Autumn Salon is more than ever one for the initiated. 'Art for Art's sake' must now be written 'Art for the Artist's sake.' For while this Salon teems with interest for the artist and the amateur, it is caviare or something worse to the general. The Paris newspaper critics, as usual nowadays, have had to choose between the bad form of not laughing at all and the probability of having to recognise a few years hence that they have laughed on the wrong side of their faces. Once again have the stouter ones disinterred that spectral word 'bariolage,' which one had rashly thought to be lying forgotten and at peace in the critical tomb of Paul Mantz. There is, of course, the frankly philistine point of view of the frequenters of the Magic City or of Luna Park. Next year perhaps the poster of the Salon will bear the words: 'Visitez la Salle Post-Cubiste: on y rit follement!' But we have not yet come to that. Certainly I found many 'horrors' in the Grand Palais. But these were not the so-called extreme works, but the many 'œuvres calmes, saines et harmonieuses – gracieuses et savantes' applauded by the *Matin*. Of all the modern work it was the cubist that I found most interesting. Over most of the other pictures Impressionism, or Post-Impressionism, lay like a wet cloth. Not the freshly kindled Impressionism of a Manet or a Berthe Morisot; not the fiery imaginings of a Cézanne or a Van Gogh, but the winding-sheet of a decadent school.

It is difficult to understand why such a fuss should be made about cubism. The theory, in its simplest form, is comprehensible enough, one would think. The cubists, like good Fauves, are trying to rid Art of the incubus of preconceived mental associations which have tied it so long to mere representation, of all that stale and vitiated atmosphere with which successions of painters and generations of the public have surrounded it. The Post-Impressionist, for instance, in painting a still-life, asks himself this question before he begins: 'What is there in this set of objects that interests me?' Needless to say, he is not interested in that decanter because it contains whiskey, or because it reminds him of thirst or of convivial evenings. Nor has he placed a pear next to it, and a chrysanthemum, because the one suggests the end of a dinner or a blossoming orchard, and the other a domestic scene in the garden. He will, of course, answer his own question in one of two ways. Either he is interested mainly in the colours of the objects, or in their shapes. If he is mainly interested in their shapes, one of the first things he will do perhaps will be to draw them a little incorrectly, so that the decanter may not immediately suggest whiskey, the pear an after-dinner feeling, and the chrysanthemum that sweet scene in the garden. But he will be careful to retain the characteristic shapes of all the objects and to arrange them in relation to one another. He may even draw thick black lines round their contours, to make them more emphatic.

The ordinary cubist, after all, goes about his work with much the same idea. Only instead of distorting his drawing, he makes a statement of form in two dimensions only, by simply outlining the various planes and dispensing with tone variations, which would form the third dimension in ordinary painting. Two pictures by de la Fresnaye, *Baigneurs* and *Joueurs de cartes*, in this Salon illustrate this kind of cubism very well. They strike the eye freshly as new things must always do; but, after all, the quality they possess is only appreciation of tone values, translated afterwards into a simplified form. This translation is just a technical achievement, nothing more; seen from the other side of the room, the tones reassert themselves – the third dimension reappears in terms of light and shade.

But this sort of thing is naturally not the all in all of cubism. Le Fauconnier, for instance, has a most remarkable and stimulating picture in the exhibition. It is a large canvas entitled *Les Montagnards attaqués par des ours*. Here the cubism is of a different sort, both technically and essentially. It is no longer a matter of planes. Le Fauconnier has simply disintegrated his forms, cut them into bits that may be called cubes if you like, and distributed them with a marvellous amount of skill over his composition. He has stolen the technique of the Futurists, but he has left them what they choose to call their philosophy. He disintegrates, not in order to express movement, which the cinematograph can do much better, but so that the eye, by taking hold of certain symbols, may transmit them to the mind and thus evoke a new spiritual image.

343

Thus it is not a jigsaw puzzle that we are asked to sift and reassemble in a familiar form, but a challenge to new vision, which is a vastly different thing. Superficially alone, it is a conspicuously able picture. To begin with, it holds together. Heterogeneous though it is in its composition, it has unity. It would be exciting as a pattern, had it no other claims. I shall come back to this picture later, because with it the whole cubist movement must stand or fall.

In the same room are other works, far less interesting, in which disintegration of form seems to be an end in itself. *La Source* and *Danses à Source*, by Francis Picabia, are the most prominent. Unlike Le Fauconnier's picture, they contain no key rhythms or shapes capable of evoking a new mental image. They are no doubt quite sincere works, and are very properly hung in an experimental Salon, but to me at least they carry no significance. Much more interesting is a very curious design by François Kupka, called *Amorpha, fugue à deux couleurs*. Every line and curve suggests rhythmic motion and virility. This picture stands quite by itself in the Salon, and is a very striking study in abstract design.

Dunoyer de Segonzac shows what is in some ways the most notable picture in the exhibition. In *Bucolique* he has made a searching study of forms, simplified to their essentials and set in astonishingly vivid relationship. Even to those who ask of a picture, 'Mais qu'est-ce que c'est que ça représente?' the canvas will be intelligible because it contains cows and shepherdesses and a host of pastoral back-memories. The painter simply has not troubled to strip his work of its sentimental associations; they are there for those who look for sentiment in a picture, and anyone who appreciates the suave lines of the cows' backs, interlaced in exquisite harmony with the medley of nude limbs and bodies, will not trouble about them either. The whole picture is held together by a strongly drawn background of trees and grassy slopes, very summarily suggested. The composition is clinched in a low and even tonality by the red backs of the cows and the yellow, sunburnt human bodies. De Segonzac shows another very beautiful picture: a study of low tree forms in evening light. Here is a subject which most painters would have felt themselves unable to express, for there is no differentiation of tone to guide the eye; the imagination is pricked by an arrangement of shapes, intensely characteristic of a wood near nightfall.

J.D. Fergusson, who exhibits two works, is a host of painters in himself, for his mind seems to be too alert and questioning ever to be content for long with one sort of achievement, however exciting it may be. His larger canvas, badly hung though it is, is stimulating in a very fine way. It is a study of a nude dancing figure, painted in strong outlines in a slightly oblique light, against a figured wall. In attempting such a subject the painter set himself the baffling problem of suggesting movement by outline, which necessitates almost full lighting, without sacrificing the solidity of the figure by merging it with a background of the same tone value. He has found a most satisfying solution of the difficulty by letting the light fall on the figure ever so slightly from the side. This has enabled him, by carefully subordinated modelling, to give the figure substance, by which it is held firmly in relation to the background, and thus the vividly drawn contours tell splendidly. It is a fine painter's picture.

Maurice de Vlaminck, who shows several pictures, seems to me to vacillate in his form of expression between realism and suggestion. He paints the forms of trees and hills very much in the abstract manner of Cézanne, while water, which figures conspicuously in his pictures, is full of painted reflections and imitative colouring. The result is a complete lack of cohesion.

The influence of Cézanne, of course, is manifest in the majority of the pictures in this Salon. But naturally enough, where it is most apparent, there seems to be little originality besides; which is another way of saying that a man is a great artist because he has something to express, and not because he uses certain forms of expression.

The large wall decoration by Anne Estelle Rice, which faces the main entrance, is a very skilful piece of work. Like Puvis de Chavannes, the painter has borne in mind throughout that she is decorating a wall space, and scarcely any stone structure would be weakened by this big canvas. The design itself is moderate enough to please the critics; it would furnish a complete answer to the stupid taunt of incapacity which the wilfully distorted drawing of some of our moderns invariably provokes from the ignorant. I miss in this Salon any of this very brilliant painter's more important and characteristic work; other members of the jury seem to have been less modest with less cause.

Othon Friesz is well represented by a picture which, though it is dirty in colour, gains significance by a repetition of interesting forms. Matisse has several canvases, but he always seems to me to be shallow in feeling, despite his vivid colouring and crude design.

In another part of the Salon is a large retrospective section, where Cézanne, Van Gogh and Gauguin rub elbows capriciously with Carrière, Emile Blanche and Boldini! And there is a rather incomprehensible room devoted to Albert Braut.

Now let me come back to Le Fauconnier and the speculations that arise from a consideration of his picture. Is it possible to evoke in the mind of the beholder, by such means as this painter uses, a new and complete mental image? It is much easier to ask this question than to answer it and indeed it is impossible to answer it for any one but oneself. This picture by Le Fauconnier is nearly allied in aim to some of Picasso's most abstract work. Picasso, by the bye, has no pictures in the Salon; it is a pity, for the two would have helped to explain each other. Le Fauconnier, in this picture at least, has used easily recognisable forms, though they are split up and scattered. It must be obvious, to begin with, that a great deal must be sacrificed to this process of disintegration. Though a number of telling rhythms may be repeated, the greater harmonies must necessarily be lost. But the greatest sacrifice of all is more fundamental. The appeal of such a picture is solely to the intellect. Of course, it may be said that the senses and the intellect overlap mysteriously. Naturally I only speak for myself when I say that the contemplation of this picture would seem to be merely a matter for the intellect. The mind strains to form an impression – to seize a symbol here and there – to co-ordinate them and weld them together. And it seems to me, too, that increased familiarity with this kind of painting would not make it any more easily or more quickly comprehensible. I may be quite wrong. I may be old-fashioned even in preferring an impression at white-heat, quickly kindled – spontaneous. I admit that Picasso, with his new abstract symbolism of line, baffles me completely. And unlike René Ghil, the symbolist poet, Picasso has not written a library of explanatory volumes for the benefit of the meek.

THE SECOND POST-IMPRESSIONIST EXHIBITION, 5 OCTOBER–31 DECEMBER 1912

When the second Post-Impressionist exhibition opened in October 1912 the press response was as overwhelming as that for the first exhibition – reviews, reports and critical articles appeared in almost every newspaper, journal and periodical, both London and provincial. Yet the tone of the critical response was completely different from that in 1910. In the two years since 1910 the British had been coming to terms with a wide range of modern art from the Continent and had seen some of the new ways of painting naturalized in English art. In this second Post-Impressionist exhibition, recent British painting was shown together with recent French and Russian art, so that Duncan Grant and Wyndham Lewis appeared with Derain and Vlaminck but most critical attention was given to the work of Matisse and Picasso.

Whereas this time Cézanne was represented by only five oil paintings, there were thirteen analytical cubist paintings and three drawings by Picasso including his *Buffalo Bill* (1911), *Tête de Femme* (gouache, 1909) and *Femme au Pot de Moutarde* (1910). Matisse was represented by nineteen oil paintings, seven bronzes and other items. The two which attracted most attention were *Le Luxe – I* (1907–8: Statens Museum for Kunst, Copenhagen) and a large design for *La Danse – I* (1910: Museum of Modern Art, New York). This last picture was painted for the Russian Prince Sergei I. Shchukin (1854–1937) and its presence at the second Post-Impressionist exhibition was recorded in a photograph of the Grafton Galleries reproduced in Benedict Nicolson, 'Post-Impressionism and Roger Fry', *Burlington Magazine*, January 1951, xciii, opp. p. 13. Braque had four *fauve* and cubist pictures in the exhibition including *Anvers* (1906), *La Calangue* (1907) and *Kubelick* (1912).

Duncan Grant showed *The Seated Woman*, together with a portrait of Henri Doucet, *The Dancers* (1910–11: Tate Gallery,

347

London) and *The Queen of Sheba* (1912: Tate Gallery, London) amongst others, and Wyndham Lewis was strongly represented by his *Timon of Athens* drawings together with other recent works.

69. Roger Fry, Introduction to the catalogue of the second Post-Impressionist exhibition

1912, pp. 7–8

The scope of the present Exhibition differs somewhat from that of two years ago. Then the main idea was to show the work of the 'Old Masters' of the new movement, to which the somewhat negative label of Post-Impressionism was attached for the sake of convenience. Now the idea has been to show it in its contemporary development not only in France, its native place, but in England where it is of very recent growth, and in Russia where it has liberated and revived an old native tradition. It would of course have been possible to extend the geographical area immensely. Post-Impressionist schools are flourishing, one might almost say raging, in Switzerland, Austro-Hungary and most of all in Germany. But so far as I have discovered these have not yet added any positive element to the general stock of ideas.

In Italy the Futurists have succeeded in developing a whole system of æsthetics out of a misapprehension of some of Picasso's recondite and difficult works. England, France and Russia were therefore chosen to give a general summary of the results up to date.

Mr. Clive Bell is responsible for the selection of the English works and Mr. Boris von Anrep for the Russian. The selection of the French works fell to my lot.

70. Clive Bell,
'The English Group'

Catalogue of the second Post-Impressionist
exhibition, 1912, pp. 9–12

Bell (1881–1964), who met Fry in 1910, had become familiar
with modern French painting during his trips to Paris in the
early years of the century.

In this introduction to the English painters in the second Post-
Impressionist exhibition Bell used, probably for the first time,
the famous – or notorious – term 'significant form'; the Post-
Impressionist, he said, regards the humble coal-scuttle as 'an
end in itself, as a significant form related on terms of equality
with other significant forms'.

For the Second Post-Impressionist Exhibition I have been asked to
choose a few English pictures, and to say something about them.
Happily, there is no need to be defensive. The battle is won. We all
agree, now, that any form in which an artist can express himself is
legitimate, and the more sensitive perceive that there are things
worth expressing that could never have been expressed in
traditional forms. We have ceased to ask, 'What does this picture
represent?' and ask instead, 'What does it make us feel?' We expect
a work of plastic art to have more in common with a piece of music
than with a coloured photograph.

The first thing to be considered is the relation of these English
artists to the movement. That such a revolutionary movement was
needed is proved, I think, by the fact that every one of them has
something to say which could not have been said in any other
form. New wine abounded and the old bottles were found
wanting. These artists are of the movement because, in choice of
subject, they recognise no authority but the truth that is in them; in
choice of form, none but the need of expressing it. That is Post-
Impressionism.

Their debt to the French is enormous. I believe it could be

computed and stated with some precision. For instance, it could be shown that each owes something, directly or indirectly, to Cézanne. But detective-work of this sort would be as profitless here as elsewhere. I am concerned only to discover in the work of these English painters some vestige of those qualities that distinguish Post-Impressionists from the mass – qualities that can be seen to advantage in the work of the French masters here exhibited, and to perfection in those of their master, Cézanne. These qualities I will call simplification and plastic design.

What I mean by 'simplification' is obvious. A literary artist who wishes to express what he feels for a forest thinks himself under no obligation to give an account of its flora and fauna. The Post-Impressionist claims similar privileges: those facts that any one can observe for himself or discover in a text-book he leaves to the makers of Christmas-cards and diagrams. He simplifies, omits details, that is to say, to concentrate on something more important – on the significance of form.

We can regard an object solely as a means and feel emotion for it as such. It is possible to contemplate emotionally a coal-scuttle as the friend of man. We can consider it in relation to the toes of the family circle and the paws of the watch-dog. And, certainly, this emotion can be suggested in line and colour. But the artist who would do so can but describe the coal-scuttle and its patrons, trusting that his forms will remind the spectator of a moving situation. His description may interest, but, at best, it will move us far less than that of a capable writer. Yet most English painters have attempted nothing more serious. Their drawing and design have been merely descriptive; their art, at best, romantic.

How, then, does the Post-Impressionist regard a coal-scuttle? He regards it as an end in itself, as a significant form related on terms of equality with other significant forms. Thus have all great artists regarded objects. Forms and the relation of forms have been for them, not means of suggesting emotion but objects of emotion. It is this emotion they have expressed. Their drawing and design have been plastic and not descriptive. That is the supreme virtue of modern French art: of nothing does English stand in greater need.

If, bearing in mind the difference between the treatment of form as an object of emotion and the treatment of form as a means of description, we turn, now, to these pictures an important distinction will become apparent. We shall notice that the art of

Mr. Wyndham Lewis, whatever else may be said of it, is certainly not descriptive. Hardly at all does it depend for its effect on association or suggestion. There is no reason why a mind sensitive to form and colour, though it inhabit another solar system, and a body altogether unlike our own, should fail to appreciate it. On the other hand, fully to appreciate some pictures by Mr. Fry or Mr. Duncan Grant it is necessary to be a human being, perhaps, even, an educated European of the twentieth century 'Fully,' I say, because both Mr. Fry and Mr. Grant – and, for that matter, all the painters here represented – are true plastic artists; wherefore the most important qualities in their work are quite independent of place or time, or a particular civilisation or point of view. Theirs is an art that stands on its own feet instead of leaning upon life; and herein it differs from traditional English art, which, robbed of historical and literary interest, would cease to exist. It is just because these Englishmen have expelled or reduced to servitude those romantic and irrelevant qualities that for two centuries have made our art the laughing-stock of Europe, that they deserve as much respect and almost as much attention as superior French artists who have had no such traditional difficulties to surmount.

No one of understanding, I suppose, will deny the superiority of the Frenchmen. They, however, have no call to be ashamed of their allies. For the essential virtue is common to both. Looking at these pictures every visitor will be struck by the fact that they are neither pieces of handsome furniture, nor pretty knick-knacks, nor tasteful souvenirs, but passionate attempts to express profound emotions. All are manifestations of a spiritual revolution which proclaims art a religion, and forbids its degradation to the level of a trade. They are intended neither to please, to flatter, nor to shock, but to express great emotions and to provoke them.

71. Roger Fry,
'The French Group'

Catalogue of the second Post-Impressionist
exhibition, 1912, pp. 13–17

Fry reprinted the text of this important essay in his collection
Vision and Design of 1920 (ed. J.B. Bullen, Oxford University
Press, 1981, pp. 166–70).

When the first Post-Impressionist Exhibition was held in these
Galleries two years ago the English public became for the first time
fully aware of the existence of a new movement in art, a
movement which was the more disconcerting in that it was no
mere variation upon accepted themes but implied a reconsideration of the very purpose and aim as well as the methods of
pictorial and plastic art. It was not surprising therefore that a public
which had come to admire above everything in a picture the skill
with which the artist produced illusion should have resented an art
in which such skill was completely subordinated to the direct
expression of feeling. Accusations of clumsiness and incapacity
were freely made, even against so singularly accomplished an artist
as Cézanne. Such darts, however, fall wide of the mark, since it is
not the object of these artists to exhibit their skill or proclaim their
knowledge, but only to attempt to express by pictorial and plastic
form certain spiritual experiences; and in conveying these, ostent-
ation of skill is likely to be even more fatal than downright
incapacity.

Indeed, one may fairly admit that the accusation of want of skill
and knowledge, while ridiculous in the case of Cézanne is perfectly
justified as regards one artist represented (for the first time in
England) in the present Exhibition, namely, Rousseau. Rousseau
was a custom-house officer who painted without any training in
the art. His pretentions to paint made him the butt of a great deal of
ironic wit, but scarcely any one now would deny the authentic
quality of his inspiration or the certainty of his imaginative
conviction. Here then is one case where want of skill and

352

knowledge do not completely obscure, though they may mar, expression. And this is true of all perfectly naïve and primitive art. But most of the art here seen is neither naïve nor primitive. It is the work of highly civilised and modern men trying to find a pictorial language appropriate to the sensibilities of the modern outlook.

Another charge that is frequently made against these artists is that they allow what is merely capricious, or even what is extravagant and eccentric, in their work – that it is not serious, but an attempt to impose on the good-natured tolerance of the public. This charge of insincerity and extravagance is invariably made against any new manifestation of creative art. It does not of course follow that it is always wrong. The desire to impose by such means certainly occurs, and is sometimes temporarily successful. But the feeling on the part of the public may, and I think in this case does, arise from a simple misunderstanding of what these artists set out to do. The difficulty springs from a deep-rooted conviction, due to long-established custom, that the aim of painting is the descriptive imitation of natural forms. Now, these artists do not seek to give what can, after all, be but a pale reflex of actual appearance, but to arouse the conviction of a new and definite reality. They do not seek to imitate form, but to create form; not to imitate life, but to find an equivalent for life. By that I mean that they wish to make images which by the clearness of their logical structure, and by their closely-knit unity of texture, shall appeal to our disinterested and contemplative imagination with something of the same vividness as the things of actual life appeal to our practical activities. In fact, they aim not at illusion but at reality.

The logical extreme of such a method would undoubtedly be the attempt to give up all resemblance to natural form, and to create a purely abstract language of form – a visual music; and the later works of Picasso show this clearly enough. They may or may not be successful in their attempt. It is too early to be dogmatic on the point, which can only be decided when our sensibilities to such abstract form have been more practised than they are at present. But I would suggest that there is nothing ridiculous in the attempt to do this. Such a picture as Picasso's *Head of a Man* would undoubtedly be ridiculous if, having set out to make a direct imitation of the actual model, he had been incapable of getting a better likeness. But Picasso did nothing of the sort. He has shown in his *Portrait of Mlle. L. B.* that he could do so at least as well as any one if he wished, but he is here attempting to do something quite different.

No such extreme abstraction marks the work of Matisse. The actual objects which stimulated his creative invention are recognisable enough. But here, too, it is an equivalence, not a likeness, of nature that is sought. In opposition to Picasso, who is pre-eminently plastic, Matisse aims at convincing us of the reality of his forms by the continuity and flow of his rhythmic line, by the logic of his space relations, and, above all, by an entirely new use of colour. In this, as in his markedly rythmic design, he approaches more than any other European to the ideals of Chinese art. His work has to an extraordinary degree that decorative unity of design which distinguishes all the artists of this school.

Between these two extremes we may find ranged almost all the remaining artists. On the whole the influence of Picasso on the younger men is more evident than that of Matisse. With the exception of Braque none of them push their attempts at abstraction of form so far as Picasso, but simplification along these lines is apparent in the work of Derain, Herbin, Marchand and L'Hote. Other artists, such as Doucet and Asselin, are content with the ideas of simplification of form as existing in the general tradition of the Post-Impressionist movement, and instead of feeling for new methods of expression devote themselves to expressing what is most poignant and moving in contemporary life. But however various the directions in which different groups are exploring the newly-found regions of expressive form they all alike derive in some measure from the great originator of the whole idea, Cézanne. And since one must always refer to him to understand the origin of these ideas, it has been thought well to include a few examples of his work in the present Exhibition, although this year it is mainly the moderns, and not the old masters that are represented. To some extent, also, the absence of the earlier masters in the exhibition itself is made up for by the retrospective exhibition of Monsieur Druet's admirable photographs. Here Cézanne, Gauguin and Van Gogh can be studied at least in the main phases of their development.

Finally, I should like to call attention to a distinguishing characteristic of the French artists seen here, namely, the markedly Classic spirit of their work. This will be noted as distinguishing them to some extent from the English, even more perhaps from the Russians, and most of all from the great mass of modern painting in every country. I do not mean by Classic, dull, pedantic, traditional, reserved, or any of those similar things which the word is often made to imply. Still less do I mean by calling them Classic that they paint *Visits to Æsculapius* or *Nero at the Colosseum*. I mean that they do not rely for their effect upon associated ideas, as I believe Romantic and Realistic artists invariably do.

All art depends upon cutting off the practical responses to sensations of ordinary life, thereby setting free a pure and as it were disembodied functioning of the spirit; but in so far as the artist relies on the associated ideas of the objects which he represents, his work is not completely free and pure, since romantic associations imply at least an imagined practical activity. The disadvantage of such an art of associated ideas is that its effect really depends on what we bring with us: it adds no entirely new factor to our experience. Consequently, when the first shock of wonder or delight is exhausted the work produces an ever lessening reaction. Classic art, on the other hand, records a positive and disinterestedly passionate state of mind. It communicates a new and otherwise unattainable experience. Its effect, therefore, is likely to increase with familiarity. Such a classic spirit is common to the best French work of all periods from the twelfth century onwards, and though no one could find direct reminiscences of a Nicholas Poussin here, his spirit seems to revive in the work of artists like Derain. It is natural enough that the intensity and singleness of aim with which these artists yield themselves to certain experiences in the face of nature may make their work appear odd to those who have not the habit of contemplative vision, but it would be rash for us, who as a nation are in the habit of treating our emotions, especially our aesthetic emotions, with a certain levity, to accuse them of caprice or insincerity. It is because of this classic concentration of feeling (which by no means implies abandonment) that the French merit our serious attention. It is this that makes their art so difficult on a first approach but gives it its lasting hold on the imagination.

72. Boris von Anrep, 'The Russian Group'

Catalogue of the second Post-Impressionist
exhibition, 1912, 18–21

Anrep (1886–1969) was a mosaicist born in St Petersburg. He
settled for a time in Paris in 1908.

Russian spiritual culture has formed itself on the basis of a mixture
of its original Slavonic character with Byzantine culture and with
the cultures of various Asiatic nations. In later times European
influence has impressed itself on Russian life, but does not take
hold of the Russian heart, that continues to stream the Eastern
blood through the flesh of the Slavonic people. One of the
peculiarities of Eastern art is a great disposition for decorative
translations of life, an ideographical representation of it, and an
imaginative design. Romanesque and Gothic art of Western-
Europe had much of the same character, but European art inclined
towards naturalism, the Russian persisted in its archaic traditions.
The Byzantine influence was of the utmost importance to Russia,
as from there came the light of Christianity. With the religious
beliefs and rites were introduced the Byzantine symbolical
representations of the Divinity as they were realised in the religious
images, called "ikones," made for devout purposes. The conventions of the ancient ikone-painters remained the only pictorial
language till the end of the seventeenth century, the art being
purely religious and under canonical regulations. In the eighteenth
century the Russian pictorial forms undergo a strong European
influence, and since then they follow European ideals. At the
present day Western influence is regarded by the nationalists as
incompatible with the deepest aspirations of the Russian soul.
Artists filled with admiration before the beauty and expressivity of
Russian ancient art aim to continue it, passing by the Western
influence, which is considered foreign and noxious to the growth
of the Eastern elements of the Russian art. The principal trait of
their personal art is a decorative and symbolical treatment of

nature combined with an imaginative colouring, that they feel answers the most to their Russian soul. Only during the last fifteen years artists of note worked for the revival of the national art. Mr. Stelletzky approaches the closest to the ancient forms. His works are not copies of the ikones but are the result of his extreme knowledge of all the possibilities that the ancient art gives; he uses the archaic alphabet which he finds the best medium for the exercise of his pictorial imagination. Count Komarovsky is not less accomplished but his colouring and forms are more tender and sensitive. Mr. Roerich belongs to the same new Byzantine group though he does not appropriate entirely the forms of the ikones, he succeeds, may be, more than others, to translate in his own manner the essence of the Russian religious and fantastical spirit. His imagination carries him further to the dawn of the Russian life, and he gives an emotional feeling of the prehistoric Slavonian Pagans.

Madame Goncharova does not realise in her art the mastership and the decorative calligraphic qualities of the ikones, but she aims for a true representation of the ancient Russian God, who is her own, and His saints. That is why sweetness, joy, tenderness and voluptuousness are far from her art as they are far from the Russian conception of the Divinity. Her saints are stern, severe and austere, hard and bitter. The revival of the Russian national art brought forth the interest of some artists to the modern popular art, the art of the unlearned lads who find their sport in painting and show by that medium their simple-natured, fresh and naïve spirit. Those artists assimilated themselves to the popular art and rejoice in its sincere directness. Their art is welcome as a counterweight to the over-refined and effeminate tastes of an influential group of aesthetical 'gourmands' of St. Petersburg. Mr. Larionoff is at the head of those 'rustical' artists. The naïve and awkward russific-ations of European forms remain as a special epoch in the history of the Russian art. Some young artists aiming for the same emotions that those simple rural imitations give, chose to use their shapes as their pictorial language, Mr. Soudejkin for instance.

Another group of artists does not exploit the national forms; their means seem to be more explicit to a modern European artist's mind: Petroff-Wodkin, Bogaevsky and Chourlianis being thoroughly different in their personalities possess the same valuable quality of keeping their art in close connection with their philosophical substance. Petroff-Wodkin gives a great spiritual meaning to the gestures of his figures, naturalistically comprehended, but coloured in a fantastical and decorative way, Bogaevsky is a landscape painter; but the *morne* cliffs, the dead cities, the desolate shores of a leaden sea are not earthly landscapes; they terrify the Russian soul as if they were terrible omens. The innermost recesses of the Russian heart are filled with mystical passions. The painter Chourlianis was overpowered by them, he was devoted to the mysteries of the Cosmos and to the music of the empyrean æther. *Rex* is one of his most important pictures. The fire, that burns in the centre of it, is surrounded by the horizon of an occult world, by the mounting spheres and by the shadows of angels. Chourlianis prematurely died last year.

As for the realistic art, the young gifted artists in Russia do not manifest any great energy in practising it, and there are but few interesting representatives of that art. Among the artists whose works are exhibited here, Mr. Sarian and Miss Joukova give the largest quantity of realistic sensations. Mr. Sarian is represented by his energetic illustrations of the Turkish life. Miss Joukova's portrait of an old woman shows a studious and sincere research for the characteristic of human nature.

It is to be noticed that both of them are still much inclined to a decorative interpretation of their feelings; that is the dominant tendency of the most interesting part of the modern Russian art.

73. Leonard Woolf,
Beginning Again: An Autobiography of the Years 1911–1918
1964, pp. 93–4

Woolf (1880–1969) was secretary of the second Post-Impressionist exhibition. He was in Ceylon during the first Post-Impressionist exhibition.

The first job which I took was a curious one. The second Post-Impressionist Exhibition, organized by Roger Fry, opened in the Grafton Galleries in the autumn of 1912. In Spain on our honeymoon I got an urgent message from Roger asking me whether I would act as secretary of the show on our return. I agreed to do so until, I think, the end of the year. It was a strange and for me new experience. The first room was filled with Cézanne water-colours. The highlights in the second room were two enormous pictures of more than life-size figures by Matisse and three or four Picassos. There was also a Bonnard and a good picture by Marchand. Large numbers of people came to the exhibition, and nine out of ten of them either roared with laughter at the pictures or were enraged by them. The British middle class – and, as far as that goes, the aristocracy and working class – are incorrigibly philistine, and their taste is impeccably bad. Anything new in the arts, particularly if it is good, infuriates them and they condemn it as either immoral or ridiculous or both. As secretary I sat at my table in the large second room of the galleries prepared to deal with enquiries from possible purchasers or answer any questions about the pictures. I was kept busy all the time. The whole business gave me a lamentable view of human nature, its rank stupidity and uncharitableness. I used to think, as I sat there, how much nicer were the Tamil or Sinhalese villagers who crowded into the veranda of my Ceylon kachcheri than these smug, well dressed, ill-mannered, well-to-do Londoners. Hardly any of them made the slightest attempt to look at, let alone understand, the pictures, and the same inane questions or remarks were repeated to me all day long. And every now and then some well groomed, red faced gentlemen, oozing the undercut of the best beef and the most succulent of chops, carrying his top hat and grey suede gloves, would come up to my table and abuse the pictures and me with the greatest rudeness.

There were, of course, consolations. Dealing with possible purchasers was always amusing and sometimes exciting. Occasionally one had an interesting conversation with a stranger. Sometimes it was amusing to go round the rooms with Roger and a distinguished visitor. I have described in *Sowing* Henry James's visit. Roger came to the gallery every day and spent quite a lot of time there. We used to go down into the bowels of the earth about 4 o'clock and have tea with Miss Wotherston, the secretary, who inhabited the vast basement, and we were often joined by Herbert Cook who owned Doughty House, Richmond, and a superb collection of pictures. I saw so much of Roger that at the end of my time at the Grafton Galleries I knew him much better than when I first went there. His character was more full of contradictions even than that of most human beings. He was one of the most charming and gentle of men; born a double dyed Quaker, he had in many respects revolted against the beliefs and morals of The Friends, and yet deep down in his mind and character he remained profoundly, and I think unconsciously, influenced by them. Like his six remarkable sisters, he had a Quaker's uncompromising sense of public duty and responsibility and, though he would have indignantly repudiated this, ultimately the Quaker's ethical austerity. And yet there were elements in his psychology which contradicted all these characteristics. I was more than once surprised by his ruthlessness and what to me seemed to be almost unscrupulousness in business. For instance, we discovered, shortly after I took on the secretaryship, that when Roger had been preparing the exhibition and asking people to exhibit, owing to a mistake of his, they had been offered much too favourable terms – the figure for the Exhibition's commission on sales was much too low. When the time came to pay artists their share of the purchase amounts of pictures sold, Roger insisted upon deducting a higher commission without any explanation or apology to the painters. Most of them meekly accepted what they were given, but Wyndham Lewis, at best of times a bilious and cantankerous man, protested violently. Roger was adamant in ignoring him and his demands; Lewis never forgave Roger, and, as I was a kind of buffer between them, he also never forgave me.

---

74. Unsigned review (probably by Robert Ross), 'A Post-Impressionist Exhibition: Matisse and Picasso'

The Times, 4 October 1912, 9

This exhibition, held like the first at the Grafton Galleries, contains a few beautiful works by Cézanne; but otherwise it is entirely given up to living artists, French, English, and Russian. It contains a large number of pictures by H. Matisse and a good many by M. Picasso, and as these are the most notorious and the most abused artists in the movement it is worth while paying particular attention to their works.

M. Matisse has been freely called a charlatan, which means that he is an incompetent or a mediocrity affecting a wilful eccentricity in the hopes that he may be mistaken for a genius. Now, his nude in the first room (9) proves that he is not incompetent. You may think it ugly; but it certainly would not be easy to paint. It is curiously like some of Mr. Walter Sickert's works, but more sharply defined and even freer from all sentimentality. Matisse has here painted a woman as if she had no more human associations for him than an animal. He has painted, in fact, as if he were not a human being himself, but a different kind of creature with a human power of representation. That is what disconcerts us in this picture, and still more in his later works in the next room. He is an artist whose vision is not controlled at all by the ordinary human interests or by the ordinary human notions of the visible world. Certain elements in reality interest him intensely; and his art has developed with a remorseless logic in the effort to insist upon these elements at the expense of all others. Take, for instance, his *Portrait au madras rouge* (31). Here again a woman is painted as if she were an animal. The artist does not seem to express any human relation of his own whatever with her. He has simplified her face and her form just as ruthlessly as if she were a piece of still life. But by means of this simplification, by means of the flatness and the rough, strong containing lines, he makes her face as completely a part of his design as the patterned handkerchief on her head. If he did nothing but this he would be merely a second-rate decorative artist without even the grace of prettiness. But he does a great deal more. With all his extreme simplification he succeeds in giving a very strong vitality and character to the face of the figure. You may say that the woman is very ugly, but the whole picture is not ugly, considered simply as an object apart from what is represents. It has the beauty of an Oriental design in pure fierce colour, while the forms, simple as they are, are intensely expressive. One can see that the design is not imposed upon the facts, but is, as it were, drawn out of them; is, in fact, the result of the emphasis which the artist's peculiar interest in reality leads him to lay upon it.

Take, again, the goldfish (37). Here the nude figure is treated as a mere accessory, and the artist plays tricks with it which disconcert the eye and look like wanton ignorance or perversity. We should not resent these tricks so much if they were played upon a piece of still life, and to Matisse this nude is merely still life. He is not interested in it enough to explain it to us. He is much more interested in the goldfish in the bowl and the nasturtiums in the green vase. The picture as a whole strikes the eye as a very novel and brilliant piece of colour, if once the mind can overcome its resentment at the manner in which the nude is treated. But this colour gets its peculiar vividness from the intensity of the artist's interest in what he represents. It is not mere abstract colour invention, which is always meaningless and tiresome in a picture; but colour expressing just what he has to say about everything he has painted. The goldfish are simple masses of red, but at the same time they quiver in the water; and it is their movement which gives significance to their colour. The vase holding the nasturtiums seems a perfectly flat shape; and its colour has the value and force that can only be given by flatness. But the enclosing line makes it seem to swim in light and has an abstract beauty of its own. Indeed, the quality of the paint in this picture is beautiful throughout; and one finds it beautiful in many other pictures when one has got over the strangeness of the method of representation.

But most people look for beauty in a picture only in the objects represented, which they expect to remind them of beautiful real things. They have no notion of beauty created, not imitated, in a work of art, and created by the effort for expression. They will therefore probably think the design for a decoration called *Les Danseuses* (185) merely hideous; and, indeed, the individual dancers do not remind one of beautiful real women at all. Here the peculiar unhumanity of Matisse is most obvious. He is not interested in his dancers as women. He is interested in the rhythm which they make together. But here again he does not impose this rhythm on them so as to make a pretty picture, nor has he found it at second-hand in other works of art. He has wrung it, as it were, out of the figures themselves. It is what he has to tell us about them; and he lays stress upon it with complete disregard of all the facts which we expect him to state. Thus he does not define his figures as we expect him to define them. The rhythmical lines are emphasis rather than definition. They express movement rather than substance, and what substance there is exists only to make the movement visible. But, as we expect those rhythmical lines to define, they disconcert us. They seem to be telling lies about form, when they are really telling truths about motion.

One may say, if one likes, that Matisse is attempting things impossible to his art, that he is trying to turn painting into music. But one need not therefore fall into a rage and accuse him of incompetence or wilful perversity or a brutal love of ugliness. He is not incompetent, but an artist of great powers, however he may use them. Nor, we believe, is he wilfully perverse, but, rather, very intense and simple in his interest in certain aspects of reality. And this interest, being itself rather unhuman, has carried his art very far away from the ordinary human understanding. But, as it seems to us, there is no cynical brutality in his unhumanity. He does not seem to hate human beings or to wish to degrade them; still less does he wish to shock us. But there is a curious asceticism in his complete detachment from all human interests. He seems to see human beings, and everything that has associations of pleasure and use for them, as if he were himself a being from another planet, watching everything with a very intense interest of his own, but an interest quite empty of all associations.

No doubt Rembrandt's art, with all its humanity, is far richer in content, as Rembrandt is a far greater artist; but Rembrandt and Matisse are not alternatives, any more than Shakespeare and Shelley. The enraged Academic puts a pistol to your head and cries, 'Matisse or Michelangelo – Choose between them.' He implies that if you get any pleasure from Matisse or see any good in him you must despise all the great artists of the past. But it would be impossible to understand or to enjoy him without some knowledge and enjoyment of the art of the past. For his art is based upon facts which past artists have stated so often that he thinks it safe to imply rather than state them; and upon the basis of implied fact he makes new statements expressing his own new and intense, if narrow, interest in reality. There is not ignorance in his very simplified forms, but a knowledge so familiar that he does not care to show it. And there is not sheer incompetence in his failures, but the rashness of an artist who experiments without fear of consequences.

The art of M. Picasso is a very different matter. He, too, is not a charlatan, but we do not believe that he is an artist of narrow intense originality like M. Matisse. Rather he seems to us to be by nature extremely inventive, and to have endeavoured to preserve himself from imitation by the pursuit of a theory scientific rather than artistic in its origin. We see him in an early portrait (65) an imitator of Goya, but without Goya's wit or spontaneity. In his large composition (46) we see him produce a work as cleverly eclectic and as sophisticated as some Italian pictures of the 17th century. And lastly, we have his purely theoretic experiments which are unintelligible to the eye and mind. Forgetting that these are meant to represent anything; we see very little abstract beauty of colour or design in most of them, though the still life (63) is an exception. They depress us as if they were diagrams of a science about which we know nothing; and where, as in the La Femme au Pot de Moutarde, a human form is obscurely discernible, it seems, but for the obscurity to be commonplace. He has every right to make his experiments, and they may perhaps prove useful to other artists in the future. He is, in fact, such a scientific experimenter as Paolo Ucelli might have been if he had had no original talent of his own, or if in him a slight original talent had been overlaid by intellectual curiosity.

75. C. Lewis Hind, 'Ideals of Post Impressionism'

Daily Chronicle, 5 October 1912, 6

Hind, who had championed Matisse in 1910 (see headnote to no. 34), was extremely hostile to Picasso's cubist work in 1911 and expressed similar reservations about Picasso's work in the second Post-Impressionist exhibition.

The camera of the photographer was aligned to a picture by Matisse. The camera was calm, but the face of the photographer wore a vast smile. He caught my eye; his look invited me to share his laughter. I declined. I was as grave as a statue by Eric Gill. Yet I was elated. Matisse elates and stimulates me. I passed on, through the Grafton Galleries, which had just been hung with a second collection of Post Impressionist works. Matisse dominates the rooms. Matisse is a serious painter. Let us pay him the compliment of treating him seriously, of trying to understand him. I admit that he is a trouble – disturbances always are a trouble; but troubles have a way, if you take them properly, of leading to serenity.

By degrees I surrendered myself to the witchery of Matisse – his colour and his rhythm. He does exercise a spell, if you can rid yourself of prejudice, and eradicate from your mind the nonsense that a new way of looking at things, and a new way of painting, is 'an insult to your intelligence,' and makes for, to quote the words of the outraged critic of the *Morning Post*, 'a deplorable and degrading show.'

I do not find the second Post Impressionist exhibition at the Grafton Galleries deplorable or degrading. It is stimulating and refreshing, but I fear that the public will find it more difficult to understand and appreciate than the first Post Impressionist exhibition. The unconscious pioneers of the movement – Cézanne, Van Gogh, and Gauguin – giants, masters who have fulfilled themselves, run their turbulent courses, and entered into their rest, dominated the exhibition of 1910. The present exhibition, again gathered together by Mr. Roger Fry, exploits the work of living men, Frenchmen, Russians, Englishmen, who are pushing the new way of seeing and painting into new glimmering or glistening paths.

OVERRUN EUROPE

A few beautiful works by Cézanne are shown, he who was the spiritual father of these moderns, who, whether they like it or not, are grouped together under the title of Post Impressionists. The meshes of the net are wide, but these revolutionaries must be called something, and the term Post Impressionist will serve, must serve, to describe the army of ardent painters – expressionists, synthesists, eliminationists, rhythmists, intensive colourists, cubists – who have overrun Europe.

Let me say again that one can be intensely interested in Post Impressionism and preserve all one's reverence and admiration for the art of the past. My respect for the Great Western Railway does not lessen because I am enthralled by aeroplanes, and I am sure that Mr. Roger Fry's fealty to early Italian art is not slighter because it delights him to express his feeling of a cascade or a terrace in Post Impressionist language. He, as a painter, has passed through the phase of eclectic classicism, but the younger Englishmen represented at the Grafton Galleries have sprung, at a bound, into the uncharted, experimental country peopled by those who try to paint, not the imitation of things, but their psychological feeling or sensation. It was Rodin who said, speaking of artists: 'Our eyes plunge beneath the surface to the meaning of things, and when afterwards we reproduce the form, we endow it with the spiritual meaning which it covers.'

INDWELLING SPIRIT

In this new movement the aim is to pierce through externals and to arrive at the significance, the spiritual indwelling. Because it is so much more comfortable to stultify one's spiritual growth, and to pretend that there is nothing in life but externals, many will be inclined to take the side of the gentleman who bawled in my ear at the Grafton Galleries, 'It's all d—d nonsense,' I led him to the end of the centre gallery, turned his stubborn head to where, isolated on the wall, hangs Matisse's decorative panel, *Capucines* – nudes and flowers and pure, refreshing colour – and said; 'Now forget yourself and your prejudices, and confess, "Isn't it beautiful?"' He answered 'No!'

Again I turned his stubborn head to the End Gallery, where Matisse's joyous, vivid *Les Danseuses* fills the wall. 'Forget.' I said, 'the pictures-frescoes in the Royal Exchange, and acknowledge that this rhythmic decoration, akin to music, makes you glad, raises your spirits.'

'That thing make me glad!' he cried. 'It depresses me. It's hideous.'

I turned away. It is silly to try to convince people. Then I perceived another gentleman, and in his eye there was something like a tear. 'Well,' I said, 'what is it?' And he answered, 'How am I to explain Picasso to my aunt?' In his hear I whispered. 'Don't,' and to cheer him led him to Mr. Spencer's entrancing *John Dome arriving in Heaven*, and then on to M. Giriend's exciting *Homage to Gauguin*.

I returned to Matisse and Picasso. The first Post-Impressionist exhibition was designed as a homage to Cézanne, Van Gogh, and Gaugin. This is a homage to Matisse and Picasso. By Matisse there are 42 paintings, sculptures, and drawings; by Picasso 17. Matisse is an artist, Picasso logician. Matisse expresses emotion. Picasso states science, and when science swallows art, good-bye art.

MATISSE HAS ARRIVED

I find Picasso's cubist portraits as unsympathetic as they are unintelligible. But he is an originator. The Futurists owe much to him, and I do not suppose that our Mr. Wyndham Lewis and others would ever have evolved Cubism had not Picasso started this geometrical game. Picasso is agonising towards a goal that I cannot think he will ever reach. He is and will be an experimenter. Matisse has arrived.

Yes, Matisse has arrived. This lonely, serious artist, this inhuman personality who, in his art, uses women, gold fish, and flowers as if they were fabrics, this man who has discarded all tradition and who paints as if there were nothing in the world but himself and the effect upon his untroubled, sensitized self of what his eyes see, and the sensation they transmit to his soul – has arrived. You may accept him or reject him, but it is certain that he is consistent, that he is travelling on his own line, and that he is expressing on canvas upon canvas magnificently unconventional, strangely beautiful, the flat symbols of his pageant of feeling. He is liberating our eyes to a new kind of beauty which, if you wait for it, will sweep from his designs like music, and which rests the heavy-laden like music.

Once I said to an ardent admirer and collector of Matisse's pictures – 'What is the effect upon you of living with them day by day?' She answered 'Tranquillity.' I gasped with astonishment. Now, I understand – and agree.

76. P.G. Konody, 'Art and Artists – More Post-Impressionism at the Grafton'

Observer, 6 October 1912, 6

For Konody see headnote to no. 45.

The first days of October have brought us a very flood of art exhibitions, large and small, but it is not likely that any of them – including even the important inaugural show of the new Grosvenor Gallery – will receive the attention they deserve whilst passions run high over the second Post-Impressionist exhibition at the Grafton Galleries. Everybody will remember the sensation created by the London début of the French P.-I.'s (the name is too long and clumsy for continual reiteration) – how people 'came to scoff and remained to pray'; how certain rash critics fulminated against the Bedlamite art of Cézanne, Gauguin and Van Gogh, and after a few weeks had to confess themselves beaten and to swallow their own words. Are we going to witness a similar spectacle this year, when the organisers of the P.-I. exhibition, elated by their success, start from the assumption that the three founders of the movement – serious and passionate, though often misguided, artists of very incomplete achievement – are now enthroned in the painters' Parnassus as 'old masters,' who have shown a younger generation the way to greater and more daring deeds? I think not. The three old masters of Post-Impressionism, in searching for a more intense and more significant expressiveness than could be found in the most brilliant realistic representation, never severed all links that connect pictorial art with the visual appearance of things. Picasso, in his later 'cubist' works, of which there is a representative selection at the Grafton Galleries, has given up – to quote Mr. Roger Fry – 'all resemblance to natural form in the attempt to create a purely abstract language of form – a visual music.'

Let us for the moment assume that Picasso is really serious, and has not adopted the new method merely to hide the mediocre skill and utter lack of technical or spiritual distinction revealed by him when he adopts a normal manner, as in the very commonplace *Mademoiselle L.B.* Let us assume the amazing and utterly unintelligible tangle of interesting straight and curved lines and planes, labelled *Tête d'homme, Livres et flacons, Buffalo Bill,* which bear not the remotest resemblance to anything on earth, are really, as has been claimed, the outcome of a deeply felt emotion. Would this justify their existence as works of art? Surely the function of a work of art is not the mere statement of an emotion, but the communication of this same emotion to others. And to profess that these jig-saw puzzles produce any emotion except hilarity could only be hypocrisy. The incoherent ravings of a lunatic may be caused by profound emotion, but they are as far removed from real poetry as Picasso's cubic pictures are from real pictorial art. Far be it from me to accuse this highly intellectual painter of insanity. In his work everything is deliberate and calculated; but, I repeat, these experiments have not the remotest connection with art. They are not even decorative as patterns. The best that can be said for them is that they are inoffensive in colour.

The question remains now: Are Picasso and his followers really as serious as their literary champions would have us believe? There are certain indications at the Grafton Galleries which justify a certain suspicion on this point. Picasso's amazing *Tête d'homme* is a geometrical tangle in which nothing is clear except a coat button and the cryptic letters GR and KOU. Near it hangs an alleged portrait or spiritual vision of "Kubelik," by Picasso's close imitator Bragne. Again, it is impossible to distinguish anything but the admirably drawn letters MOZART and KUBELICK (sic!). In the first room is another medley by Picasso, labelled *Le Bouillon Kub*, and introducing a packet marked in block letters 'KUB.' Note the significant connection between these 'CUBist' masterpieces with the lettering 'KUB,' 'KUBelick,' and 'KOU.' Can it be that Kubelik was chosen as a subject merely because his name begins with these three letters? Were it not for the lettering, the picture might as well be called *A View from a Garret Window*, or *Still Life*, or *The Dream of an Opium-eater*. And does not *Le Bouillon Kub* suggest that Picasso is merely laughing at his admirers and telling his admirers: 'These are the ingredients with which I stew my cubist bouillon'?

Among Picasso's less extreme pictures are a dull still life, introducing a typical P.-I. collapsible jar, a *Tête de femme* which suggests a wood-carving in the first stage of chipping out the broad planes, and the amusing *Femme au Pot de Moutarde*, in which it is possible not only to recognize the woman and the mustard pot, but also the clearly stated fact that the woman has just swallowed the contents of the pot.

The art of Henri Matisse, which has never before been so fully represented at a London exhibition, is scarcely less disconcerting than that of Picasso, although Matisse, a real master of superb decorative colour, rhythmic line and expressive simplification, never departs so far from visual truth that his intention could be disguised by an impenetrable veil. Take him at his best – he is to be seen at his best in *Les danseuses: design for a decoration in Prince*

Tschonkine's Palace at Moscow – and it is impossible to deny him that admiration which the indiscriminating lavish upon his most repulsively contortionate caricatures of nature and humanity. This group of dancing women, pointed in a flat pattern of blue, green, rose and black, has all the significant movement, the broad simplification, the swinging rhythm, but passionately intensified, that are to be found in the figures painted on Greek vases. This panel is a decoration that could probably not be rivalled by any living painter. But it stands practically alone among Matisse's works at this exhibition. His inventions of rare schemes of boldly, yet subtly, contrasted colours are almost without exception pure delight to the eye. There is, however, throughout his work what appears to be a research for the repulsive in form and movement, and a certain coarseness of statement that are the opposite pole to the beauty and refinement of that Far Eastern art by which he seems to be so strongly influenced. Matisse, the painter, is immensely stimulating at an exhibition, but I should think it impossible to live with any of the pictures from his brush that have been brought together in the large gallery.

In the work of Matisse, the sculptor, I can see nothing but unredeemed, deliberate repulsiveness. That it is deliberate, the artist himself proves by showing a *Buste de Femme* in four consecutive stages of completion, in which the approximate resemblance to a human being of the first stage is gradually eliminated until the final head in bronze assumes an aspect of inexpressible loathsomeness, such as could not be even distantly approached by the most debased and depraved type of humanity, or even of the simian tribe. The female nude, *L'arraignée*, is constructed on the same principle as the excessively slender and long-necked grotesque *Liberty cats*, which, at the time of the aesthetic craze, were so popular a decoration of the overmantel at the suburban villa.

Not all the French P.I.'s go to the same extremes as Picasso and Matisse. Flandrin, for instance, though striving for expressive simplification, and at times somewhat arbitrary in patterning his masses of colour, is more Impressionist than 'Post,' in so far as he has not cut loose altogether from that much reviled 'represent-ation' of nature which the new school professes to despise and altogether to discard. Herbin is a mild *cubist* of not very firm conviction, who becomes almost photographically precise in a large Still Life, and scales the height of absurdity in the grotesque jumble of leaning houses, which he chooes to label *Le Pont Neuf*.

The eccentricities of the new school, which cause so much merriment among the general public are, of course, not due to incompetence. It is so easy to copy a jar or a vase fairly correctly, or to draw simple houses with perpendicular walls, that it would be absurd to charge artists, of frequently very striking talent and of vast antecedent experience, with inability to do what is within the power of any schoolboy. The fact is that the P.-I.'s are too scientific; that they try to give pictorial expression to abstruse theories. If De Vlaminck and Othon Friest carefully avoid to make their walls plumb, it is probably due to their conviction that the perspective law of the vanishing point may be applied to vertical lines. But the reiteration of the new formulæ is apt to become terribly wearisome. The painting of a vase, for instance, from two points of view, so that the spectator can see the unforeshortened body as well as look into the round mouth, may amuse the first time; but if a dozen artists or more adopt or continually repeat the same device, it is uncommonly like having continually dinned into your ears the very obvious truth: 'This vase has a mouth, although, unless you look down upon it, you can't see it.'

Some pictures of truly remarkable beauty are to be found among the less extreme attempts. Among these are a presumably early nude figure by Matisse – a searching study of values and significant planes; a flower-piece by Girieud, that it would be difficult to rival for sheer loveliness of luscious colour; Chabaud's sombre and dignified *Chemin dans la Montagnette*; Derain's *La Fenêtre sur le Parc*, painted in pleasantly archaistic imitation of a Pierdei Franceschi background; and, above all, the *Maissonneure* and *Gennevilliers* by Cézanne. But, then, Cézanne can scarcely be called a P.-I., if the same term applies to Picasso and Matisse. He is an Impressionist more occupied with the accentuation of structural form than with the play of light on the surfaces.

Of the English and Russian contributors I shall have to speak on a later occasion.

77. Arnold Bennett, diary entry

8 October 1912

For Bennett see headnote to no. 22.

2nd Post-Impressionist Exhibition. Self-satisfied smiles of most people as they entered. One large woman of ruling classes with a large voice and *face-à-mains*, in front of a mediocre picture: 'Now no one will ever persuade me that the man who painted that was serious. He was just pulling our legs.' Self-satisfied smiles all over the place all the time. One reason of the popularity of these shows is that they give the grossly inartistic leisured class an opportunity to feel artistically superior. A slight undercurrent of appreciation here and there. A woman to whom a young man pointed out a pencil drawing by Matisse said: 'That's what I call beautiful.' (It was.)

I met Frank Harris. He was prepared on principle to admire everything, though there was a large proportion of absolutely uninteresting work. When I said I had seen much better Picassos than there were there he hardened at once. 'I find it all interesting,' he said grimly. The photograph room, where photos of Gauguin, Van Gogh, etc. supposed to be on sale, was in charge of an ignorant young ass who had all the worst qualities, from the languishing drawl to the *non possumus* attitude, of the English salesman.

78. Desmond MacCarthy, 'Kant and Post-Impressionism'

Eye Witness, 10 October 1912, 533–4

MacCarthy takes up Bell's term 'significant form' which he relates to Fry's ideas on 'aesthetic emotion' in the context of Post-Impressionist painting.

'Of all the cants that are canted in this canting world the cant of criticism is the worst.' This sentence occurs somewhere in the writings of Sterne. I remember it because it pleased Schopenhauer hugely. He was fond of English quotations and he hated critics, and if he spelt the last half of the quotation so – 'the Kant of Criticism is the worst,' it expressed not only his general attitude towards critics, but also one of his most cherished convictions.

Posterity, at least that section of it which tries at all obstinately to think what's what, rates Kant a great deal higher as a philosopher than it does Schopenhauer. But for all that Schopenhauer's theories have had much more influence upon artists, poets, and musicians, in so far as they have attempted to formulate æsthetic theories of their own. Indeed, his influence upon artists has been so constant, that it is something of a surprise to find the preface to the Catalogue of the Post-Impressionist Exhibition harking back to his predecessor for its point of view. I do not think myself that Kant would have cared for the pictures on the walls of the Grafton Galleries, but that is neither here nor there. The interesting thing is that Mr. Fry, wishing to prepare the public for what they will see, and to state the æsthetic case for these painters, falls back on Kant's definition of the proper object of æsthetic emotion. It is not my purpose here to criticise the pictures, but to review the preface, or rather these prefaces, for there are three, one by Mr. Fry, one by Mr. Clive Bell, and one by M. Anrep.

The characteristic of these pictures, which Mr. Fry knows will disconcert or infuriate people most, is the complete disregard which Post-Impressionists show of the usual attempt to produce illusion in pictures, and he knows that it will not placate spectators, when it is borne in upon them that this disregard has no excuse in lack of skill, to find that it is deliberate. 'But why, *why*, WHY,' shouts the exasperated spectator, 'do they falsify or simplify the forms of things till they are either absolutely unrecognisable – just look at these Picassos, I beg you, LOOK! – or till they lose all power to remind me of the beautiful qualities the things in reality possess.' It is at this point Mr. Fry and Mr. Bell (I seem to see them closing in upon an excited gesticulating figure) ever so gently come to the rescue. It will not do to whisper in his ear, 'I thought you admitted that an artist need not, indeed *must* not copy nature, and if so why do you object,' for this will only excite the patient more. He can reply with perfect truth that when admitting that art was not photography, or that the business of the artist was not to produce the greatest degree of delusion, he did not mean that the merit of a work of art was entirely independent of its representative qualities. Yet this notion is what Mr. Fry and Mr. Bell must induce him, if not to believe, at least to accept on approval, while he is going round the pictures.

I can imagine them arguing thus: 'Before you dash your umbrella through that canvas, answer me one question, what quality is it in picture or statue that *ought* to give you pleasure?' (This is a difficult question to answer and time will be gained.) 'You know you despise as sentimental and deplorably un-æsthetic those people who enjoy a picture because the subject interests them; who admire, say, Sir Luke Fildes's picture of *The Doctor*, because they feel so sorry for the poor parents, and the doctor looks a kind man who will sit by the little one's side all night till the crisis is past. You despise such people's opinion of art, don't you? You are sure that the æsthetic merit of a picture does not depend principally on the moving ideas which its subject is capable of calling up – in fact, that its æsthetic merit is independent of those ideas?'

This step is easy, and Mr. Fry may be sure of taking at any rate a cultivated visitor to the Grafton Galleries with him, so far. The next step is more ticklish; it consists in making the suggestion that perhaps the principal reason why the lack of representative qualities in the pictures is so deeply resented, may be the same at bottom as that which so wrongly induces an un-æsthetic person to prefer *The Doctor* to, say, a sketch of a ballet-girl by Degas; namely, that even the cultured picture-lover has derived more pleasure from irrelevant qualities than from these which really make a picture a work of art, as distinguished from a mere copy of some scene or object. The sly suggestion is that we miss the representation of nature so much only because all the time our attention has been fixed more upon the ideas which the picture calls up, than upon the actual plastic beauty of the lines, masses, and colours which should be the sources of our emotion. Let us, therefore, says the Post-Impressionist artist, get rid of representation as much as possible. It is only a source of distraction. The less a picture imitates natural forms the more likely are we to derive a purely *æsthetic* pleasure from contemplating it, instead of one derived from the associated ideas called up by forms reminding us of real things.

Here Kant's theory of æsthetics comes in. Kant laid great stress on the immediacy of the æsthetic judgement and its disinterestedness. By immediacy he meant that beauty was a quality perceived as directly as a colour itself; and that no analysis could reconstruct or explain that impression. Æsthetic judgements were therefore not susceptible of proof, they could only be evoked; and therefore there could be no such thing as scientific criticism. Art criticism in the last resort could only point.

By disinterestedness he meant that the æsthetic emotion is one entirely detached from a sense of the qualities of things as they appeal to the imagination, or to the moral or practical judgment. He distinguished between 'free or disinterested beauty' and 'secondary beauty,' which is felt through the medium of associated ideas. He refused to call 'secondary' beauty, beauty – why I cannot think. An arabesque or a melody is an example of 'free' beauty according to Kant; but he denied that the human face (he had not, of course, seen Picasso's portrait of Buffalo Bill) could be beautiful in art, because the beauty of the human face must depend upon ideas, the idea of human qualities.

In Mr. Bell's preface Kant's theory of æsthetic emotion takes this form:

> We can regard [he writes] an object solely as a means and feel emotion for it as such. It is possible to contemplate emotionally a coal-scuttle as the friend of man. We can consider it in relation to the toes of the family circle and the paws of the watch-dog. And, certainly, this emotion can be suggested in line and colour. But the artist who would do so can but describe the coal-scuttle and its patrons, trusting his forms will remind the spectator of a moving situation. His description may

interest, but, at best, it will move us far less than that of a capable writer. Yet most English painters have attempted nothing more serious. Their drawing and design have been merely descriptive; their art, at best, romantic.

How, then, does the Post-Impressionist regard a coal-scuttle? He regards it as an end in itself, as a significant form related on terms of equality with other significant forms. Thus have all great artists regarded objects. Forms and the relation of forms have been for them, not means of suggesting emotion but objects of emotion. It is this emotion they have expressed. Their drawing and design have been plastic and not descriptive. That is the supreme virtue of modern French art; of nothing does English stand in greater need.

What Mr. Bell means by 'significant form' is what Kant meant by 'free' beauty.

Mr. Fry makes Kant's distinction between 'free' beauty and 'secondary' beauty equivalent to the distinction between classic and romantic art:

> All art depends upon cutting off the practical responses to sensations of ordinary life, thereby setting free a pure and as it were disembodied functioning of the spirit; but in so far as the artist relies on the associated ideas of the objects which he represents, his work is not completely free and pure, since romantic associations imply at least an imagined practical activity. The disadvantage of such an art of associated ideas is that its effect really depends on what we bring with us: it adds no entirely new factor to our experience.

Now in this passage Mr. Fry does not deny that 'secondary' or 'romantic' beauty is a proper object for æsthetic emotion; but he gives it much less importance. I do not propose to criticise the pictures at the Grafton Galleries now. It is true that the beauty which some of them possess is of the first kind, which Mr. Fry calls classic, and that, owing to the lack of representative qualities in them, the second kind of beauty is almost totally absent. Personally, I think that its presence is of great importance, and that a picture should rouse both kinds of emotion if it is to rank as a magnificent work of art.

79. Unsigned review (probably by Robert Ross), 'The Post Impressionists: Some French and English Work'

The Times, 21 October 1912, 10

People in England at present are more inclined to discuss Post-Impressionism in general than the merits of any particular Post-Impressionist. They ask what is the value and intention of the movement, and condemn or approve the artists wholesale. But, of course, no movement can provide an artist with talent who has not got it by nature; and some Post-Impressionists are gifted, while some are not. The best a movement can do is to teach an artist a method of using his natural gifts, suited both to those gifts and to the circumstances of his time. Now, Post-Impressionism, as a movement, has obvious dangers, and has already produced a good many absurdities; but it does enable an artist of modest gifts to exercise them upon tasks which do not fatally overstrain them.

In the Grafton Galleries there are a good many pictures which will give real pleasure to any one who does not demand of them more than they offer. M. Marchand, for instance, is not a great artist; but his *Vue de Ville* (89) is a pleasant picture, because in it he has represented certain facts very simply and clearly, and has combined them so that his design is lucid and, at the same time, expressive of the character of the place. In all pictures we expect, or ought to expect, a greater lucidity than we find in reality. This lucidity does not necessarily explain to us things which we may want to know about the objects represented, for the artist's interest may not be identical with ours. But it assists our vision to grasp what he has to show us. It is, in fact, the result of his emphasis, which expresses his own interest in reality; and that emphasis produces a lucid and coherent design, if it is itself produced by a real and coherent interest. For most of us M. Marchand's town, if we saw it in reality, would be a mere confusion to the eye, however much it might interest us. And many skilful artists would represent it as a mere confusion, struggling desperately with all kinds of facts of light and colour and form, giving us, as it were, a number of

378

breathless reminders of reality, which could only satisfy us if we expected a picture to be less lucid than our own vision of reality. M. Marchand aims at no particular passages of illusion. He makes his picture by simplifying and emphasizing the different planes of the buildings, so that the eye grasps them easily and is led just where he wants to lead it through the picture. But the picture is more than a mere geometrical exercise, because he does manage to convey the character of the place by means of these simplified planes and of a colour also very much simplified but quite free from a sentimental or merely decorative prettiness. It is colour restrained and consistently restrained rather than falsified, and it invites our confidence like a quiet and homely style in a writer. M. L'Hote is a more fantastic artist than M. Marchand. His *Port de Bordeaux* (86) is further away from reality, and looks like a memory of a scene heightened in memory. But here again the simplicity and homeliness of the method give us confidence in the artist. He seems to represent no more than he does remember. The shapes in the picture are rather blunt and crude; but they do not make the picture itself ugly, and their bluntness and crudeness is the result of omission rather than of misrepresentation. No doubt a very great artist could have made more of the scene than this, for he would have known more about it and would have been able to make use of all his knowledge in his art. But M. L'Hote has succeeded modestly because he has made the best use of what knowledge and skill he does possess. His picture may not contain enough to satisfy some tastes, but at any rate it does not contain too much.

M. Derain is an artist both more ambitious and more austere. His *La fenêtre sur le parc* (13) is a picture in the great classical French tradition. It looks Post-Impressionist mainly because the artist has simplified all his forms further than any artist of the 17th century and because he has tried to keep his masses of paint alive rather than smooth. In his very serious treatment of still life he resembles some of the great Chinese artists. Like them, he tries to give it an abstract grandeur of design while preserving the character of the objects represented. He is, of course, far behind them in technical skill, but he is an accomplished artist by modern standards. In his landscape *L'eglise* (19) he combines an Impressionist force of light with a representation of space and masses not often found in Impressionist pictures. This picture looks easy, but it is the ease of a clear conception simply executed. M. Vlaminck is more uncertain and more various in his aims. His *Viaduct St. Germain* (222) is splendid and not unreal in colour; but he reminds one, in his treatment of form, of the speech of someone who has a severe cold in the head. In his pictures the shapes seem to run rather than to melt into each other. M. Herbin is the very opposite in his treatment of form. He sacrifices everything to the representation of mass, and his design in *Le Pont Neuf* (94) certainly has a startling lucidity. There never was a picture more easy to grasp. Indeed, his pictures seem to be as pointed as a good joke or epigram. But as soon as one has seen the point of them one loses interest in them. They seem to exploit reality like many clever Japanese pictures, though in a very different way; and they are amusing rather than satisfying. M. Bonnard's *Cascade* (157) is wonderfully skilful and charming; but his design lacks lucidity. This picture has a general air of untidiness owing to its lack of relief, and bewilders the eye instead of guiding it. M. Girieud is very charming in his small pictures, but rather futile in his larger ones.

The examples of the new Russian school, in which there is an attempt to revive the ancient Byzantine art of Russia, do not seem to us to promise much. M. Stelletzky's *Yard of the Kremlin in the time of John the Terrible* (138) reminds one of Gentile Bellini pleasantly enough, and his *Stag Hunt* (235) is spirited and amusing. Mr. Komarovsky's very Byzantine *Gabriel* (227) makes a novel and delightful pattern of colour. Mr. von Anrep's allegorical composition has more originality than the other Russian works; but even here we suspect that the quaintness is archaistic, and we doubt whether any of the artists really feel Byzantine. They are right, no doubt, in their desire to escape from the cosmopolitan art taught in Paris; but Byzantinism may be just as foreign to them as that art. What they have to discover is how Russians of to-day would naturally paint; and they will not necessarily do that by painting like Russians of some centuries ago.

Among the English Post-Impressionists Mr. Duncan Grant shows the most talent; but his charming *Pamela* (102) is not at all Post-Impressionist in its treatment of form. He seems to be a colourist by nature who experiments with different methods of execution to see which will express his sense of colour best. His *Queen of Sheba* (74) is a very amusing illustration, which would have told better, as an illustration, in black and white. For here the colour, good as it is, seems irrelevant, and the spotty execution of the background is incongruous with the painting of the Queen's face. We hope, however, that Mr. Grant will continue to produce illustrations, for he shows a delicate sense of character in this work; and, if he can develop that, it will counteract a dangerous tendency to be too artistic, which we notice in pictures like his *Dancers* (81). Mr. Etchells has some pleasant little landscapes. His large picture, *The Dead Mole* (103), leaves us with a sense of discomfort partly because it lays an awkward stress on a trivial incident, partly because the man is represented, here in three dimensions, and there in two. But Mr. Etchells is an artist of real promise, though very uncertain accomplishment; and promise is more healthy than accomplishment in an artist of his youth. Mr. Adeney's landscapes, especially the *Saw Mill* (118), are very pleasant, unaffected pictures. He expresses the character of a scene by the simplest means, and succeeds where many more skilful painters have failed. Mrs. Bell's *Asheham* (77) is equally simple and pleasant. Mr. Fry's *Angles sur Langlin* (120) is rather Impressionist than Post-Impressionist. It looks like a landscape painted, so to speak, straight from the model, with only so much simplification as was needed to make a lucid design. In his *Cascade* (82) he attempts to do with simple masses of paint what Chinese artists did so well with delicate outlines and washes. His paint renders the force and movement of the water very powerfully, but seems heavy and dead in the masses of the rock.

On the whole, the English pictures are inferior to the French just where we should expect them to be, in their rendering of mass and form. Without a strong underlying sense of mass and form a Post-Impressionist picture is empty rather than simple; and emptiness betrays itself more quickly in Post-Impressionist pictures than in others. Whatever we may think of M. Matisse's method, his pictures are not empty. Facts are implied rather than stated in them, but he has a thorough grasp of the facts which he implies. It would be easy to produce colour fantasies in the manner of M. Matisse; but if they were mere fantasies they would be like nonsense verses. If Post-Impressionist art becomes popular in England we shall have hundreds of nonsense pictures, mere caprices in colour with no more meaning in them than a Turkey carpet. The pictures of MM. Derain, Marchand, and L'Hote have meaning because they are conceived and designed in masses; but already in the English pictures we see a tendency to sacrifice mass to colour; and this must be checked if the movement is not to end, as so many movements have ended in England, in an empty decorative convention.

---

80. C.H. Collins Baker, 'Post-Impressionist Prefaces'

Saturday Review, 9 November 1912, 577–8

For Collins Baker see headnote to no. 55.

If no better case than Messrs. Roger Fry's and Clive Bell's can be made out for Post-Impressionism we cannot be blamed for dismissing the whole business as a mysterious and rather boring conspiracy to fool the public. How many Post-Impressionists can solemnly swear to themselves, on their private altars, that nothing but pure faith sends them to the stake of much advertised notoriety is a question that no sceptics have a right to answer. But we shall not wonder, after reading the prefaces to the Grafton Gallery exhibition catalogue, if we find our sceptics undiminished.

Mr. Clive Bell's contribution is a muddle of clouded thought, loose argument, historical inaccuracy and phrases, such as that all the pictures in the show are 'manifestations of spiritual revolution which proclaims art a religion and forbids its degradation to the level of a trade', that in American are described as 'hot air'. Mr. Fry, whose conviction none I think would question, appears to be somewhat in difficulties as to more or less fundamental laws of physical and psychical processes. (Art nowadays, you see, has become an abstruse kind of metaphysic that seems to stand in need of extraordinary ingeniousness.)

Mr. Bell assures us that Mr. Wyndham Lewis' art is practically independent of 'association or suggestion'. Obviously if this means anything it is that this gifted painter's art means nothing. Humanly speaking, it is impossible for a man to picture things that are associated with nothing, suggested by nothing, and which suggest nothing; conceptualist diagrams are another matter. Surely Mr. Bell might reflect a little on the conditions of consciousness and perception, and so spare us these inconceivable ideas. He then engagingly concedes that fully to appreciate Mr. Fry's art and Mr. Grant's 'it is necessary to be a human being'. Presumably it is a sine qua non that we be non-human in order to apprehend this Mr. Lewis. What guarantee, though, have we that rabbits and clams can apprehend him, or that he has established communication with the hypothetical inhabitants of another solar system? Surely we are not exigent in demanding more satisfying proof than Mr. Bell's mere implication and Mr. Lewis' indefinable pictures. I need quote the former but a little further, for then we shall have reached not only his climax but also, if I read Mr. Fry correctly, a curious point where both prefaces coincide. 'All the painters here', says Mr. Bell, 'are true *plastic* artists; wherefore the most important qualities in their work are quite independent of place or time or a particular civilisation or point of view. Theirs is an art that stands on its own feet instead of leaning upon life; and herein it differs from traditional English art.' Which, to cut the sentence shorter, 'has been for two centuries the laughing-stock of Europe'.

Now why is true *plastic* art independent of time or point of view; indeed, how can any thinking person hold that any expression of consciousness is independent of point of view and chronology? Pictorial expression is consciousness visualised, and consciousness must surely depend on time and 'point of view'. Mantegna, for example, is Mantegna; the inevitable result of his conditions. In the same way the least plastic art is rigidly dependent on conditions. If these loose meaningless statements of Mr. Bell are irritating, what are we to think of his next assertion that Post-Impressionists are independent of life, the presumable meaning of his epigram about 'standing on its feet instead of leaning upon life'? Mr. Fry chimes in with his explanation of French Post-Impressionists who 'do not seek to imitate form but to create form; not to imitate life but to find an equivalent for life'.

As everybody knows, the function of pictorial art is to transmit to our consciousness the visualised consciousness of others, who are more or less fitted by patient mastery of obvious external appearances to reveal more spiritual qualities. Vital art is that which by communion gains ever more the confidence of life (or nature); academic art is that which rests content with superficial knowledge and falls back upon a dummy life whereby an effect well within the understanding of popular ignorance can be repeated with facility. It is obvious that as visualised consciousness is wholly dependent for sustenance upon perception of external life, so the shutting off of life-perception affects art as starvation affects the body. 'Not leaning upon life', then, and 'creating an equivalent for life' are nothing more than euphemism; on this clause Alma Tadema and Mr. Marcus Stone could be smuggled into the rare heaven we are asked to accept as harbouring the Post-Impressionists. As for Mr. Fry's 'creation' of form (his apology for Picasso's *Head of a Man* being something totally unlike a human head), in the case of a starving body that would be called living upon its own waste, unreplenished matter.

I said something about Mr. Clive Bell's historical inaccuracy, referring to his assertion that English art has been Europe's laughing-stock for two centuries. Surely Mr. Fry might have edited this queer nonsense, for he has a deep knowledge of art history. Wherein may I ask Mr. Bell did English art in 1712 differ from French or German or Italian, save in degree? Does he think Hogarth, Reynolds and Gainsborough ever were ridiculous to foreign eyes; or that Constable and Turner were not vital forces in the development of French art? This feeble special pleading is poor service to a cause. A truer sense of history and a clear-thinking mind would have landed Mr. Bell in a more logical extravaganza; had he recognised the fundamental unity of aim in European art he then could have assured us that Europe had been the laughing-stock of Europe for six hundred years. I don't say that this assurance would better the chances of Post-Impressionism; but it would be logical.

The theory that these pictures are not decorative properties ('not pieces of handsome furniture') seems to me to broach a possible solution, to indicate an easy way out; for indeed they are not pictorial. Pictorial art is clearly conditioned; anything is not a picture, as one might say, nor is everything pictorial. You cannot paint the time or a creed; and coloured charts, house elevations, maps of blood-vessels or mathematical diagrams are not pictorial expression. A reasonable definition of Post-Impressionism is an abstruse science for the propagation of passionately emotional diagrams. Surely we can find a place for this new movement without incorporating it with art. Supposing the special war correspondents at the front took to reporting Balkan battles in the manner of M. Picasso's *Head of a Man*, in purely abstract language, displacing the cardinal facts of the campaign, would their feverish protestations of equivalents for truth, creation of tactics and passionate emotion convince us that their proper job was journalism? No unbigoted person would advocate a suppression of Post-Impressionism; by all means we must support religious toleration (see Mr. Bell's preface). But let it be classed not with art but with conceptualist empiric sciences.

One of these prefaces claims that Post-Impressionist art is contemplative, the work of seers who have gained communion with the mysteries of life. Are we to take this seriously – that a whole batch and school of young painters, most of them comparatively untrained and as yet capable of but a superficial perception of surface appearances, has miraculously and unanimously achieved the plane of contemplation, at which Rembrandt arrived only after half a lifetime's unparalleled 'imitation'? This looks like contemplation learned by rote, based upon fixed academic rules and easily communicated. True art, ever pursuing elusive life, is fluent; a science of cubes and triangles, suspiciously akin to kindergarten formulæ for drawing cows or cats, is fixed and exhaustible. It has taken Mr. Wilson Steer nearly thirty years to see the truths that make wonderful the sky in his *Summer Evening* at the Goupil Gallery. I don't suppose Mr. Steer advertises his artistic religion, his passionate attempts or contemplative vision; unselfconscious contemplation, however, is a more authentic 'proposition'.

81. P.G. Konody, 'Art and Artists: English Post-Impressionists'

Observer, 27 October 1912, 10

For Konody see headnote to no. 45.

'Their debt to the French is enormous' – even Mr. Clive Bell, with all his enthusiastic admiration of the English Post-Impressionists at the Grafton Galleries, is forced to make this admission. Every word of their artistic language is traceable to some French root. There is no eccentricity, no affectation, no mannerism in French that does not find a ready echo is English Post-Impressionist art. And, let it be said at once, like every echo, it is feebler than the original sound. The aims are identical, but the achievement is very often more timid, so that whilst our *fauves* are less likely to provoke derision and violent abuse than their French prototypes, they are at the same time less plastic in design, less emphatic and less exciting. Wherever the English pictures are grouped together on one wall they appear dull and almost colourless compared with the surrounding orgies in primaries.

This generalisation applies, of course, only to the pictures brought together at the Grafton Galleries, for elsewhere – at the Stafford Gallery – is to be seen the work of a group of English and Scottish Post-Impressionists, headed by Mr. Peploe and Mr. Fergusson, who apply the new principles as passionately and fearlessly as their French fellow-workers. It is difficult to understand why no place should have been found in Grafton-street for the very interesting work of Messrs. Peploe and Fergusson, Miss A.E. Rice, Miss Jessie Dómorr and Miss Ethel Wright. With all their faults – and the chief fault is a monotonous liveliness of pattern and bright colour which defeats its own and through lack of contrasting repose – they at least have made up their mind to work in two dimensions, whereas the Grafton Post-Impressionists still frequently waver between the two and the three dimensional.

Mr. Duncan Grant is the most unequal and, perhaps for this very reason, the most interesting of the Grafton band, just as

Mr. Wyndham Lewis is the most monotonously consistent, and, therefore, the most tedious of them all. Perhaps Mr. Lewis realises the hopelessly mechanical aspect of his sternal spheres and geometrical diagrams – Picasso's cubism simplified and 'standardised.' Perhaps it is for this reason, and with a desire to shock other than artistic sensibilities, that he applies his compasses and T-square to the sacred motif of the *Mother and Child*. I have actually seen a shocked parson raise his protesting hands at this 'blasphemy,' but personally do not understand how this calculated piece of geometrical design, which is but remotely connected with art, can arouse anger or emotion of any kind. His painting of *Creation* is cooked after the same recipe, and so are his black and white drawings for *Timon of Athens, The Thebaid*, and *A Feast of Overmen*.

Mr. Duncan Grant, when at his best and most personal, is scarcely a Post-Impressionist. Indeed, his exquisite painting of *Pamela* by a lily-pond, which has all the subtle charm of Vuillard's colour with more lucidity and coherence than are at the French painter's command, is quite out of place in this gathering, and would fit far better into a display of Impressionist art. His *Queen of Sheba* is a charming piece of decoration, a firmly knitted design of lovely colour in which the neo-impressionist system of detached isolated dabs of colour is employed somewhat aggressively. No matter what distance you are from the picture, these touches remain isolated and do not fuse into luminous tomes, so that the effect is more as of a mosaic than of an oil painting. But what is one to say of the same artist's *The Countess*? Here he merely emulates Matisse at his silliest. This is nursery art without the child's ingenious sincerity. The child honestly tries its best. The artist who consciously strives to acquire the child's naiveté dishonestly gives of his worst.

In *The Dancers* Mr. Grant wrestles unsuccessfully with a problem triumphantly solved by Matisse in his large decorative panel in the end room. The comparison is obvious, since motif and general disposition are identical. But Mr. Grant does not arrive anywhere near Matisse's superb rhythm of movement, which is the more telling as it is expressed with the utmost simplicity of synthetic outline and flat colour masses – as flat as the paintings on a Greek vase. Mr. Grant's dancers have no real abandon: they are all posed and stiff. Nor does he fully accept the decorative convention of flat patterning. He wavers between it and three-dimensional realism, and the result is neither the one nor the other.

Intensity of expression is certainly to be found in Mr. Etchell's *The Dead Mole*, although the thing expressed – the evil smell of putrefaction – is scarcely worth expressing. It is a picture that can scarcely be taken seriously, and yet it holds hints of unusual gifts, of which nothing is to be discovered in the same artist's *The Blue Thistle*, with its distorted jar, and bust of a woman whose head and neck have a comical effect of being detachable and only loosely stuck into the garment. Mrs. Bell's *Asheham* belongs to the inlaid linoleum type of Post-Impressionist landscape; whilst in her *Nosegay* she is so bent upon searching for non-existing angles that the flowers look for all the world as if they had been badly cut out of paper. Another picture by the same lady, *The Spanish Model*, which is a good beginning of a sketch, has apparently found favour with the Contemporary Art Society, by whom it has been lent to the exhibition.

In the majority of these pictures the beauty and value of pigment for its own sake are altogether disregarded. Cézanne had a keen sense of it. Van Gogh's frenzied brushwork gave every touch a strangely vital quality. In English Post-Impressionism this appreciation of pigment per se is rare. It enters largely into the work of Mr. Roger Fry, who in *The Terrace* happily applies the new principles – emphatic structural design, pure colour, insistence upon the permanent and elimination of the transitory elements in Nature – with a reverence for the classic tradition of his earlier training.

A curiously fascinating picture is Mr. Spencer Gore's *Letchworth Station*. It is about the last subject that any artist with the old-fashioned sense of the 'picturesque' would have chosen for representation. But this is far more than a representation of uninviting facts. It may or may not be a 'portrait' of Letchworth Railway Station. What is suggests is the silent protest of a lover of the green countryside against the intrusion of unbending iron and black smoke. An almost cruel stress is laid on all that is hard and stiff and graceless, dingy and unpleasant in and around a railway station; everything is concentrated on that 'spiritual significance' that has entered so largely into the jargon of Post-Impressionist criticism.

And finally there is the sculpture of Mr. Eric Gill, primitive

Egyptian in its austere massiveness, broadly simplified planes and lovingly wrought and polished surfaces. I have had occassion to say harsh words about what appeared to be sensationalism and lack of sincerity in certain works shown elsewhere – by the way what has the Contemporary Art Society done with them? – but here Mr. Gill's accomplished carvings afford welcome relief after M. Matisse's indescribable outrages. Mr. Gill is, above all, a stone carver, and he treats his material with respect. His *Garden Statue*, his *Contortionist* (which would make an admirable paper-weight), and *The Poser* cannot possibly give offence, because they are so highly conventionalised that they cannot suggest either a wilfully childish or an incompetent attempt at representing reality. They are frank imitations of archaic Egyptian sculpture, and as such above reproach. The little *Golden Calf* – the sign of the cabaret of that name in Heddon-street – perched high on its square pedestal, is a masterpiece of expressive simplification. It is big in feeling and wrought with the loving care that distinguishes all Mr. Gill's works.

If the contributions sent by the Russian artists are truly representative of the form taken by the Post-Impressionist movement in the Tsar's domains, they can only lead to the conclusion that in Russia, as in all other countries, the revolt against modern realism his led to a new start from archaic forms. And since Russian art in its infancy was Byzantine art, Russian Post-Impressionism is throughout tinged with Byzantinism. Thus Stelletzky's four decorative panels are Byzantine pure and simple. To the same tendency Roerich adds something of that Eastern mysticism which entirely dominates the abstruse and unintelligible art of Chourlianis. Apart from their undeniable decorative effectiveness, these Russian paintings are too far removed from the Western European conception of art to arouse a deeper interest than that of passing curiosity.

82. Roger Fry, 'Art: The Grafton Gallery: an Apologia'

Nation, 9 November 1912, 249–51

Fry focuses his defence of the second Post-Impressionist exhibition on the work of Matisse and Picasso.

However well-fitted to criticise the present exhibition at the Grafton Gallery I may consider myself, I can hardly suppose that my claim to do so would be accepted. This, then, must be taken as a speech for the defence, not a judicial summing up.

The prosecution has had time to develop its ideas with volume and vehemence. There is something admirable in the reckless courage with which a large section of the press has damned the Post-Impressionists. It shows that British Philistinism is as strong and self-confident and as unwilling to learn by past experience as ever it was, and doubtless these are among the characteristics which have made us so proudly and satisfactorily what we are. For in spite of the fact that one or two of those critics whose learning and reputation give them something of a position of leadership have been either favorable, or at least respectful – Sir Claude Phillips, for instance, with a candor and courage worthy of his sincere devotion to art has withdrawn the suggestion of charlatanry made against Matisse – in spite of facts such as these which might give a less expert critic pause, the generality of critics have given vent to their dislike and contempt in unequivocal terms. One gentleman is so put to it to account for his own inability to understand these pictures that he is driven to the conclusion that it is all a colossal hoax on the part of the organisers of the exhibition and myself in particular. However flattering to my powers of persuasion such a theory is, I fear I must decline the honour of going one better than Captain Köpenick.

One feature of the attacks is of peculiar interest. Two years ago Cézanne's works drew down the most violent denunciation. He was 'a butcher who had mistaken his vocation,' a bungler who could never finish a picture, an impostor, he was everything and anything that heated feelings and a rich vocabulary could devise. This year Cézanne is always excepted from abuse. He, at least, is a great master; but whatever advantage might be given by this concession is instantly taken back by the statement that he is not a Post-Impressionist, and has nothing to do with the rest. It is an old and wellworn device, but I doubt if it will do Matisse any more harm than the recent hurried canonisation of Gladstone has done to Mr. Lloyd George.

In any case, as to Cézanne, we are not happily all agreed, and I can only rejoice at the rapidity of the conversion. Perhaps two years more will see Matisse and Picasso on the same pedestal.

Bur however imposing this vigorous attack of the general run of critics may be as a moral spectacle, it shows, I think, a curious intellectual and æsthetic weakness. The exhibition provokes a number of very interesting and difficult questions in Æthetics, and yet no writer for the prosecution has taken the trouble to discuss them or to give reasons for his dislike. Almost without exception, they tacitly assume that the aim of art is imitative representation, yet none of them has tried to show any reason for such a curious proposition. A great deal has been said about these artists searching for the ugly instead of consoling us with beauty. They forget that every new work of creative design is ugly until it becomes beautiful; that we usually apply the word beautiful to those works of art in which familiarity has enabled us to grasp the unity easily, and that we find ugly those works in which we still perceive the unity only by an effort.

Many critics, too, have exaggerated the destructive and negative aspect of this art, affecting to find in it a complete repudiation of all past tradition. I certainly should like to hope that it will be destructive of the great mass of pseudo-art, but it is destructive not by reason of its denials but of its affirmations. By affirming the paramount importance of design, it necessarily places the imitative side of art in a secondary place. And since it is true that the demand for mere imitation and likeness has, in the last five hundred years or so, gradually encroached upon the claims of design, this art appears to be revolutionary. But in its essentials it is in line with the older and longer and more universal tradition, with the art of all countries and periods that has used form for its expressive, not for its descriptive, qualities. So far from this art being lawless and anarchic, it is revolutionary only in the vehemence of its return to the strict laws of design. If it is not too rash to try to coin a single phrase to explain a very varied movement, I should say that it is marked by the desire for organic unity in a work of art, as opposed to that search for casual and factual unity which attempted, but unsuccessfully, to satisfy the public of the last century.

There is much work of immature or minor artists in the Grafton Gallery, work which has, I think, great promise for the future, but I must confine myself in the present article to the work of two men who stand out at the present moment as leaders – Henri-Matisse and Picasso. No sharper contrast can be imagined than exists between these two men, and it is, indeed, one of the hopeful signs of the present movement that it allows of such striking diversity. It is easier to speak of Matisse, for he has achieved something like a definitive form, something complete in its way, and his whole development has proceeded by such clear and logically related steps that one need not forecast any striking or bewildering change in his methods. He is indeed a singularly precise and methodical artist – one whose intelligence keeps pace with his sensibility, making clear to him at each point the next position to be gained. There is absolutely nothing fantastical or whimsical about Matisse, nothing, when once one has seized his method of expression, that is bewildering or disconcerting. All proceeds by singularly clear and deliberate steps towards a definite end. As an illustration of his method the four busts of a woman are peculiarly instructive. In the first state he has rendered the head more or less naturalistically. In each successive state he has amplified the forms, working always towards a more complete and inevitable plastic unity; one in which the relations, only dimly apprehended in the first study, become entirely explicit. The final result is from a purely descriptive point of view monstrous and repellent. I mean that, if taken as a likeness of an actual woman, we should speak so of the model from which it had been copied, but judged as pure form it has an intensity, a compactness, an inevitability which gives it the same kind of reality as life itself.

In his painting, no less than in his sculpture, we find the idea of equivalence by means of amplification. In order that each form may have its full significance in the whole, may hold its own in the equilibrium of all the forms, it must be as ample and as simple as possible. It is because he has followed out this scheme so fearlessly that his designs have their singular compelling power. This is particularly observable in the great decoration of the *Dance* where the rhythm is at once so persuasive and so intense that figures that pass in front of it seem to become part of the rhythmic whole. The rhythm passes out of the picture and imposes itself on its surroundings. Matisse has himself noticed this, and again and again in subsequent compositions, parts of the *Dance* are woven into the background; the pattern of actual things and the pattern of the painted figures fusing to form a new synthesis.

Matisse is essentially a realistic painter: that is to say, his design is not the result of invention, but almost always comes out of some definite thing seen. In a sense he is always trying to make his works like the thing seen, but not like in the literal sense. What he does is to draw out to its last and final explicitness the effect of things seen upon his sensibility. In his earlier work he still modelled with light and shade, and how vigorously one may judge from the *Pose du Nu* (No. 9), but he soon found that color and line expressed more clearly his conception. He began then to translate contrasts of shade into contrasts of pure color, as in *Le Madras Rouge* (No. 31). Finally, he has learnt to dispense even with this, and to give to each figure its volume and mass without any perceptible color contrasts. Thus in his latest work the same identical color may be used to express a number of distinct planes and different objects, and that without confusion or loss or spacial definition, thanks to the increased amplitude and simplicity of the design. This use of pure flat masses of color without degradations or transitions enables him to give to color a purity and force which has scarcely ever been equalled in European art except by some of the French glass designers of the thirteenth century.

Matisse's art is singularly aloof, singularly withdrawn from the immediate issues and passions of life. Even in his *Dance* there is nothing Dionysiac, but rather a perfect equilibrium of motion. In his *Conversation* there is not dramatic tension. This commonplace event is seen with epic generalisation. It becomes placid, monumental, and sedate, like some early Assyrian sculpture. At first sight it is grotesque. That is because of our inveterate habit of translating images back into life instead of regarding them simply and passively. In looking at early art we have learned this passive attention because the fact of translation is difficult to us; we know too little of the actual life which gave rise to the image. It needs some familiarity with such a decoration as the *Conversation* to do this, but when once it is done, the strange impressiveness of the design, the perfect rightness of the relations, becomes apparent, and in the end one is inclined to agree with Matisse that the mood his art inspires is one of serenity and repose.

It is dangerous and difficult to speak of Picasso, for he is changing with kaleidoscopic rapidity. There have been moments when his art seemed stable, when he seemed to have established a definite form, but instantly the balance has broken down and a new conception has begun to emerge. It is difficult then to judge of his achievement, though it is easy to show the fertility of his work in its influence on other, less restless, less adventurous spirits. He is the most gifted, the most incredibly facile of modern artists. No *tour de force* of imitative art would have been difficult to such an eye and hand. But the very facility that might have made him the darling of the Academies has stood in his way in the line of advance he has chosen. Again and again he seems to have dreaded its effect on him, and to have deliberately countered it by adopting some more abstract and unrepresentative idea of form.

Because the latest developments of Picasso's art have in this way come to take on geometric form, some have supposed him to be a pure theorist, working out abstract intellectual problems of design. This seems to me to be a mistaken view; one has only to look at the quality of his work, to mark the nervous sensitiveness and delicacy, the rare distinction of his touch, to see that his sensibility is his most salient characteristic. And he is unlike Matisse in that this sensibility is not controlled by a clear and methodical intellect. Hence his continual experimenting. He reaches out in any direction which his instinct dictates for the possible expression of his sensibility to actual objects. As I have said, his art is rarely complete; generally it seems to be in labor with a new idea, and almost always before the new idea has been completely realised, another possibility is beginning to dawn. It may be doubted if such is the character of the greatest artists, but it is typical of great originators and inventors. In Picasso's early work there is more than a trace of sentimentality. As though conscious of this danger he threw aside all those means by which the associated ideas of a picture may interfere with perceptions of pure form, gradually reducing his shapes to a geometrical abstract. But the quality of his temperament comes through, even in such pieces as the *Têtes de Femmes* (Nos. 64 and 66). In the still-life pieces of this period (Nos. 60 and 63) Picasso seems to me to come nearest to complete realisation. He has the power of building up out of the simplest objects designs of compelling unity and precision. And more than in his other work we feel in these the concentrated passion, the almost tragic intensity of his mood. For all the remoteness from natural form, the abstract and musical quality of his designs, Picasso's temperament is less serenely remote than Matisse's. It has the gloomy force and intensity of the Spanish genius.

As to the latest works of all, those in which Picasso frankly abandons all direct reference to natural appearance, I confess that I take them to some extent on trust, a trust which is surely justified by his previous work. They certainly have the beauty of intensely organised wholes, but I apprehend the unity almost dispassionately and intellectually. I find that they move me only by the charm and distinction which is inalienable from everything that Picasso does. The idea seems to me intelligible enough, namely, the construction of a fugal arrangement of forms out of the elements given in any natural object, without taking those elements in the same order or relation that they have in actual life. It is, of course, possible that we are not yet sufficiently accustomed to interpreting the meaning of such purely abstract form for us to feel to the full the effect of these compositions. It is also possible that they are but an intermediate stage on the way to a clearer and more explicit form.

83. Frank Rutter, 'An Art Causerie'

Sunday Times, 10 November 1912, 19

Rutter continues to urge his revolutionary view of Post-Impressionism. For Rutter see headnote to no. 1.

POST-IMPRESSIONISM OR —

Whether we approve or disapprove, whether or no we understand what is meant by the term, there can be no denying that public interest in the work of the so-called 'post-impressionist' painters is exceedingly keen and widespread. During the last few months I have lectured among various districts and classes in the West Riding of Yorkshire, but wherever I have been, of whomsoever my audience was composed, I have rarely escaped the question, 'What do you think of Post-Impressionism?' Controversy in the Press and the exhibitions of the Contemporary Art Society at Leeds, Bradford, and elsewhere seem to have stirred up the whole country to an unusual pitch of excitement about paintings which few have seen and fewer still can understand. Indeed, my own experience in the provinces is that everybody wants to know what post-impressionism is and means, but nobody is quite certain who is a post-impressionist painter and who not. When I have been asked if Mr. Wilson Steer is a post-impressionist I have ventured to reply in the negative, but when the same question has been asked me about Mr. Sickert I have evaded the real issue by saying that it all depended on what you meant by post-impressionism. To my thinking, the collection of works now on view at the Grafton Galleries avails us little in solving this problem. If, as the title suggests, all the exhibitors are post-impressionist painters, then I ask myself how it is that Lamb and Spencer Gore are post-impressionists when John, Sickert, and Ginner apparently are not. Under the classification adopted by the organisers of the Grafton Galleries exhibition the Camden Town Group is rent asunder, and nobody but Mr. Roger Fry and Mr. Clive Bell can tell who is a post-impressionist and who not.

NEO-GOTHIC ART

The utter confusion which at present exists owing to the coining of the word 'post-impressionist' by a few English writers might be partially cleared up if the public could be brought to realise that the term as used in England covers some half-a-dozen distinct and separate art movements which in France are given separate names. In the last fifty years the pseudo-classic academic art of Europe has been placed in a position similar to that occupied by Rome at the beginning of the fifth century. It has been overwhelmed by wave after wave of fresh, young, vigorous invaders; and if we accept the parallel we may liken the impressionist movement led by Manet and Monet to the first Visigothic invasion, the neo-impressionist movement led by Seurat and Signac to the second Visigothic invasion, and after these come the Huns, the Ostrogoths, and the Lombards in the form of the Fauves, the Cubistes, and the Futurists. Under the circumstances, I think the phrase 'Neo-Gothic' art would be far more appropriate and more intelligible than the term 'post-impressionism.' My only doubt is whether the Lombards have yet arrived on the scene. The complicated events of the fifth century in Italy would not be made easier to understand if historians jumbled up together the Visigoths, the Huns, and the Ostrogoths. This, I am afraid, is what has been done at both the 'post-impressionist' exhibitions held in the Grafton Galleries. Those who during the past ten years have been patiently endeavouring to follow waves of invaders and to discriminate between them will not be surprised at the confusion that has resulted.

AN HISTORICAL EXHIBITION

I can imagine an exhibition which might have shown the course of events to the public without confusing it. The first room would have shown the older impressionists, with the exception of Cézanne and Van Gogh, whom history has decided to tear from this group and use as links with a later generation. The second room would show the neo-impressionist movement represented by Seurat, Signac, Théo van Rysselburg, etc., all of whose art was based on luminist ideals. The third room would only contain works of Cézanne, Van Gogh, and Gauguin, while a note in the catalogue would call attention to their connection with the earlier impressionists. If the first room contained Manet's *Zacharie Astruc* the connection would hardly need explaining. The fourth room, which would need to be a large gallery, would be devoted to the Fauves, grouped round their chief brigadiers, Matisse, Van Dongen, Othon Friesz, and De Vlaminck. Their connection with the Cézanne room would be obvious, the ferocity of their work would explain why the painters were nicknamed the 'wild beasts,' and the difference between their aim and that of the neo-impressionists could hardly fail to be recognised. Then would follow a small room devoted to Picasso, who is a tribe of invaders in himself, and the last two rooms would be occupied by those who have stumbled after him, the Cubistes and the Futurists. It will be noted that I have found no place for Bonnard, who is so strangely included in the present Grafton Galleries exhibition. If, however, it is thought advisable that the *intimiste* movement should be represented, then Bonnard, Vuillard, and others might be found a small room somewhere between the neo-impressionists and the Fauves. Artists like Braque, who have suddenly turned a somer-sault in their practice, might, if included at all, be represented in each of the movements to which they have successively adhered. I do not suppose an exhibition on the lines I have sketched would be much more popular than the collection in Grafton-street, but at least it would be shown systematically and avoid creating a confusion of thought as unnecessary as it is undesirable.

A FOREGONE CONCLUSION

That all these different tribes of invaders, so confusingly lumped together as 'post-impressionists,' should have spread dismay among the old-fashioned academic painters and raised the ire of their admirers is perfectly natural. What Roman could welcome the Gothic invaders of Italy in the fifth century? To him the coming of these barbarians must have seemed the end of all things, and how could he be expected to foresee the masterpieces of Gothic art which these barbarians were one day destined to achieve? We, knowing that history repeats itself, may exercise more caution in approaching the barbaric invasions of nineteenth-century art. Some of these invasions may prove to be mere raids, like that of the Vandals in 455, and pass away without leaving any lasting results. But signs are not wanting that the Fauves, the strongest of all these tribes of modern art, are destined to make a permanent settlement, and perhaps it is to these we shall look for an equivalent to the Gothic cathedrals. At all events, their work has many Gothic characteristics in its rudeness, its strength, and its summary expressiveness. So far it does not show the grace that we associate with classic art, but as the descendants of the 'wild beasts' of yesterday and to-day become tamer and more domesticated they may absorb grace from their defeated foes as did the Goths in Italy.

84. Anthony M. Ludovici, 'Art: the Pot-Boiler Paramount'

New Age, 21 November 1912, 66–7

Ludovici attacks the fashion for Post-Impressionist technique as the proliferation of mediocrity. For Ludovici see the headnote to no. 61.

Some years ago there were two kinds of pot-boiler in the picture world. There was the pot-boiler painted by the truly tasteful but impecunious artist who could do better things, but who was compelled at least once per annum to lay aside his more inspired work and to paint a picture for merely trade purposes; and there was the pot-boiler which was produced, not in a moment of cupidity or of lust for mere gain, but normally, continually, perpetually, by the kind of painter who could not rise above the pot-boiling standard. The producer of the first kind of pot-boiler was generally a very gifted and very estimable fellow, who honestly admitted that his pot-boilers were wretched stuff, but who pleaded poverty as an excuse for his annual, or sometimes biennial, deflections from the path of high art. The producer of the second kind of pot-boiler was merely a variety of the ordinary, honest craftsman who produced his picture just as his fellow craftsman produced brown and black boots, and who, while being aware of his limitations in art, still plumed himself on possessing some taste and higher culture, because he had chosen the palette rather than the last.

No tasteful purchaser of pictures was deceived by either of these two kinds of pot-boilers, and they both went either into the channels of real trade in the form of advertisements, almanacs, covers of chocolate boxes, etc., or were hung in some bourgeois home where boots and pictures were purchased in accordance with the same utilitarian point of view. A boot had to fit, a picture had to tell an obvious tale of interest, or, better still, of sweet sentiment, which could be understood immediately by all.

Since the good old days when these two pot-boilers reigned supreme, however, many changes have come over the world of pictorial art. Started by earnest and gifted pioneers, movements have been set on foot which have proved as revolutionary as they were unprecedented. Extraordinary, original techniques have come into being, like those of the pointillistes (Monet) and a-chiaroscurists (Manet), each of which has been taken up by hosts of admiring followers or mere imitators. Nevertheless, in the early days of these movements, the person of modest powers was still unable to pretend that he could go very far beyond the ordinary recognised pot-boiling standard, because some serious schooling and great original gifts continued to be required in order to produce the Monet and Manet type of picture. To have the pretensions of genius in Monet's or Manet's style was extremely difficult.

Soon, however, the sharp line of demarcation between the genuine painter and the mere painter of pot-boilers was to be bridged, at least so far as a certain portion of the public was concerned. And with the assistance of a fair modicum of blindness, both among the people who are devotees of art and the critics, a curious thing happened in the art world. A sort of vanishing trick was performed under the very eyes of all those in whose best interest it would have been to allow nothing of the sort to happen.

The mere pot-boiler, the painter of pot-boilers vanished! By a curious trick of sleight of hand, he was merged into the exalted company of the painters whose work was high art. Or, if you would like it put in another way, a new kind of pot-boiler was discovered for the pot-boiler painter. He could now paint masterpieces! – misunderstood strokes of genius!

Tricks, moments of carelessness, moments of depression, are easily emulated. Gauguin and Van Gogh – to mention the greatest of the Post-Impressionists – like all men, had little knacks, moments of carelessness, and moments of depression which could be imitated with ease. Very quickly, therefore, the worst examples of their work became a sort of canon for a legion of mediocre people who saw fame, or at least a higher level of appreciation than mere pot-boiling would bring, if only they could imitate, not the highest achievements, but rather the vagaries of genius.

Nonentity after nonentity arose, who could now scornfully laugh at the career of painting pot-boilers, and could conceal his incompetence and vulgarity beneath a deceiving mantle composed of the ostensible eccentricities of great minds.

With the appearance of the Futurists, this exulting band of 'emancipated' painters of pot-boilers saw yet another chance of ascending the ladder to 'high artistic achievement' without possessing the necessary gifts thereto, and very quickly the market was flooded with the 'inspired' work of a legion of Post-Impressionists and Futurists, whose true business thirty years ago would have been the trade pot-boiler, provided of course that they had been able to reach even that standard – a question which gives rise to a good deal of doubt. Before long, perhaps, a still further reduction in the standard of what is supposed to be a great painting will enable every man Jack of us to be 'artists' and the producers of masterpieces, and then 'art' will be general and we shall all feel what a great age is ours.

These are some of the thoughts that came to me on my second visit to the present Post-Impressionist Exhibition at the Grafton Galleries, and the more I studied the exhibits the more convinced I became that the vanishing trick above described had actually been performed.

It must not be supposed, however, that the man who knows will be actually taken in by this feat of legerdemain, although I cannot help thinking that large numbers of the public are. The very colours these people use in their work are in most cases self-revelatory and betraying. Look at Marchand's *Nature Morte* (No. 10), for instance, or Derain's *Le Rideau* and *La Forêt* (Nos. 11 and 12) or De Vlaminck's *Les Figues* (No. 17). These men are in my opinion heralds of the decay and dissolution of art, and their colour is the colour of decomposed tissues and of putrefying corpses.

I will not enter into the subject of the content of their canvases, because from that point of view there would scarcely be a single picture worth saving in the whole exhibition. But even from the point of view of manner alone, how few could one choose, and how small would be the reward of one's search in the end! Cézanne's *Le Dauphin* (No. 4); Chabaud's *Chemin dans la Montagnette* (No. 41) for its design; Van Dongen's *Portrait de Madame Dongen* (No. 43); Marquet's *Le nue à contre-jour* (No. 55); Flandrin's *Porte de la Cuisine* (No. 57); Asselin's *Anticoli* (No. 80); Grant's *Pamela* (No. 10); Mrs. Bell's *Nosegay* (No. 109); Fry's *Angles sur Langlin* (No. 120); Flandrin's *Pivoines* (No. 158); at a pinch one might have been tempted to carry off one of these as trophies; but, in a show of 242 pictures, the number is small, not more than four per cent., and for the life of me I could not add to it.

This is the heyday of the mediocre person. Let him profit while he may from the confusion and doubt that prevail about him. But do not let him try to convince us that his work is anything more than the pot-boiler paramount.

85. Rupert Brooke,
Cambridge Magazine

23 November 1912, ii, 125–6 and 30 November 1912, ii, 158–9

Brooke (1887–1915), who had been educated at King's College, Cambridge, published his *Poems* in 1911 and died as a commissioned officer on the Greek island of Scyros in 1915.

Two years ago, such English people as look at pictures were startled and shocked by an exhibition of 'Post-Impressionist' art in London. Names that have since grown familiar – Cézanne, Van Gogh, Gauguin – began to crop up in conversation and discussion. Some praised the novelties, some wondered, many laughed. In France, Germany, and Russia, of course, these pictures had been known for many years. We get things late in England. But by now the first shock, even with us, has worn off. And the 'Second Post-Impressionist Exhibition' in the Grafton Galleries, off Bond Street, finds a more critical, and less shockedly hostile, public. It is almost unnecessary in the *Cambridge Magazine* to say that everybody who is interested in pictures must go to this exhibition. Most people of that kind who read this page will already have been. For those who have not been, and those who are not certain if they should go, I want in this article to suggest shortly what the 'new movement' in art is, what this exhibition is, and what its merits and faults are. And I must defend myself beforehand by saying that I am neither painter nor art-critic, but write as an ignorant and common-place person, with a mild liking for good pictures.

In the first place, 'Post-Impressionist' is rather a silly name. It has the negative advantage of covering a great many different schools and tendencies – all, in fact, that come after the Impressionists. But if in the various currents of modern art there *is* one general stream – and there probably is – this name does not help to recognise it. In France the modernists in art are usually known as *Les Fauves*, in Germany as *Die Wilden* or *Die Expressionisten*. *Expressionism* is, on the whole, the best name that has been found. It will probably spread. It recognises what is, roughly, the main reason of this modern art – a very sensible one – namely, that the *chief* object of a good picture is to convey the expression of an emotion of the artist, and *not*, as most people have been supposing, his impression of something he sees. In other words, the goodness of a good picture does not consist in its resemblance to 'nature.' It is not true that the better a picture is, the more like reality it is. It is not true that the less like reality a picture is, the worse it is. Giotto is better than Andrea del Sarto. Passion before Perspective.

This is, of course, the merest common sense. But the feature of this new movement that has swept in most of the best artists of modern Europe, and has begun to touch England, is that it has begun to act on these common-sense truths. I have tried to give the simplest theoretical essence of the 'Expressionists.' Their theories are pushed much further in the most divergent directions. Some declare their object to be to portray 'the treeness of the tree.' Some talk only of 'design.' Some would bring reality to you through the brain, reducing it to geometrical shapes. And, of course, the general aesthetic tone of most of these new pictures (for there is one, in the same way as there is in Renaissance sculpture, or Elizabethan plays, or, I am told, Russian literature), which is the important thing and necessarily the justification of the whole movement, would take far more space to explain than there is room for here.

The present exhibition professes to represent the 'Post-Impressionists' of three countries – France, Russia, and England – and contains only the work of living artists, except for a few pictures by Cézanne. It is a pity that the committee could not have included works by, at any rate, Erbslöh, Jawlensky, and Kandinsky of Munich, Pechstein of Berlin, and Kokoschka of Vienna, who paint pictures at least as good and as interesting as most of those here. To dwell further on this would be ingratitude to the energy and love of art that have given us the chance of seeing such an exciting and lovely exhibition. But one grumble must be permitted, at the air of slight incompetency that hangs over the whole exhibition. When I last visited it, for instance, on November 5th, it had been open just a month. And still some of the best Russian pictures, referred to in the preface, were not hung. And look at the catalogue! For the immense sum of a shilling you get three little prefaces, a list of the pictures and owners, and an index, which occasionally gives the date and place of the artist's birth; sometimes, as with Mr. Gill, merely mentions the heartening fact that he *has* been born; and sometimes gives no information at all. A decent catalogue would have contained small reproductions of some of the better pictures, and, most certainly, the date of painting of each exhibit. The latter omission is particularly scandalous in the case of Picasso.

The three prefaces, by Mr. Clive Bell on 'The English Group,' by Mr. Roger Fry on 'The French Group,' and by M. von Anrep on 'The Russian Group,' are worth reading. M. von Anrep gives information, the other two write mostly on theory. Mr. Fry is the more helpful. But the innocent spectator will do best if he approaches the pictures with as willing and unsuspicious a mind as possible. Modern art can make its own defence against anything except prejudice.

The collection is rather a hotch-potch. Some of the pictures ought rather to be in the Royal Academy, some in the New English Art Club, a few in the muck-heap. There are, for instance, some sickening soapy pictures by a M. van Dongen which would disgrace even the New English Art Club. I shall only touch briefly on some of the artists who, being both good and 'Post-Impressionist,' are rightly in this show. To begin with there are the French, headed by Matisse. The great glory of this exhibition is that it gives us at length a chance of judging and appreciating Matisse. Some twenty pictures and nearly as many drawings. The pure bright and generally light colour, and the stern simplicity and unity of design, fascinate the beholder. Look at *Les Capucines, Les poissons rouges, Coucous sur le tapis bleu et rose*, and the great *La danse*. There are moments in the life of most of us when some sight suddenly takes on an inexplicable and overwhelming importance – a group of objects, a figure or two, a gesture, seem in their light and position and colour to be seen in naked reality, through some rent in the grotesque veil of accidental form and hue – for a passing minute. Matisse seems to move among such realities; but lightly and dispassionately. He is entirely, almost too purposefully, free from the emotions they bring. His world is clean, lovely and inhuman as a douche of cold water. He paints dancing; and it is the essential rhythm of dance that, with a careless precision, he gets. God, certainly, does not paint like Matisse; but it is probable that the archangels do.

Picasso, the other most famous modern master represented, is very different and distinctly inferior. His method opens – as far as one can judge at present – but a narrow field. It is he who extracts from the real object he paints a pattern of lines, angles, whorls, surfaces, curves, and shadings, which is unintelligible to the spectator who would connect it with the reality. You can see his progress in this logicalisation of the real world, from pictures with an obvious relation to the object portrayed, to arrangements of lines which seem to have no connection at all with those human or material forms, the names of which supply their titles. The pictures he painted *during* this progress are unsuccessful. They tend to have an uncomfortable, unhelpful similarity with the real. But his journey has landed him in an absolute art which grows better as it disentangles itself from that unholy hybrid condition. Unfortunately Picasso seems not to care about colour. His paintings have a drab ugliness about them for this reason; and so his merits can best be seen in the two little drawings in the end gallery. 'Patterns,' they are, if you like; but patterns, queerly, in three dimensions; lovely and self-sufficient and inexplicable as a fugue. They do not 'represent' anything. Why should they? But they express that slight, real, half-romantic pathos – almost verging on sentimentality – which pervades all Picasso's work. A 'minor' artist . . . .

Several other good painters of the younger generation in France are well represented here. There are certainly a dozen to twenty pictures of exhilarating strength and beauty. One, perhaps the best picture in the exhibition, certainly the most excitingly beautiful, is by M. Simon Bussy; his only exhibit. A fortnight ago it was not hung. When it is it will be worth a trip to London to see that picture alone. But M. Bussy is a school to himself, and stands outside the ordinary modern movement. Cross and Signac, who are as good as most of these artists, are – perhaps justifiably – not represented here. But you find the gloomy passion of Derain, some admirable still-life by Herbin, and attractive, earnest work by Flandrin, Vlaminck, and Girieud – though the latter is not very well represented. L'Hote and Braque (when he is not following Picasso) have a lovely light way of treating landscape. It is, indeed, in landscape that these younger French painters achieve their greatest successes. They paint it with an individual simplifying inspiration that starts originally from Cézanne. There is, alas, only one picture by the extraordinary and gifted Rousseau.

The Russians in this exhibition have hitherto been unknown to England; almost, I should think, to Western Europe. They have almost no connection with the French, German and English 'Post-Impressionists.' In his preface, indeed, M. von Anrep especially disclaims Western influence. Their work is difficult to appreciate at first. It is heavy with soul, packed with a religious romanticism. The first attitude an ignorant Western mind takes up towards it is one of suspicious awe. The pictures seem to vary between pomposity and real mysticism, and to be of a more 'literary' nature than the French and English exhibits. At first sight M. Stelletsky's seem the most convincing. Even his Byzantinism is affected by the desperate and gloomy heat of unquiet that tinges all these Russian pictures.

One turns to the English section with the most lively interest. Can we hold our own yet in modern art? The answer is 'No!' But it is certainly refreshing to see that 'advanced' English art has moved beyond the stage of a few years ago, when a simple recipe for producing a picture throbbing with 'lyrical beauty' was to depict a human figure (preferably female) with one or both arms uplifted in unusual attitudes. There is not much cohesion about the English Post-Impressionists – perhaps it is as well. A not inconsiderable section of them is not represented here. Of those that are, Mr. Stanley Spenser exhibits the most remarkable picture – especially remarkable as it was painted in his eighteenth year – *John Donne arriving in Heaven*. It has a passion of design and form sadly absent from many of the others; a crude and moving nobility. Mr. Wyndham Lewis is more or less alone in resembling Picasso in method. But he gives an angular geometrical representation of reality which has an unexpected amount of emotional appeal – of a different kind from Picasso's, stern and rather simple. His *Mother and child*, and some of his black and white work, are powerful.

Messrs. Fry, Grant, and Etchells form more of a group. Mr. Fry has the advantage of knowing what a good picture is far better than the others. He aims at a pure, neither grandiose nor twisted, beauty; and achieves it with a delightful certainty. There are several pictures in this exhibition which may or may not be distinctly better than his, but few that are so certainly and definitely beautiful pictures as at least two by him. Mr. Grant and

Mr. Etchells, being younger, are declared to show more 'promise.' Mr. Grant has painted better things, perhaps, than any he shows here. But several of these are lovely. He is always a trifle disappointing. One always feels there ought to be more body in his work, somehow. Even his best pictures here are rather thin. But there is beauty in *The Seated Woman*, an exquisite wit and invention in the delightful *The Queen of Sheba*, and grave loveliness in *The Dancers*. His genius is an elusive and faithless sprite. He may do anything or nothing. Also, he is roaming at present between different styles and methods. What an eye for beauty! Why aren't his pictures better? But it's absurd to suppose they won't be when he has 'found himself.' Both he and Mr. Etchells are unfortunately fond of a spotty way of laying their paint on, which, as they employ it, seems to the inexpert eye not to advantage them. Mr. Etchells' danger also lies in a lack of fervour; but he inclines to stolidity, Mr. Grant to prettiness. It is a pity Mr. Etchells is largely represented by landscape, as his portraits are better. His *The Dead Mole* has honesty and some power.

There is also a small amount of sculpture in the exhibition. Some by Matisse, which verges on the commonplace. It is a pity there is nothing by Lehmbruck; but it is a pleasure to find several things of Mr. Gill's. Mr. Gill is as certainly better than any Continental 'Post-Impressionist' sculptor, as the Continental 'Post-Impressionist' painters are better than ours. His genius is not represented by any of its greatest triumphs, except, perhaps, the *Garden Statue*. But there is enough here for even those unacquainted with his work to begin to suspect that he is the greatest living English artist, and one of the three or four great sculptors of the past hundred years.

86. Virginia Woolf, letter to Violet Dickinson

24 December 1912

Virginia Woolf (1882–1941) took a rather sceptical view of her friends' enthusiasm for Post-Impressionism. In this letter she refers to an idea of Fry's which later developed into the Omega Workshops.

The Grafton, thank God, is over; artists are an abominable race. The furious excitement of these people over their pieces of canvas coloured green and blue, is odious. Roger is now turning them upon chairs and tables: there's to be a shop and a warehouse next month.

87. Unsigned review (probably by Robert Ross), 'Cézanne and the Post-Impressionists'

The Times, 8 January 1913, 10

The composition of the second Post-Impressionist exhibition was changed somewhat at the beginning of 1913 by the addition of thirty watercolours by Cézanne.

A number of pictures have been added to the Post-Impressionist Exhibition in the Grafton Galleries, and among them about 30 watercolours by Cézanne. Many of these are slight and perhaps mere projects; but they are all of great interest both for their own sake and because of their relation to the later Post-Impressionist movement. In them, more clearly, perhaps, than in most of Cézanne's oil paintings, one can see the new tendency which he brought into painting and which distinguishes his art so sharply from that of the Impressionists.

Like the Impressionists, Cézanne was constantly occupied with reality and tried to see it without any artistic prejudice; but he also incessantly tried to discover in it designs of abstract grandeur to which he would sacrifice all irrelevant fact, as the Impressionists sacrificed irrelevant fact to the facts which interested them. Both he and they maintained the freedom of the artist; but he made a different use of it. He was not so much interested in any fact for its own sake as in a kind of music for the eye which he sought for in all facts and to which he subordinated them all. In this respect he was like many artists whom we call decorative; but he differed also from most of these in one important particular. The tendency of most decorative painting is to reduce everything to two dimensions. This is carried to its furthest point in pure pattern, as in a

Persian carpet, where there is no representation at all, but only an abstract music for the eye.

But Cézanne's designs, however abstract, were always conceived in three dimensions; his music was a music of masses, not of lines or flat spaces. That is what makes his art original and at the same time difficult. For we are used to think of all the means by which mass is represented as means of pure illusion. We can see a harmony of pattern easily enough, because we are used to the sacrifices of fact necessary to produce it. But we are not used either to harmonies of mass or to the sacrifices of fact necessary to produce them; and so pictures like Cézanne's *Rocks* (7 and 9) may seem to us both meaningless and unbeautiful. They do not remind us either of rocks themselves or of a Persian carpet; they have neither illusion nor music for the eye. But Cézanne had learnt to compose in abstract masses as Cimabue could compose in lines and flat spaces, and he drew these masses without any design of producing an illusion of three dimensions, but only so as to reveal the new kind of music which he found in them. If, then, we are to understand his works and to see their beauty, we must not look for flat pattern in them nor must we look for the kind of abstract design that goes with flat pattern. We must accustom ourselves to abstractions in three dimensions, to a new music of masses, which at first is very disconcerting to the eye.

It is disconcerting, because, in a work like *Le Chemin Tournant* (21), it seems to be merely inadequate illusion. We are not used to abstract grandeur of design in landscapes of this kind, but only to an illusion of pleasant realities. Cézanne does not aim at that illusion at all, but he happens, in insisting upon his new kind of design, to use some of the means which we associate with an old illusion. He thus leads us to expect more of it than we get, and we are tantalized rather than satisfied. But if we can rid ourselves of this expectation of illusion we shall see the grandeur of design which Cézanne has introduced into modern painting, and with which he has invested every kind of object. For in modern art there has been a growing divorce between beauty of design and the illusion of three dimensions very harmful to all kinds of painting. It has made illusive painting merely illusive and decorative painting merely decorative. Cézanne, like another Giotto, found a new music in masses, and he has, perhaps, given a new start to European painting. At any rate, he has inspired nearly all the most interesting works of the Post-Impressionist movement: and if we study his water-colours we shall be able to understand them. Then we shall see that the Cubists, whether they succeed or fail, are not mere charlatans, but are aiming at something intelligible. They, too, are trying to design in masses; and their sharp distinction of planes, though it often becomes a mechanical device, is an insistence upon mass just as legitimate as a rhythmical artist's insistence upon outline.

We can see the beginning of Cubism in Cézanne; and we can see it more consciously but not mechanically employed in several works newly hung in the Gallery; as, for instance in the *Allée d'Arbres* (88) of M. Marchand. Here the masses of foliage are simplified into sharply contrasted planes very agreeable to the eye: but there is some incongruity between the convention here employed and the illusive treatment of the ground, and especially of the shadow. M. Marchand's other new work, *Baigneuses* (86), is more completely successful. It is, indeed, a new kind of romantic picture, not great, but very amusing and delightful. M. Picart le Doux in *La Fête* and *Le Village* (41 and 47) makes some sacrifice of mass to gaiety of colour; but both pictures are very easy to enjoy. M. Thiesson's *Paysage aux Pommiers* (152) is a quiet but distinguished little work. Mr. Adeney's *London Square* and Mr. Wadsworth's *Viaduct* (122 and 182) are novel only in their unaffected simplicity; and Mr. Gore's *Cinder Path* (116) is very accomplished.

88. P.G. Konody, 'Art and Artists: More Post-Impressionists'

Observer, 19 January 1913, 9

For Konody see headnote to no. 45.

The new and revised edition of the Second Post-Impressionist Exhibition at the Grafton Galleries is a very distinct improvement upon its predecessor, in so far as quite a number of the very debatable eccentricities by Matisse and Picasso have been removed and replaced by more desirable things, among which a very comprehensive collection of water-colour drawings by Cézanne lent by M.M. Bernheim-Jeune and Co., will prove of supreme interest to all serious students. To many, indeed, they will be a perfect revelation that is likely to dispel certain misconceptions concerning the master's art which have gained wide currency.

There is about many of his paintings, and especially about the often mediocre examples shown in London during recent years, a curious air of an unsuccessful struggle with material difficulties which at times almost savours of incompetence. They impressed one as the work of a passionate amateur who had never mastered the language of his craft, and who often remained inarticulate with all his craving for intensified expression – an artist who would attempt to run and jump before he had learnt to walk. The last thing they would have led one to expect him to have been is a patient, painstaking, methodical student of nature who would go out day by day with a sketchbook to record his impressions in his own personal way, and yet at times with a precision that would have delighted Ruskin himself – his impressions of Nature's vastness as well as of her details.

It is just this side of Cézanne's great personality that is revealed by the sketches and studies at the Grafton Galleries. It is difficult to describe or to explain their haunting beauty and extraordinary fascination. They are utterly unlike any other artist's water-colours, although superficially some of them, like *Le Réflet dans l'eau*, recall the delicate caligraphic method of the early Chinese painters. In these drawings may be found the explanation of the reverence with which Cézanne is regarded by the present generation of French artists, for they are the nearest approach to the realisation of a new artistic aim which is but imperfectly realised in most of his finished paintings.

Western painting has hitherto been mainly occupied with finding a perfect method of transferring the reflection of Nature from the retina of the human eye on to canvas or paper. The picture on the retina is, of course, flat or two-dimensional; and it is a well-known fact that to the infant, until it has learnt by the constant use of its fingers to associate certain surface appearances, such as the light and shade of objects, with depth and distance, everything appears to be two-dimensional only. European art has been wedded to the representation of these surface appearances and to the creation of an illusion of depth by means of light and shade; whereas Eastern art, contemptuous of 'mere representatism,' and aiming at the expression of the spiritual significance of all objects in Nature, has eliminated light and shade altogether, and has arrived at a decorative convention of flat pattern and caligraphic line that is peculiarly favourable to concentration on the qualities aimed at.

Cézanne's drawings constitute, as it were, a link between the Western and the Eastern conception of art. They are neither an exact representation of facts nor are they designed in pursuance of the Eastern ideal. The Chinese and Japanese artist, it must be understood, does not work direct from Nature. He is taught to use his eyes and to observe; but his actual work is done from memory. And since the memory retains the significant and essential features of form, movement and character, rather than insignificant and disturbing accessories, he is able to concentrate upon the inner meaning of whatever subject he may be treating.

Cézanne has chosen the more difficult path of working direct from Nature, whilst aiming at an æsthetic ideal that has more in common with Chinese than with European art. His selection of the essentials and rejection of everything else is deliberate, and every touch of his brush is guided by the passionate desire to achieve something more significant than a mere *trompe l'œil* of reality. He finds this significance in the volume of objects, and his whole work is a search for means of expressing this volume with greater intensity than it can be expressed by ordinary methods of literal representation.

It is all very well to jeer at that quality which the new post-impressionist jargon has described as the 'treeness of the tree' and the 'wallness of the wall' – in other words, the 'volume' of tree or wall – but the very essence of the æsthetic enjoyment of a picture lies in that intensity of expression which makes the beholder realise the peculiar qualities and form and volume of an object more quickly and directly than does the object itself. The means by which Cézanne achieves this end in the drawings at the Grafton Galleries are too subtle to analyse; but the result is there, and is such as to justify the inclusion of this rare artist among the great masters of the nineteenth century. Quite apart from the vital qualities, from the astounding expressiveness of these drawings, they have a daintiness and lightness of touch and a delicate loveliness of pale colour that one would never have expected from the men whose still life and portrait painting in oils so often betrays such brutal and uncouth vigour.

Another most acceptable addition to the exhibition is a superb flower piece by Van Gogh – a picture that surely should silence that artist's many revilers. Here, again, by a magic that it is impossible to describe or explain the bunch of flowers, painted in a mosaic-like thick impasto of pure pigment, have an almost uncanny, passionate vitality that is distinctly more stimulating to the Æthetic sense than any real flowers in all their dewy morning freshness.

The Russian pictures which have been added to the collection are only interesting as curiosities and as evidence of the firm hold that post-impressionism has gained on Muscovite art. Most of them are full of intolerable affectations. Von Anrep's ambitious *Composition*, for instance, can scarcely be taken seriously. Subject matter and method of expression are here grotesquely at variance: a group of female football players of tragic aspect and of Giottesque design with a good seasoning of Matisse!

89. O. Raymond Drey, 'Post-Impressionism: the Character of the Movement'

Rhythm, January 1913, 363–9

It is generally the fate of new movements to bring down upon themselves the derision of the majority, and in painting this is more universally true than in the case of the other Arts. Literature of quality appeals first to its own cultured audience and music is judged more by the emotions than by the intellect. Merely by submitting his work to public exhibition the painter lays himself open to attack by the motley host of ignorant people who drift into picture galleries and exhibitions much as they drift into a race-course enclosure or a fashionable restaurant. Even the critics, who ought to know better, usually judge new work by old standards, which often do not apply at all. The vulgar and the instructed thus join forces to decry what they do not understand, so that to-day it is counted almost a public offence to show work which is not easily within the general comprehension. Two years ago, if the early days of Post-Impressionism as a self-conscious movement may be ascribed to so recent a date, public indignation vented itself not on one man only but on the work of a body of painters who were striving, each in his own way, to gain a common end. The whole movement was promptly denounced as either insanity or char-latanism by an overwhelming majority of the critics who thus expressed the feeling of the public and avoided, at the same time, the irksome process of setting their minds to the understanding of a new artistic idea. It was obvious at least that the well-worn phraseology of journalistic criticism was inadequate to the occasion; the old stereotyped phrases could no longer be made to serve their turn, and new 'clichés' were yet to be made. Criticism and public opinion formed a kind of unholy alliance and did their best to kill the new painting with abuse and ridicule. Even to-day there are not more than two or three critics in England who make any attempt to do justice to Post-Impressionism. And yet this despised development of painting is increasing its influence day by day, in spite of the newspaper critics and in spite of the hostility and laughter of the crowd. Its influence is slowly making its way into public opinion, removing gradually the barriers of ignorance and prejudice. More than anywhere else it is making itself felt directly in the art of painting. Post-Impressionist works are hung at all the chief exhibitions on the Continent: names of Post-Impressionist painters are becoming as well known as those of the Academicians. It is becoming increasingly evident that youth is with the new movement; that the vital work of the day is being done by young men who are under the Post-Impressionist influence; that their work draws its sustenance from life and not from imitation of the great pictures of the past. On the other hand it is becoming known that Post-Impressionist painters have studied the old masters ardently; that they have copied old pictures reverently in the galleries of Europe and that they do not despise the past because they are fired with the spirit of the present. But where the artist is concerned there comes a time when book-learning must be left behind and a fresher impulse sought in the actual experience of life which reveals itself in new forms to each generation of men. The old charge of notoriety-hunting is still heard, it is true, but it has become much less general. Increasing familiarity with the new mode of expression has softened the shock which the public mind always suffers when grappling with the unknown.

Now all art labels are necessarily misleading unless we take them simply. The term Post-Impressionism will serve well enough if we interpret it as denoting the first new movement of importance which has followed Impressionism. Like all great movements it is partly a revolt and partly the manifestation of a new creative impulse. Just in the same way the work of the Impressionist painters was not only a defiance of the sterile practice of an academic formula grown old and meaningless; it was also the outcome of a new artistic consciousness. The Impressionist painters, disgusted with official Napoleonic art, left the vitiated air of the studio and sought, by going straight to Nature, to wrest her secrets from her. The vivid beauty of Monet's work derives from a true perception of atmospheric values and of the quality of light, gained in the open air. He and his fellow painters saw that shadows in strong sunlight are seldom black or brown, but nearly always blue. They saw, because they used their eyes and trusted them, that a tree is not always green, that water is not always green or blue, that the colour of everything is continually changing under changing conditions of light, and that every colour is modified more or less appreciably by the proximity of another. In these discoveries, indeed, was matter enough to inspire the labours of the new artists. Just at this time Helmholtz and Chevreul published their chromatic theories, and the more scientific of the Impressionists were quick to apply them to their art. Signac and Seurat, the first Pointillists, made a new science of the palette, breaking up their colour into small spots or patches which corresponded nearly to the colour divisions of the spectrum. The early Impressionism of Manet and his friends, which was really a revival of the robust painting of the great Dutchmen and Spaniards and owed its strength to a new appreciation of tone-values, led the way to a purely intellectual consideration of colour. More and more colour and light monopolised the attention of the young painters. Monet to this day paints the self-same scene a score of times in different conditions of light and colour. A new academy was born, with rules more stringent and elaborate than any that had gone before. Painting was in danger of becoming a method, to be learnt by rote.

But already in the heart of Impressionism a new impulse was stirring. Manet, the greatest painter of the movement, reverted late in life to tendencies which the optical discoveries of his friends had concealed rather than overthrown. Manet had always had a great sense of design, stimulated, no doubt, by the interest which amateurs were then taking in Japanese and Chinese art. *Olympia*, which is, perhaps, his finest picture, is sufficient evidence of this. The others, in their enthusiasm for light and colour, forgot design or chose to ignore it in favour of tone and colour arrangements. In the Caillebotte collection at the Luxembourg hangs a small head by Manet which is the true parent of Post-Impressionism. It ought, in its abstraction and in its lineal quality, to be as shocking to the vulgar as a painting by Herbin.

But Manet was too great a re-discoverer himself to be the standard-bearer to yet another movement, and this work was left to Cezanne, Gauguin and Van Gogh, the three great founders of a new school – the first 'Post-Impressionists.' The new movement began, like the old, with rebellion against a mode and with the statement of fresh and vital principles. The Impressionists, preoccupied with the study of effects of light, had lost the power to create, which is the child of imaginative sensibility. They had become enslaved to theory and to a new imitative convention. Impressionism, in a word, had become Realism.

It was left to the so-called Post-Impressionists to bring back to painting the qualities of rhythm and plastic design. The followers of a great leader are generally the first to discover that the work of their master is built upon a substructure of ideas, intimately associated and inevitable, it is true, yet not completely realized by the master himself in the spate of his genius. His artistic egotism made it hard for him to admit that the convictions which he saw plainly enough in the work of Gauguin and Van Gogh were his too. Creative passion often obscures the critical faculties in great artists. It was so with Cézanne, whose artistic hunger was too unappeasable, too insistent for him patiently to analyse its characteristics. But his canvases tell their own tale plainly. His landscapes, his figure pieces and above all his studies of still-life reveal a new conception of visual relationships, the relationships of line, of colour and of mass. In 1904, the year before his death, he wrote to his friend Emile Bernard:

'Permettez-moi de vous répéter ce que je vous disais ici: traiter la nature par le cylindre, la sphère, le cône, le tout mis en perspective soit que chaque côté d'un objet, d'un plan, se dirige vers un point central.'

Here, concealed beneath the trite phraseology of the art-school, is the statement of a new faith, of a new vision. It is a new conception of form, the first promise of a new art of rhythmic simplification and design. So far at least was Cézanne conscious of theory, for by this simple avowal he disclosed the whole emotional content of his art. His landscapes with their broadly treated masses, supremely coherent and inter-related, where every part of the composition leads inevitably to a climax – not to a climax of tone but to a climax of lineal form – beckoned imperiously to those artists who read in them the indication of a new truth. His pictures of still-life bore a yet plainer message, where he expressed form almost nakedly as an end in itself, disdaining the tricks of a representative realism.

What is this distinction between interpretative and representative painting? All art, of course, whether it be good or bad, is expressive of the artist's personality in some degree. To speak of exact imitation of nature in painting implies an equivocal use of the term, an artistic evasion, a shirking of the point at issue. Our vision is reformed and modified unceasingly by physical and psychical phenomena over which we have no control—not two men out of a thousand see nature in the same way. On the other hand, the very act of painting involves the acceptance of an artificial expedient for producing on a flat surface, in two dimensions, the similitude of something that exists in three. But between the subconscious self-expressiveness of imitative painting and the conscious striving after interpretation of the Post-Impressionist the distinction is obvious. The aim of the Post-Impressionist is to express by every means in his power the emotion evoked in his mind by the subject he is painting. Everything in his picture must help to build up a mental image; everything must serve to strengthen the dominant idea which he is trying to express. Is it not true that every landscape forms a mass composed of numerous smaller masses, all knit together by a natural rhythm in which alone lies the secret of its compelling mood? The mood is the painter's, of course; in the painter's conception of nature nothing exists beyond the magic circle of his own vision.

Cézanne never spoke of rhythm; probably it never struck him that his pictures were rhythmic. But painters who studied his work with enthusiasm and compared it with the work of others saw in its rhythmic quality the means to a new æsthetic excitement, more powerful than any that painting had known before, because it touched the strings of a universal capacity for response. In the paintings by Gauguin and Van Gogh the message was obscured by an abnormal psychological sensitiveness in the presence of nature. But beneath their overwhelming virility lay the same conscious-ness of new values, of a new means of expression. A new movement was founded – a new train of ideas was laid. It manifested itself at first as a revolt against a tyrannical convention, which tied painters to the consideration of effects of light. But it lives because it is inspired by a new artistic consciousness, by a new creative spirit.

90. Clive Bell,
'Post-Impressionism and Aesthetics'

Burlington Magazine, January 1913, xxii, 226–230

For Bell see the headnote to no. 70.

In discussing æsthetics the first question to be asked is: 'How do we distinguish works of art from all other objects?' If it be conceded that the characteristic of a work of art is its power of raising a peculiar emotion, called æsthetic, we can pass forthwith to the fundamental problem and inquire, 'What quality is common to all objects that do raise it?' Clearly, the answer of each to this question will depend upon his particular æsthetic experience, for each will seek this quality only in those objects that have moved him. Unless they have done so, he will not rate them works of art, for a man can be sure of the nature of no one's feelings except his own. One who was moved exclusively by red objects would conclude that redness was an essential quality in a work of art; nor would he have any certain means of discovering whether the emotions of others for blue objects were precisely the same as his for red. I lay some stress on this small point to escape an imputation of arrogance. In elaborating a theory of æsthetics my only data are my own æsthetic experiences. Since I am seeking a quality common to works that move me, I should merely court disaster by considering those that do not. The degree in which my conclusions commend themselves to others will depend upon the degree in which my experience tallies with theirs. A good critic can make me see in a picture what I had formerly overlooked, but if what he makes me see still leaves me cold, there is no way of forcing my æsthetic emotions. The doctrine that about tastes there is no disputing is no novelty; and, ultimately, all systems of æsthetics must be based on the tastes of the individuals who devise them. Nevertheless if, after examining a number of works about the excellence of which there is general agreement, we find some quality common to all and absent from none, we shall have gone a good way towards formulating, if not a true, at least an acceptable, æsthetic hypothesis. In this essay, having discovered such a quality in a number of old and universally admired works, I shall attempt to show that it is the same quality that moves me in certain modern and more disputable achievements.

Sensitive people seem to agree that there is a peculiar emotion provoked by successful works of art. I do not mean, of course, that all such works provoke the same emotion. On the contrary, every work produces a different emotion. But all these emotions are recognizably the same in kind. So far, at any rate, opinion is not much divided: that there is a particular kind of emotion provoked by works of visual art, and that this emotion is provoked by every kind of visual art, by pictures, sculpture, buildings, pots, carvings, textiles, etc., etc., is not disputed, I think, by anyone capable of feeling it.

Now, fixing my attention on painting and sculpture, the two kinds of visual art with which I am immediately concerned, I find this peculiar emotion stimulated in me, in varying degrees, by works of all ages and all countries: and, fixing it on what is rather arbitrarily called modern European painting, by which is meant, I gather from the historians, almost anything painted anywhere in the West and Centre of Europe since 1150, I find it stimulated by many French, Italian and Spanish primitives, by Giotto, by Piero della Francesca and by Nicolas Poussin intensely; moderately by a small percentage of the others; hardly at all by the mass – until I come to the Post-Impressionists. Amongst the Post-Impressionists I find one giant who moves me supremely, Cézanne; several masters who stir my æsthetic emotions considerably; many good artists who stir them; and, of course, many camp-followers whose works, not stirring them at all, cannot, on my hypothesis, be reckoned art. For it is just this power of raising æsthetic emotion that makes a thing a work of art. And to discover the quality common to all works that raise it is, I conceive, the problem of æsthetics.

There is, then, a peculiar emotion provoked by some quality common to all works of visual art. This quality is the essential quality. It is often found in company with other qualities, no doubt; but they are adventitious, it is essential. Without it a work of art cannot exist; and no work that possesses it, in the least degree, is altogether worthless. Such a quality there must be; otherwise our classification of certain objects as works of art is senseless. Either all works of art have something in common, or, when we speak of 'works of art' we gibber. What is this quality? What quality is shared by all works that stir our Æthetic emotions? What quality is common to S. Sophia and the windows at Chartres, Mexican sculpture, a Persian bowl, Chinese carpets, Giotto's frescoes at Padua, the masterpieces of Poussin, of Cézanne, and of Henri Matisse? Only one answer seems possible – significant form. In each, forms and the relations of forms stir our æsthetic emotions. Form is the one quality common to all works of visual art.

At this point one cannot suppress an irrelevent query: 'Why are we so profoundly moved by forms related in a particular way'? The question is extremely interesting, but irrelevant to æsthetics. If, later, I make an attempt to answer it, I shall do so by way of postscript and not under the impression that I am rounding off my theory. For a discussion of æsthetics it need be agreed only that forms arranged and combined according to certain unknown and mysterious laws do move us profoundly, and that it is the business of an artist so to combine and arrange them that they shall move us. These moving combinations and arrangements I have called, for the sake of convenience, and for the reason that will appear hereafter, significant form.

My hypothesis has at least one merit denied to many more famous and more striking – it does help to explain things. We are all familiar with pictures that interest us, that excite our admiration, but do not move us as works of art. To this class belongs what I call 'Descriptive Painting', that is, painting in which forms are used not as objects of emotion but as means of suggesting emotion or conveying information. Portraits of psychological and historical value, topographical works, pictures that tell stories and suggest situations, illustrations of all sorts, belong to this class. That we all recognize this distinction is clear; for who has not said that such and such a drawing was excellent as illustration, but as a work of art worthless? Of course, many descriptive pictures possess, amongst other qualities, formal significance, and are therefore works of art; but many more do not. They interest us; they may move us, too, in a hundred different ways, but they do not move us æsthetically. According to my hypothesis, they are not works of art. They leave untouched our æsthetic emotions because it is not their forms, but the ideas or information suggested or conveyed by their forms, that affect us.

The latest works of Picasso (*Buffalo Bill, Tête d'homme, Le Bouillon Kub*) leave me cold. I suspect them of being descriptive, of using form as the Royal Academicians use it, to convey information and ideas. I suspect Picasso of having come for a moment under the spell of the Futurists. The theories of the Futurists have nothing whatever to do with plastic art. A Futurist picture aims at presenting in line and colour the chaos of the mind at a particular moment; its forms are not intended to promote æsthetic emotion, but to convey information. These forms, by the way, whatever may be the nature of the ideas they suggest, are themselves anything but revolutionary. In such Futurist pictures as I have seen – I except those of Severini – the drawing, wherever the painter has been so indiscreet as to leave it visible, is in that curiously soft and common convention brought into fashion by Besnard some thirty years ago, and much affected by Beaux-Arts students ever since. As works of art the Futurist pictures are negligible. But they are not to be judged as works of art. A good Futurist picture would succeed as a good piece of psychology succeeds; it would reveal, through line and colour, the complexities of an interesting state of mind. If Futurist pictures seem to fail, we must seek an explanation, not in a lack of plastic qualities that they never were intended to possess, but rather in the minds the states of which they are intended to reveal.

Most people who care much about art find that of the work that moves them most the greater part is what is called 'primitive'. Of course there are bad primitives. For instance, I remember going, full of enthusiasm, to see one of the earliest Romanesque churches in Poitiers (Notre-Dame-la-Grande) and finding it as ill-proportioned, over-decorated, coarse, fat and heavy as any better-class building by one of those highly civilized architects who flourished a thousand years earlier or eight hundred later. But such exceptions are rare. As a rule primitive art is good – and here again my hypothesis is helpful – for, as a rule, it is also free from descriptive qualities. In primitive art you will find no accurate representation; you will find only significant form. Yet no other art moves us so profoundly. Whether we consider Sumerian sculpture or pre-dynastic Egyptian art or archaic Greek, or the T'ang masterpieces, or those early Japanese works of which we had the luck to see a few superb examples (especially two wooden Bodhisattvas) at the Shepherd's Bush Exhibition in 1910, or whether, coming nearer home, we consider the primitive Byzantine art of the 6th century and its primitive developments amongst the Western barbarians, or, turning far afield, we consider that mysterious and majestic art that flourished in central and south America before the coming of the white men, in every case we observe three common characteristics, – absence of representation, absence of technical swagger, sublimely impressive form. Nor is it hard to discover the connexion between these three. Formal significance is incompatible with exact representation and ostentatious cunning.

Naturally, it is said that if there is little representation and less saltimbancery in primitive art, that is because the primitives were unable to catch a likeness or cut intellectual capers. The contention is beside the point. There is truth in it no doubt, though, were I a critic who desired to impress by a display of knowledge, I should be more cautious about urging it than such people generally are. For to suppose that the Byzantine masters wanted skill, or could not have created an illusion had they wished to do so, seems to imply ignorance of the amazingly dexterous realism of the few notoriously bad works of that age. Very often, I fear, the misrepresentations of the primitives must be attributed to what the critics call 'wilful distortion.' Be that as it may, the point is that, either from want of skill or want of will, primitives neither create illusions, nor make display of extravagant accomplishment, but concentrate their energies on the one thing needful – the creation of form. Thus have they created the finest works of art that we possess.

By the light of my hypothesis it seems to me possible to read more clearly than before the history of art, and see in that history the place of Post-Impressionism. Primitives produce art because they must: they have no other motive than a passionate desire to express their sense of form. Untempted, or incompetent, to create illusions, they devote themselves entirely to the creation of form. Presently, however, the artist is joined by a patron and a public, and soon there grows up a demand for 'speaking likenesses'. While the gross herd still clamour for likeness, the choicer spirits begin to affect an admiration for cleverness and skill. The end is now in sight. In Europe we watch art sinking, by slow degrees, from the thrilling design of Ravenna to the tedious portraiture of Holland, while the grand proportion of Romanesque and Norman architecture becomes Gothic juggling in stone and glass. Before the high noon of the renaissance art was almost extinct. Only nice illusionists and masters of craft abounded. That was the moment for a Post-Impressionist revival.

For various reasons, not to be set out here, there was no revolution. The tradition of art remained comatose. Here and there a great genius appeared and wrestled with the coils of convention and created significant form. Nicolas Poussin, Claude, El Greco, Chardin, Cotman; these men move as Giotto and Cézanne move. But the bulk of those who flourished between the high renaissance and the Post-Impressionist movement may be divided into two classes, virtuosi and dunces. The clever fellows, who might perhaps have produced a little art if painting had not absorbed all their energies, were for ever setting themselves technical acrostics and solving them. The dunces continued to elaborate chromophotographs, and continue.

Like all sound revolutions, Post-Impressionism is nothing more than a return to first principles. Into a world where the painter was expected to be either a photographer or an acrobat burst the Post-Impressionist claiming that above all things he should be an artist. Never mind, said he, about representation or accomplishment, mind about creating significant form, mind about art. Creating a work of art is so tremendous a business that it leaves no leisure for catching a likeness or displaying address. Every sacrifice made to representation is something stolen from art. Far from being the insolent kind of revolution it is vulgarly supposed to be, Post-Impressionism is, in fact, a return, not, indeed, to any particular tradition of painting, but to the great tradition of plastic art. It sets before every artist the ideal set before themselves by the Primitives, an ideal which, since the 12th century, has been cherished only by exceptional men of genius. Post-Impressionism is nothing but the reassertion of the first commandment of art – Thou shalt create form. By this assertion it shakes hands across the ages with the Byzantine primitives and with every vital movement that has struggled into existence since the arts began.

Post-Impressionism is not a matter of technique. Certainly, Cézanne invented a technique, admirably suited to his purpose, which has been adopted and elaborated, more or less, by the majority of his followers. The important thing about a picture, however, is not how it is painted, but whether it provokes æsthetic emotion. Essentially a good Post-Impressionist picture resembles all other good works of art, and only differs from some, superficially, by a conscious and deliberate rejection of those technical and sentimental irrelevancies that have been imposed on painting by a bad tradition. This becomes obvious when one visits an exhibition such as the Salon d'Automne of Les Indépendants, where there are hundreds of pictures in the Post-Impressionist manner many of which are quite worthless. These, one realizes, are bad in precisely the same way as any other picture is bad; their forms are insignificant and compel no æsthetic reaction. In truth, it was an unfortunate necessity that obliged us to speak of 'Post-Impressionist pictures', and now, I think, the moment is at hand when we shall be able to return to the older and more adequate nomenclature and speak of good pictures and bad. Only we must not forget that the great movement of which Cézanne is the earliest manifestation, and which has borne so amazing a crop of creative art, owes much, if not everything, to the liberating and revolutionary doctrines of Post-Impressionism.

The silliest things said about the pictures at the Grafton Gallery are said by people who regard Post-Impressionism as an isolated movement, whereas, in fact, it takes its place as part of one of those huge curves into which we can divide the spiritual history of mankind. I believe it to be the first upward stroke in a new curve to which it will stand in the same relation as 6th-century Byzantine art stands to the old. We may compare Post-Impressionism with that vital spirit which, towards the end of the 5th century, flickered into life amidst the ruins of Græco-Roman realism. Post-Impressionism has a great future; but when that future is present Cézanne and Matisse will no longer be called Post-Impressionists. They will be the primitive masters of a movement destined to be as vast, perhaps, as that which lies between Cézanne and the masters of S. Vitale.

Post-Impressionism is accused of being a negative and destructive creed. In art no creed is healthy that is anything else. You cannot give men genius; you can only give them freedom – freedom from superstition. Post-Impressionism can no more make good artists than good laws can make good men. Doubtless, with its increasing popularity, an annually increasing horde of incapable painters will employ the so-called 'Post-Impressionist technique' for presenting insignificant patterns and recounting foolish anecdotes. Their pictures will be dubbed 'Post-Impressionist', but only by gross injustice will they be excluded from Burlington House. Post-Impressionism is no specific against human folly and incompetence. All it can do for painters is to bring before them the claims of art. To the man of genius and to the student of talent it can say, 'Don't waste your time and energy on things that don't matter: concentrate on what does'. Only thus can either give the best that is in him. Formerly because both felt bound to strike a compromise between art and what the public had been taught to expect, the work of one was grievously disfigured, that of the other ruined. Tradition ordered the painter to be photographer, acrobat, archæologist, and litterateur: Post-Impressionism invites him to become an artist.

So much for Post-Impressionism and Æsthetics. May I, now, take leave, for a moment, of solid ground and venture a short flight of hazardous speculation? There remains the metaphysical question. Why do certain arrangements and combinations of form move us so strangely? For æsthetics it suffices that they do move us; to all further inquisition of the tedious and stupid it can be replied that, however queer these things may be, they are no queerer than anything else in this incredibly queer universe. But to those for whom my æsthetic theory seems to open a vista of possibilities I willingly offer, for what they are worth, my dreams.

The wisest philosophers of all ages have believed in the existence of a reality of which the physical universe is but the appearance; and even modern men of science are beginning to see that no other hypothesis will explain many of those things that most need explanation. Let us suppose, then, that behind the world of appearance lies a world of reality. Consider art. Consider music and visual art, both of which depend for their effect on pure form. One by combinations and arrangements of sounds, the other by lines and masses, both purely by form, provoke emotions that exalt us to a peculiar state of mind. We are transported to a fabulous and unfamiliar world with laws and logic of its own, where certain combinations and relations are perceived to be right and necessary, although by the rules of the world we have left they are nothing of the sort. We bow to a new order because we are inhabiting a new universe. As the saying goes, we have been carried out of ourselves. Whither have we been carried?

I believe we have been carried into the world of reality. Form is the bridge between two worlds. When the artist contemplates objects or visualizes his own conceptions he beholds form and behind form reality. Only in form can he express and communicate the emotion that he feels. It is because his forms express this emotion that they are significant; because they fit and envelop it they are coherent; because they communicate it they exalt us to ecstasy. Form is the boat in which artists ferry us to the shores of another world.

There is an experience common to children and not unknown, I think, to men and women. We come upon some scene or object, a tree, a field, a wall, a landscape, and, suddenly, we find ourselves in that world to which great art transports us. The scene or object before us is charged with an extraordinary significance, we, ourselves, are filled with an unreasonable delight. I am persuaded that at such moments we see things as great artists see them. Our vision is disinterested: we see things as ends instead of thinking of them as means. Therefore we see them as pure forms related on terms of equality with other forms. For a moment the world has become a work of art: we see it as form: and behind form we catch a glimpse of reality.

91. Anthony M. Ludovici on Van Gogh

Introduction to *The Letters of a Post-Impressionist: Being the Familiar Correspondence of Vincent Van Gogh*, 1912, pp. xxiii and xxxiv–vii

Some of Van Gogh's letters appeared in a German edition in 1911 and were reviewed by Michael T.H. Sadleir in *Rhythm* (Autumn 1911, 16–19). Ludovici translated a number of them into English and added an introduction for this selection which was published in 1912.

I have myself seen pictures which I could not help thinking must have been painted in Van Gogh's academic period; Meier Graefe even thinks that Van Gogh's work of this period is likely to rise in public esteem; I have little doubt, therefore, that Van Gogh did go through an academic stage, however short or however undistinguished it may have been. . . .

Now all that he has acquired – art-forms, technique, stored experience, practised observation – is but a means, a formidable equipment which he is deep enough, artist enough, human enough, to wish to lay at the feet of something higher. Now his storehouse of knowledge becomes an arsenal which he consecrates solemnly to the service of a higher cause and a higher aim than the mere immortalizing of 'decorative pages of colour' – 'interesting and strong colour-schemes' and 'exteriorisations of more or less striking impressions.' When these things are pursued as ends in themselves, as they were by the Impressionists and the Whistlerites, they are the signs of poverty, both of instinct and intelligence. They are also signs of the fact that the mere craftsmen, the simple hand-workmen, or the mere mechanic – in other words, the proletariat of the workshop, has been promoted to the rank of artist, and that matters of decoration, technique and treatment (which are fit subjects for carpenters, scene-painters, and illustrators to love and to regard as the end of their mediocre lives) have ursurped the place of higher and holier aims.

In about as many years as it takes some painters to learn their palette, Van Gogh had learnt the great and depressing truth at the bottom of all the art of his age – the truth that it was bankrupt, impoverished, democratized, and futile. Divorced from life, divorced from man, and degraded by the great majority of its votaries, art was rapidly becoming the least respected and least respectable of all human functions.

He realized that art was an expression of life itself, that pictorial art was an expression of life's satisfaction at her passions become incarnate. All expression is self-revelatory. Pictorial art, then, is the self-revelation of life herself looking into her soul and upon her forms. It is life pronouncing her judgment on herself. Alas! it is less than that: it is a certain kind of life pronouncing its judgment on all life. Where life is sick and impoverished, her voice speaking through the inferior man condemns herself, and paints herself bloodless and dreary, probably with a sky above depicted in a lurid and mysteriously fascinating fashion, calculated to make the earth seem gray and gloomy in comparison. Where life is sound and exuberant, her voice, speaking through the sound man, extols herself and paints herself in bright, brave colours, which include even bright and brave nuances for pain and the like.

The sound, healthy artist, then, once he has attained to proficiency in his *métier* – a result which, if he be really wise and proud, he will not attempt to accomplish before the public eye as every one is doing at present – naturally looks about him for that higher thing in life to which he can consecrate his power. His passion is to speak of life itself, and life in its highest manifestation – Man. But, alas, whither on earth must the poor artist turn to-day in order to find that type which would be worthy of his love and of his pictorial advocacy?

Is the hotch-potch, democratic, democratized, hard-working, woman-ridden European a subject to inspire such an artist? True, he can turn to the peasant, as many artists, and even Van Gogh himself, did. At least the peasant is a more fragrant and nobler type than the undersized, hunted-rat type of town-man, with his wild eyes that can see only the main chance, with his moist finger-tips always feeling their way tremblingly into another's hoard, and with his womenfolk all trying to drown their dissatisfaction with him by an endless round of pleasure and repletion; but, surely there is something higher than the peasant, something greater and nobler than the horny-handed son of toil?

Gauguin and Van Gogh knew that there was someone nobler than the peasant. But the tragedy of their existence was that they did not know where to find him.

Fortunately for himself Van Gogh died on the very eve of this discovery. Gauguin suffered a more bitter fate than death; he went searching the globe for a nobler type than his fellow-continentals, at whose feet he might lay the wonderful powers that nature, study, and meditation had given him. But in doing this he was only doing what the whole of Europe will soon be doing. The parallel is an exact one. The prophecy of the artist will be seen to have been true. And Gauguin's search for a better type of humanity is only one proof the more, if such were needed, of the intimate relationship of art to life, and of the miraculous regularity with which art is always the first to indicate the direction life is taking.

I have shown how, from a negative and futile impressionist, Van Gogh became more and more positive and human in his content, and ever more positive, brave and masterly in his technique, and that this healthy development naturally led him to the only possible goal that lies at the end of the path he had trodden – Man himself.

In 1886 he writes to Bernard: 'I want to paint humanity, humanity and again humanity. I love nothing better than this series of bipeds, from the smallest baby in long clothes to Socrates, from the woman with black hair and a white skin to the one with golden hair and a brickred sunburnt face' (page 85).

At about the same time he writes to his brother: 'Oh, dear! It seems ever more and more clear to me that mankind is the root of all life' (page 89); and 'Men are more important than things, and the more I worry myself about pictures the colder they leave me' (page 131).

But the finest words in all these letters, words which at one stroke place Van Gogh far above his contemporaries and his predecessors, at least in aim, are the following; 'I should like to prepare myself for ten years, by means of studies, for the task of painting one or two figure pictures...' (page 152).

In his heart of hearts, however, Van Gogh was desperate. There can be little doubt about that. Not only did he feel that his was not, perhaps, the hand to paint the man with the greatest promise of life; but he was also very doubtful about the very existence of that man. Not only did he ask: 'But who is going to paint men as Claude Monet painted landscape?' (page 103); he also shared Gauguin's profound contempt of the white man of modern times.

Indeed, what is his splendid tribute to Christ as a marvellous artist, a modeller and creator of men, who scorned to immortalize himself in statues, books, or pictures (pages 65 *et seq.*) if it is not the half-realized longing that all true artists must feel nowadays for that sublime figure, the artist-legislator who is able to throw the scum and dross of decadent civilizations back into the crucible of life, in order to mould men afresh according to a more healthy and more vigorous measure? The actual merits of Christianity as a religion do not come into consideration here; for Van Gogh was not a philosopher. All he felt was simply that craving which all the world will soon be feeling – the craving for the artist-legislator, which is the direst need of modern times. For, in order that fresh life and a fresh type can be given to art, fresh vigour and a fresh type must first be given to life itself.

---

92. James Bone, 'The Tendencies of Modern Art'

Edinburgh Review, April 1913, 420–34

Ostensibly a review of Ludovici's *The Letters of a Post-Impressionist* together with Cosmos, *The Position of Landscape in Art* (London: George Allen and Co, 1912) and C. Lewis Hind, *Hercules Brabazon Brabazon* (London: George Allen and Co, 1912), this article surveys French trends in art from an English point of view.

The manifestations of the advanced art of our time are so disturbing to settled convictions derived from Victorian days, that even the larger public is beginning to ask what has happened that artists, whose calling has hitherto in England seemed so tranquil and enviable, should be engaged in a sort of holy war with one another. The battle in progress has sympathetically affected both literature and music, and is being carried on with a fury which shows that the causes at issue go far deeper than in the Whistlerian and Impressionist conflicts.

Looking back on the Victorian age, the social student finds that at the time when England was manifesting a stupendous burst of energy in scientific discovery and thought, and in the practical application of science to industry, the art of the nation had sunk to its most timid and parasitic condition. There were, of course, exceptions, and it is of these exceptions that we chiefly think when we speak of Victorian art. But in the main the reaction of art to the age was astonishingly feeble. Gifted technicians there were in plenty, but – if we except the great landscape painters in the earlier

433

years and the continuation of their work by the Impressionists at the close, the short sudden burst of angry beauty in the Pre-Raphaelites, the occasional English epic by Watts, and some tender febrile portraits that appeared mainly as the result of Whistler's influence – the token of Victorian painting seems to be a solace for the spare moments of a strenuous age. This is all the more striking, for in the eighteenth century England, through the influence of the travelled connoisseur, had maintained the continental grand manner and standard of style. But in the Victorian age the patrons of the artist were mainly the new class of wealthy manufacturers and traders, and for a time art almost ceased to be intended either for the cultured aristocracy like eighteenth-century painting, or for the poorer middle class like the prints and engravings from Hogarth to Rowlandson. The new patron, being in the main untravelled, and gentility having spoilt the racy tastes of his humble beginnings, might have been thought particularly open to take his art unquestioningly from his artist. It ought to have been possible for English artists to have imposed upon these merchants of a proud and stirring age an art as opulent and ambitious as the Venetian artists gave to the great traders of Venice. Here were men, for the most part simple and eager for the refreshing fruits which they had been told by Mr. Ruskin art had to offer, ready with large cheques and unfailing hospitality as a poor recompense to the high priests of the Temple for their great boons. But all that was given to them was – well, the Tate Gallery, the Holloway Collection, and the off days at Christie's will tell what was given to them. As a class, the railway kings, nitrate kings, pill makers, and great tradesmen bought rubbish and made bad bargains; but their ideals were good. At any rate they desired for their money higher and less material satisfaction than the freak dinners and the motor and Riviera life of our own newly rich. As it happened, curiously enough, the taste at Court corresponded with the simple external preferences of these patrons. ('I like my pictures glossy,' a very august lady is reported to have said to Sir Edwin Landseer.) In the painting of animals the patronising attitude of the time was more nakedly revealed. Wild animals were interesting, so far as human attributes could be read into them. Dogs were shown as *Dignity and Impudence*, the lion as *The King of Beasts*, and so on. What was alien, wild and menacing to the soul of man in nature and in animals was ignored. Long, Goodall, and hosts of others searched the East for subjects, and came back with a good story or two in paint and a little Byronic romance. Alma Tadema's pretty and learned reconstructions, so inadequate as a vision of the lust and terror of the antique world, were hardly challenged in their day, so little was expected of emotional truth, although the Pre-Raphaelites had come and gone, and their pale procession of camp-followers was still trailing 'greenery-yallery' lotus flowers through the galleries of London. The artists were encouraged to produce large, heavy, smirking pictures, with a trivial motive that dulled and devitalised life instead of intensifying and penetrating it. Their pictures now cover the walls of our big municipal galleries, and it is surely one of the most pathetic things in the public history of our art that the awakening of the great cities to a knowledge that art had something to give them ended in the filling of their new galleries with the topical stupidities of the time. Each silly 'picture of the year' added to these galleries, and confirmed the indolent mentality and false sentiment that beset the practice of art in Victorian England. In the twentieth century the attitude of the time, as expressed by the patron, has quite changed. Every day at Christie's sees a growing intolerance of the complacent ideals and methods of the Victorian favourites. We have other foolishness, no doubt, and are still paying thousands for second-rate Barbizon pictures and dubious old masters, and 'Cries of London' (that are not from the heart); but on the whole the instincts of the time are better.

The best indication of the difference between the Georgian and the Victorian eras is the extraordinary enlargement of the public acquaintance with Old Masters. Their art has been analysed, discussed, and illustrated with a thoroughness that we used to give only to the supernatural. Photography, that to its old-fashioned devotees (like Mr. G.B. Shaw) seemed at one time likely to supplant the Old Masters, has been tied to their chariots and made to testify to their glories. Although concern with art in any of its manifestations is visible only in a minority of people of any class, the concern for Old Masters runs vertically through the nation, and touches every degree of culture. However our new art may develop, however wild may be its extremists, the insular self-satisfaction of Victorian art is gone for ever. The artist again lives in a highly critical cosmopolitan surrounding, the patron is again the student of art, and the philosopher is always at the artists's elbow, asking 'What is art?'

It is unnecessary now to trace the pedigree of the movement for full freedom to express visual truths that was termed 'Impressionism.' In landscape the link is the work of Turner, Bonington, and Constable at the beginning of the century. If their countrymen treasured only the ashes of their art, France stole some of its heat, and its influence on the Barbizon men, who in turn influenced Monet and the Impressionists, is admitted by every Englishman. One English Impressionist there was whose long life spanned the gaps between the going and the returning of the new desire for atmosphere and luminosity, and the flowering of colour on canvas and on paper; although his first exhibition was not till 1892 when he was in his seventy-second year. The public were certainly not till then ready for Hercules Brabazon Brabazon's lovely art, shell-like in its small iridescent perfections.

At the time when Brabazon died Impressionism had been accepted in England, and the Royal Academy had assumed the aspect that the independent societies wore about ten years earlier, while in these societies signs were appearing that the desertion from Impressionism to a more synthetic and self-revealing art had already begun. The increasing seriousness of purpose and revival of draughtsmanship, that were soon to make themselves felt, owed their inspiration, not to official schools nor to a great native exemplar, but to Alphonse Legros, a French artist deeply versed in the Old Masters, who lived the main part of his long life in England and devoted his many gifts to the discovery and teaching of what is permanent and communicable in the great art of the past. He had a sort of second sight which enabled him to see Old Masters at work in the fields of Watford and Wembley Park. The new feeling for style and form, which is characteristic of our time, owes much of its quickening to him.

But in the first decade of the new century these signs were only apparent to the close student of art. The impetus of Impressionism which we had received from France began to slacken in sympathy with the turn of the tide there. Our new forces were experimenting, in strange eclectic company, for a form to express their new sensibility. Work, whose merit was its individualist character, was being done in isolated quarters, but as a whole there was a general weakening of intention and questioning of reputations that corresponded to the trend of the national temper of the time. The inquiry into the administration of the funds of the Chantrey Bequest had shaken the reputation of the established corporation of English art beyond repair in our day. This became very evident as time went on, and the Academicians themselves, being mainly old men and seeing the heavy fall in the price of their pictures in the auction market, and the steady conversion of Bond Street to Old Masters and etchings, perceived clearly that art was in a bad way and, as is customary, looked across the Channel for the cause of this distressful state of affairs. Very little observation was required to satisfy them that a new movement had arisen there and was known everywhere but in England. That was Post-Impressionism.

From the lamentations over the decline of English art that have appeared in the Press during the past twelve months it is easy to construct an approximate image. We must imagine English Art as a female Job, sitting in peace in her household and awaiting the arrival of the messengers. The first messenger comes with the tidings that the anecdote picture is dead. The oxen were ploughing and the asses were feeding together when Mr. George Moore rushed down upon them, and only Mr. John Collier was left to paint the tale. And while he was speaking there came another, beating his breast and crying that allegoric high art too had perished at the hands of critics (who were, however, incompetent and unworthy of attention), and he, Sir W.B. Richmond, alone had escaped. And while he also was yet speaking there came another and cried that Impressionism was perishing even in the house of the New English. And lo! even as he spoke behold another came who said 'Mr. Sargent has given up painting portraits.' Then English Art rose up and rent her garments, and having shaven her head with a potsherd fell down upon the ground and her friends knew her not.

Yes; the complete cessation of portrait painting by the most gifted and powerful portrait painter of the age is certainly the culminating point in the woeful calendar. It is difficult to imagine such a renunciation by a great painter of an earlier time. Have the doubts of the validity of Impressionism begun to assail even him in whose hand it was a wonderful instrument to probe into hidden truths of personality? Mr. Sargent has hinted that it is even so. A couple of years ago a picture by him appeared in the New English

Art Club's exhibition showing a landscape painter hugging a large box of paints to his heart and peering solemnly before him for a little subject. The sunshine struck his white shirtsleeves, and made little spiky lights on the latchets of his boots. Great mountains rose all around him. The little man with his big box of paints seated so self-importantly, peering around for his little effect, while Nature in its vastness stared down at him, made one think of a man hunting a rabbit while a lion stands behind him. Possibly Mr. Sargent's only intention was a portrait of a friend on the Alps, but a future generation may prize it as the declaration of the end of a faith.

The new movement that was then emerging from English art was largely influenced by the Pre-Renaissance Italian masters, by archaic Greek, Byzantine, Egyptian, and Assyrian art, and by the art of the Far East. Mr. Augustus John, its leader, already occupies a position for which there is no parallel in our history in that his art, which is supported by many of the most fastidious and erudite connoisseurs of the time, has for its content democratic and revolutionary ideals of the most uncompromising kind. The first rankness of his subject-matter has subsided with the passing of the stuffy materialism in official paintings, against which it was a protest. It is no longer spoken of as 'high art' in the sense that game is 'high.' The art of this group has much in common with that of the French Post-Impressionists, although Mr. John's development seems to have had no connexion with their experiments; but the plastic freedom of Puvis de Chavannes undoubtedly gave important hints to both schools. It is noticeable that they have sought in the first place to simplify their art by simplifying their technical method as well as their representations. They use tempera, and in their experiments with oil have often reduced their colours to a few tints prepared beforehand; and besides demanding freedom in drawing and in treatment of the subject they refuse to be bound by the accepted system of atmospheric tones. They seek to dislocalise their figures so that they belong to no class, no place, or time. By all these devices they aim at a lean athletic art to run deeper into our consciousness. For that purpose they have stripped art of much that was comfortable and informing, of many graces and charms, and of many truths that we had come to think inseparable from it, and it is natural enough that in the eyes of the older generation the result should have a naked, disquieting look. Mr. John's masterpiece, *The Girl on the Cliff*, is like nothing else in English painting in the pure keenness of its imaginative invention. The master draughtsman of his time, he has been strong enough to yield up every appearance of skill and of grace, and to limn his idea with the fresh, short-cut directness of a child.

This we see alike in his gigantic groups of gipsies arbitrarily grouped together in a cold bright transfiguration of English countryside; in his primitive matrons, sealed in knowledge, mysteriously smiling; and in strange girl figures with dilated eyes, roaming solitarily in remote places. He attains his mysticism without vagueness. His outlines are clear and hard as mountain crests, and his tones are never indeterminate. His poetry is his own, and unlike the Pre-Raphaelites he moves us without inspiration from the poets. He uses none of the usual devices to glamour you into his country. He blows his high, clear trumpet, and the curtain of our every-day mind is rent and his world opens before us. Compare his *World Elsewhere* with the anæmia and luxury of Burne-Jones's conception. This bright, clear world he possesses as definitely and fully as Blake did his; his power to render it is greater, and that power has been purified to its essence by his single-minded passion to get closer and closer to his image. This is a supreme quality of the rarer masters, one of the most lasting weapons in art's armoury, and rarer in proportion to the gifts of the artist which tempt him to demonstration of his powers.

The intense faithfulness to the creatures of the mind, that we identify most clearly in that much-loved master Fra Angelico, is the marking characteristic of John's imaginative art. In his types we see embodied ideals that have been long absent from our art, if indeed they have ever been assembled within it: brute strength, independence, and life on primitive and patriarchal terms. No weak-looking man ever finds a place in his pictures; the old men look cunning and tough, the children untamed and fierce, the women deep-breasted, large-bodied, steady-eyed, like mothers of a tribe. A bracing wind seems always blowing and the hills are darkling in the distance, but there are flowers underfoot. His people are never in an interior, except sometimes in a tent. They stand firmly on the earth and regard civilisation with eyes that have judged it and found it wanting. Unlike Brangwyn and Meunier, who have been termed the artists of democracy, John rarely shows a figure at work. Strangest of all the impressions one gets from these wild wayfarers is responsibility. He makes you see that his strong men and women in poor clothes, standing with beauty under cold skies, have chosen their part and challenge you to judge them. This is John's message. Nor is it unrelated to a spirit of the age that is reflected in other activities. The distrust of comfort, of cities, of society in its present organisation, even of civilisation, and the desire for a simple life and the recovery of the virtues that lie in a more physical communion with the earth, are all questions of the time that many writers are urging upon the people, and that many are putting to the test of experiment.

With John one classes several other artists who have set out into a new and untilled field of art through the gap that he had made. Of these the most distinguished is Henry Lamb, who with subtle gifts for colour and design in the high Italian tradition seems to be seeking in many strange ways to find pictorial expression for conceptions as far from sentient experience as is music. He has travelled farthest on his way towards the same house as the Post-Impressionists in his inventions of figures in a fantastic setting.

French Post-Impressionism first landed at Brighton in 1910, and reached London about a year later. Naturally a movement so fiercely opposed to the established practice of art in England aroused a shout of condemnation and ridicule, but at the same time artists, critics, and public were unusually ready to expose their sensibility to the action of this new, uncanny art. The works of Cézanne, Gauguin, and Van Gogh could not be dismissed by serious students as daubs by charlatans, anarchists, and self-advertisers. Cézanne was a pious Catholic, a *rentier*, and a good family man. When he laid down the difficult and original lines on which his art was to develop he said good-bye to all prospect of fame, and was sure of nothing but the hostility of dealers and patrons. He only once exhibited a painting during twenty years. He pursued his ideal till the end, when he died with a brush in his hand. Van Gogh and Gauguin, weak in body and miserably poor, found no appreciation in their lifetime, neither did they seek it. Gauguin found his chief inspiration in the people of the Tahiti and the Marquesas Islands. Van Gogh's work was all done when living in obscurity in a French town, far from the excitement of studios and exhibitions. His letters, edited by Mr. A.M. Ludovici (whose introduction is one of the most probing contributions to recent criticism), are among the very few revelations of an artist's soul that the world possesses. These three artists were Impressionists who had given up their faith. From their new point of view it followed that nearly all that Impressionism had painfully garnered was valueless and was only a lure to entice art from its strait path.

Our debt to the French Impressionists is that they gave an impression of the world infinitely more vivid and real than existed before; but had their success remained unchallenged, their worship of the illusion of reality as an art in itself would have become an intolerable tyranny, which would have forced painting to have exercised only one side of her powers and atrophied completely the side on which she claims kinship with a pure and abstract art like music. So, swiftly on the heels of the Impressionists, the Post-Impressionists were bound to come and bring redress. Their maddest things may be taken as inarticulate outcries that something was wrong. The leaders, Cézanne, Van Gogh, and Gauguin, not only reproduced a furious indictment of the whole aim of contemporary art, but discovered strange enlarging avenues for a new advance into the Unseen:

> And so bring the invisible full into play:
> Let the visible go to the dogs – what matter?

As a glimpse of the spirit animating these men this excerpt from a letter by Vincent Van Gogh is illuminating. Van Gogh imagined himself painting an artist friend – 'an artist who dreams and works as the nightingale sings songs.' He writes:

Let us imagine him a fair man. All the love I feel for him I should like to reveal in my painting of the picture. To begin with, then, I paint him just as he is, as faithfully as possible – still, this is only the beginning; the picture is by no means finished at this stage. Now I begin to apply the colour arbitrarily. I exaggerate the tone of his fair hair; I take orange, chrome, and dull lemon-yellow. Behind his head, instead of the trivial wall of the room, I paint infinity. I make a simple background out of the richest of blues, as strong as my palette will allow. And thus, owing to this simple combination, this fair and luminous head has the mysterious effect upon the rich blue background of a star suspended in dark ether.

In another letter he said: 'It is my most fervent desire to know how to achieve such diversions from reality, such inaccuracies and such transfigurations that come about by chance. Well yes, if you like, they are lies; but they are more valuable than real values.'

Transfiguration is the desire that underlies the best work of the school. Its members do not look back to Titian or to Rembrandt, or to Leonardo, in whom the perfect balance between a noble mould of design and realism of representation was struck – the equipoise of the subjective and objective – but throw the balance on the side of design. Their followers in France have, in the main, thrown all their weight on that side and, as it were, have brought the scale down heavily on the subjective foundation. So, in the hands of Picasso and his followers painting is fast passing into an abstract state, purged of any associations, and becoming something more analogous to the free art of music.

Of the three great forms of expression of the spirit of fine art we find poetry at one end of the scale and music at the other. Pictorial and plastic art lies midway, and accordingly as it has kept the balance between the quality most contained in each of the sister arts so far it has attained the highest expression of itself. With poetry the immediate appeal is the smallest of all; and although it is held that poetry is the marriage of sound and sense, sense is so far the predominating partner that we only care to read poetry in an unknown tongue for a very short time – many people, for instance, cannot tolerate more than two verses of Burns's 'Poor Mailie.' With sculpture and painting the two properties merge more intricately into one another, for in the one operation of the eye both things claim equal attention. A picture appeals to us by its design and colour before we have considered what it represents. That immediate effect, however, has an organic relation to the content of the picture. It is conceived by the artist, consciously or unconsciously, as the symbol of his subject, and is the last thing to disappear. So we find in old and faded works by good masters wonderful ghosts of form and colour after features, dress, and action have almost disappeared. In every real work of art this quality that is not representation exists, and apart from all veracity of statement, or beauty of the subject-matter, or illustration of life, it is what makes the painting a work of art. Primitive art has this first-hand expressiveness; and although we may argue that it was the result of incapacity to represent objects, that as soon as the primitive could draw his images better he did so, and that this is what we understand by the progress of art, it does not follow that this lost expressiveness is not worth much of what we have gained, or rather, that something of its spirit cannot be recovered and developed in another way than that of the Greeks and the Renaissance. Even the stoutest of us, as he has stood by the eternal-looking figures in the Assyrian and Egyptian rooms in the British Museum, must have had his moments of doubt whether all the victories won by Greek art have not cost us more than a defeat.

Mr. George Moore in one of his early essays raised this interesting speculation: what would have happened to Japanese art if a cargo of the Elgin marbles had been wrecked on the coast of Japan? A speculation that is more to the taste of our day is: what might have happened here if, before the Renaissance, a cargo of Egyptian and Assyrian figures had found its way ashore on both sides of the Channel? Each year, however, sees more and more examples of the ancient art of the East assembling in European museums, and the impact of this penetrating expressiveness upon the more sensitive minds of our generation is probably one of the causes of the dissatisfaction with the whole trend of art that is now becoming manifest all over Europe. But the more potent influence has been the new knowledge of Chinese art which has moved Western artists to the greatest heart-searchings. The whole art of the Orient is at last receiving respectful study – its significance as well as its form – and the artists are beginning to follow the students, and their studies are carrying them far in directions that seem like madness to their older contemporaries. For instance, the Indian many-limbed figures, which a generation ago were dismissed as barbaric and debased forms of art, are now thought no more strange than the centaurs and fauns of the Greeks, and it has become the focus of discussion whether they do not represent further possibilities of making sculpture more symphonic or processional, the many limbs, it is argued, having power to suggest infinity. In a word, the question is raised whether the drift of art was really Westward.

The new learning of Oriental art bulks more and more formidably every day, and the forces are gathering in all parts of Europe. The solid foundations (especially in the Grafton Gallery) are beginning to tremble underfoot, and men are asking one another whether there is any law (or by-law) on earth why there should not be more than one Renaissance. In Italian art before the Renaissance the masters exercised a plastic freedom over their designs and a power to intensify and exaggerate expressiveness that gives them often a curious kinship with Eastern art, this power fading in the fifteenth century and in the seventeenth dying away (shall we say?) in the enigmatic smile – faintly Chinese – of the Mona Lisa. It may be that some day, when Oriental learning has wrought a complete change, people will say of her (for, of course, she will be rediscovered by then) that she was smiling at the wrong Renaissance.

Since the powerful influence of that Renaissance, art has gone fast and far along the road to complete imitation of nature. But as the artist's power of representation has enlarged, the problems of this function have in an increasing measure occupied his mind, and the objects upon which he has exercised it have slackened their demand upon his power to invent and magnify. There were, of course, a host of other factors, such as the modern concern with light, which became the 'hero of the picture,' but the decline in the importance of the subject-matter and a lessening capacity for pregnant design are indisputable. The question that falls to be considered is whether the synthesists (to group together the French Post-Impressionists and the Augustus John group in England under one ugly but convenient title) do not really go far to remedy the two disabilities into which modern art has fallen. The one group, which includes most of the living Frenchmen who exhibited at the Grafton Gallery this year, have accepted and carried yet further the negation of the subject-matter, but they seek to make the composition monumental in its own right by the value of the pigment, the strength and intensity of colour, and the simplification of form to shapes that convey this sense of permanence. The other group, which includes John, Lamb, and Grant, in England, and Maurice Denis, and a number of Frenchmen, seeks to revive the importance of subject-matter, and to concentrate upon the emotional significance that arises from the subject. Whether we agree or not that they have found the remedies, we must admit that both sections are serious about serious things, and that their search for monumental form and style is all for the good. It is significant that many of the most learned and most thoughtful of our critics here and in France have given their general support to the movement; that it has attracted back to contemporary painting the more serious section of our connoisseurs who want art to be anything but a solace for tired minds; and that it has stabbed through the indifference to art into which the general public had fallen since the Pre-Raphaelites.

How far the movement is leading us, and how changed the criteria of criticism are becoming, may be gauged from an excerpt from a Post-Impressionist article in the *Burlington Magazine* by Mr. Clive Bell, a leading apologist of the school, who puts the case for perfect freedom in this way:

> Either all works of art have something in common, or when we speak of "works of art" we gibber. What is this quality? What quality is shared by all works that stir our aesthetic emotions? What quality is common to St. Sophia and the windows at Chartres, Mexican sculpture, Chinese carpets, Giotto's frescoes, the masterpieces of Poussin, of Cézanne, and of Henri Matisse? Only one answer seems possible – significant form. In each, forms and relations of forms stir our aesthetic emotions. Form is the one quality common to all works of visual art.

The claim is distinctly staked out that the plastic arts are not representative but presentative like music. The purest form of art by this theory would be art purged of its content, and reduced to cubes and patterns, for then there would be nothing but form and the relations of form. Cézanne, Van Gogh, and Gauguin had this in common, that they sought to reclaim the ancient prerogative of the artist to deal with his subject after his own law. Their transfigurations were expressed in terms of abstract form. They were synthesists above everything else. A few of their French followers seek, like John and Lamb, to transfigure their subjects in terms of significant form, but the majority are content with the aim of form without concern for the subject. England therefore, at a time when she has not made up her mind whether pictorial art can be severed from literary associations, finds herself facing the spectacle of pictorial art trying to sever itself from all associations. Can pictorial art live apart from its association content like music, or will it become gibberish, as poetry does when the poet seeks to use words for their rhythmic value apart from their meaning? Can it give up ethics and cease to have the responsibility of poetry without lowering its whole value to the human race? The answer surely is that it cannot; that although works of art have in common the language of significant form, a work of art to be great must also have a moral value that can be expressed in that form, as form in poetry rises to its heights when it is expressing its most ecstatic thoughts. That any movement can alter a principle so deeply rooted in the human race, is, at the least, unlikely, but that such an attempt is valuable as a corrective of the ills of modern art – the otiose condition of our classicalists, the indigestion of our Impressionists – is surely undeniable.

Nevertheless, England, at least, need have little fear of Post-Impressionism or any other form of imported art. England only imports what can be dealt with by her national temperament, and that she speedily transforms into a home product on which the original exporter cannot find his trade marks. How different is the wayward, dainty Impressionism of Steer, Clausen, McTaggart (who got Impressionism by wireless, for he never saw a Monet till he was over sixty), of the Glasgow School, of Brabazon, Holmes and Houston, from the Impressionism that seized and possessed Monet, Renoir and Pissarro, to whom their art was a new religion! The prevailing instinct of English art is the desire for beauty, and we pay the penalty in the national cult of prettiness, which is as far into her territory as most of us can enter. Our boast might be that we make two pretty things grow where one idea grew before. However the mandarins may rage against them, even our pioneers are never ahead of beauty. But the French can forget her in their search for truths, and it is they who must take their consolation from Whitman's lines:

> . . . The Great Masters
> Do not seek beauty, they are sought;
> For ever touching them, or close upon them, follows beauty,
> Longing, fain, lovesick.

No characteristic of the Englishman is more clearly expressed in his art than his love of an harmonious life within the walls of that much-vaunted castle of his, which is inviolate, because the authorities know perfectly well that nothing dangerous is concealed within. (Who ever heard of a really dangerous English anarchist?) We have an incurable gift (called 'Spirit of Compromise') for taking an ideal, domesticating it, and making it something with which we can live harmoniously. Life must be pleasant and seemly. The French have a gift for making life fit an ideal – or be damned. It was they who had the Revolution and the Commune. There is a shy, wild-flower quality in our English art that makes it perhaps seem fragile and accidental when seen beside the art of contemporary schools of the Continent, with their strong intensive culture. But England's spontaneous charm never altogether fails, and is ever springing up under the most unlikely hedgerows and in queer company to carry us through seasons when professional crops on the Continent have perished in the drought.

But the continuous, laborious, seriousness so characteristic of the French mind is as alien to us as is its gaiety. French genius takes *pains* in the real sense of the word. Millet, Degas, Monet, and Cézanne belong to a line of artist-explorers to whom we have no equivalent. It is this spirit that England needs most, for in our island art loses her divine fierceness and challenge, and we forget that beauty should be more than sweetness, that art at her noblest can be 'terrible as an army with banners.' Therefore let us not shut our gates to all that comes with the smoky flares of the Post-Impressionists.

What the future may hold for English art is more than ever an enigma; but of one thing we may be sure: Post-Impressionism, either as a poison or a medicine, will never be taken here in its purity. You never get in England the empty vessel. Artist, musician, writer, politician – their capacity is always nearly full: only a little can go on top and the body of the liquor remains much the same. None of our national bogeys are really dangerous. No anarchists, Jesuits, or Post-Impressionists can ever have their will of us. South Kensington and Hammersmith can sleep safe o' nights, well guarded by the Spirit of Compromise, formidable to Art as to Anarchy.

93. Albert Gleizes and Jean Metzinger, *Cubism*

1913, pp. 14–17

Gleizes (1881–1953) together with Metzinger (1883–1957) exhibited with the cubists in Paris in 1910 and 1911 but made no real contribution to the development of cubism. They were better known as publicists of the movement.

We must not regard Impressionism as a false departure. The only possible error in art is imitation; it infringes the law of time, which is the Law. Merely by the liberty displayed by their technique, or shown in the constituent elements of a tint, Monet and his disciples helped to widen the field of effort. They never attempted to render painting decorative, symbolic, or moral. If they were not great painters they were painters, and for that reason we should respect them.

People have tried to present Cézanne as a sort of genius *manqué*; they say that his knowledge is admirable, but that he cannot sing or say his ideas; he only stutters. The truth is that he was unfortunate in his friends. Cézanne is one of the greatest of those artists who constitute the landmarks of history, and it ill becomes us to compare him to Van Gogh or Gauguin. He suggests Rembrandt. Like the painter of the *Pilgrims of Emmaus*, neglecting idle applause, he has plumbed reality with a resolute eye, and if he himself has not attained those regions in which the profounder realism is insensibly transformed into a luminous spiritualism, at least he has left, for those who desire steadily to attain it, a simple and wonderful method.

He teaches us to overcome the universal dynamism. He reveals the reciprocal and mutual modifications caused by supposedly inanimate objects. From him we have learned that to alter the coloration of a body is to corrupt its structure. He prophesies that the study of primordial volume will open unknown horizons to us. His work, a homogeneous mass, shifts under the glance, contracts, expands, fades or illumines itself, irrefragably proving that painting is not – or is no longer – the art of imitating an object by means of lines and colours, but the art of giving our instinct a plastic conciousness.

To understand Cézanne is to foresee Cubism. Henceforth we are justified in saying that between this school and the previous manifestations there is only a difference of intensity, and that in order to assure ourselves of the fact we need only attentively regard the methods of this realism, which, departing from the superficial reality of Courbet, plunges, with Cézanne, into the profoundest reality, growing luminous as it forces the unknowable to retreat.

Some maintain that such a tendency distorts the traditional curve. Whence do they derive their arguments? From the past or the future? The future does not belong to them, as far as we are aware, and one must needs be singularly ingenuous to seek to measure that which exists by that which exists no longer.

94. Unsigned review of Gleizes's and Metzinger's *Cubism*

Athenaeum, 17 May 1913, 548–9

This little book (of 60 or more pages of illustrations, and as many somewhat slenderly furnished with letterpress) is difficult to review in the sense of giving any idea of its contents. A rage for condensation and a distrust for the kind of clarity which expresses a general idea under a concrete, but too rigidly binding form have resulted in an orgy of abstract nouns which may mean anything or nothing, according to the equipment which the reader brings to his task. To some extent the authors seem to have contemplated such a result. They wish to suggest rather than to convince. 'We will not attempt definitions,' they say; 'we honestly believe that we have said nothing which is not calculated to confirm the true painter in his personal predilections': a danger rather easily avoided if they cast their suggestions in so vague a form that the reader, if he does not like them taken one way, can take them another. They not only refrain from definitions, but also shirk the occasion for giving instances, even when (as on p. 43) typical examples are really necessary to consolidate the reader's apprehension of the contrasting categories previously described. It would have been well to recognize that the concrete instances of the reasoner, like the concrete images of the artist, are bound up in this compact form for purposes of transit from mind to mind.

We make these criticisms on the style of the book because it would be a pity if it were regarded as meaningless. We have here no dazzle of pyrotechnics verbally plausible, but intellectually void, like the manifestos of the Futurists, nor, in spite of our own occasional failure to hatch anything out, are there, we believe, any merely 'lapidary sentences – having the value of chalk eggs, luring the thinker to sit.' Messrs. Gleizes and Metzinger follow, we doubt not, a consistent train of thought, but, as in a Cubist picture, it takes so capricious a course, with such sudden breaks in its continuity and unexpected plunges from one plane of thought to another, that we are tempted to abandon our study of the 'integration of the (literary) consciousness' of the authors for the collection of aphoristic fragments, more striking than the argument in which they are set. 'Let the artist's function grow profounder rather than more extensive' is a saying which has strayed from its position as presenting the case against decoration. Of Impressionism we are reminded, 'here the retina predominates over the brain; but the Impressionist is conscious of this, and to justify himself he speaks of the incompatibility of the intellectual faculties and the artistic sense.' Another palpable hit in the same direction is conveyed in the remark that

> the least intelligent will quickly realize that the pretence of representing the weight of bodies and the time spent in enumerating their various aspects is as legitimate as that of imitating daylight by the collision of an orange and a blue.

Excellent, too, are the use of the word 'taste' as 'the consciousness of quality,' and the repudiation of the terms 'good' and 'bad' taste: 'A faculty is neither good nor bad, it is simply more or less developed.'

It will be observed that in each of these fragments our authors rather make contributions to general art criticism than explain the principles of Cubism. With an energy hardly necessary nowadays, they clear a way for an art of design more free from realism than any European art in the past; but, though they claim that Cubism is 'the only conception at present possible of the pictorial art,' they do not make clear the grounds of their pretension, or the nature of the tenets of the school. 'The science of design consists in instituting relations between straight lines and curves,' we are told; and again:—

> Form appears endowed with properties identical with those of colour. It is tempered or augmented by contact with another form, it is destroyed or emphasized, it is multiplied or it disappears.

These are dicta which hold good for Cubism, but do not necessarily imply Cubism, as it has in fact developed. For others besides the Cubist 'lines, surfaces, and volumes are only modifications of the notion of plentitude.' The crux of the matter is only touched in the page or two devoted to pictorial space, which 'we have negligently confounded with pure visual space or with Euclidean space.' To clear up this alleged distinction 'we should have to refer – to the non-Euclidean scientists, we should have to study at some length certain theorems of Reimann's.'

In thus shirking the one dubious part of their subject by vague reference to writers with whom neither artists nor critics are likely to be familiar, the authors disappoint us. These theorems should have been quoted or described, for the further arguments adduced cannot be accepted as sufficient to persuade us of the reality of the distinction. We are assured that 'the convergence which perspective teaches us to represent cannot evoke the idea of depth.' In the ordinary sense of words this is simply untrue, though doubtless, if the pure linear system be complicated with colour and modelling, it may 'evoke the idea of space' no longer. Obviously, in that case, 'to establish pictorial space, we must have recourse to tactile and motor sensations – indeed, to all our faculties.' This does not, however, distinguish pictorial as imaginatively differing in kind from visual space. That 'the Chinese painters evoke space, although they exhibit a strong partiality for *divergence* (i.e., inverted perspective), only indicates that success is dependent not on the elaboration of the mathematical means employed, but on the ratio between those means and the pretensions of the artist. Within the flat convention and calligraphic line of those artists, radiation – whether convergent of divergent – of lines assumed as parallel, is a sufficient symbol for space.

Neither does it follow that to respect the integrity of visual space implies great circumstantiality in its presentation – still less that it implies the 'imitation of volumes.' Forms may be assumed as boldly interpenetrating, the mathematical elements of objects may be disengaged with complete disregard to their actual surfaces, fantastic exaggerations and imaginative reactions may have free play, and the curiosities of optics be exploited for purposes of emphasis or distortion, yet the mathematical idea of pictorial space need not necessarily be other than that of visual space. Why then should all these things, by comparison so readily apprehended, be insisted on, while the fundamental thing is casually mentioned and dropped? Virtually we are told: 'These pictures escape your merely geometrical criticism, and obey other principles – as to which we need not trouble you.' With due gratitude to Messrs. Gleizes and Metzinger for a stimulating book this, we submit, is hardly playing the game.

95. Frank Rutter,
Art in My Time
1933, pp. 156–7

Rutter here refers to the sixth exhibition of the Allied Artists' Association of 1913 which developed further the trend for showing modern English and modern continental work together. For Rutter see headnote to no. 1.

Even as late as 1913 this was the only exhibition in London which welcomed work of a post-impressionist tendency. The London Salon of 1913 contained several paintings by Maurice Asselin, Henri Doucet, Picart Le Doux, Mme. Renée Finch, Per Krohg (the

Norwegian), and other Paris artists associated with the *Fauves*. The English exhibits included Mr. Roger Fry's *La Salute* and *Still Life*, and paintings by W.B. Adeney, Nina Hamnett, Wyndham Lewis and Edward Wadsworth.

But *the* feature of the Allied Artists in 1913 was not so much the painting as the sculpture. Here for the first time in England was seen the work of Brancusi – *Muse Endormie* – and of Zadkin, who sent three heads. Epstein, who was sensitively impressionist in his older *Head of a Babe* (1907), was uncompromisingly post-impressionist in his abstract *Carving in Flenite*. To these exhibits, all of which excited much discussion, must be added those of that greatly gifted sculptor, Henri Gaudier-Brzeska, who made his first appearance this year. His exhibit consisted of *Oiseau de Feu*, *Wrestler*, *Madonna* and three portrait busts. To-day Gaudier has his place in the Victoria & Albert and other museums, and is generally recognised as having been a true genius. The Allied Artists recognised him in January 1914, when he was elected Chairman of Committee. He further increased the esteem in which he was held when that year he exhibited his *Bird* in gun-metal and his *Insect*. All he did had distinction, but alike in his drawings and his sculptures he showed a special gift for interpreting animal and bird form.

96. Review signed 'X', 'Modernism at the Albert Hall'

Athenaeum, 26 July 1913, 92–3

A review of the Allied Artists' exhibition to which Rutter refers in no. 95.

Whether one is in sympathy or not with the work now on show at the Albert Hall, it would need a singular narrowness to impugn the policy of the Allied Artists' Association, of which it is the outcome. That policy should appeal to all unprejudiced and liberal-minded men. By it every artist, be he never so heterodox and disconcerting in his conception of his art, has the chance to face directly the judgment of the world at large. Nor do even questions of mere technical excellence go against him. The very pavement-artist, so long as he can contrive to pay the requisite fee, is made as free of the Association as any Futurist, Cubist, or Academician of them all. Here, if anywhere, that latent or discouraged talent, so often hinted at, has its opportunity.

The exhibition is, on the whole, pretty much what might be inferred from these conditions. It contains good, bad, and indifferent work, but one is inclined rather to wonder that there are so many works of the first-class, and so few of the second. The curse of the average modern picture show is the dull, monotonous level of more or less competent mediocrity. That charge, at least, cannot be brought against the Allied Artists. At their worst they are amusing, if nothing else; at their best, undeniably stimulating.

All the influences now at work in the art world find expression here, and it is with the very 'advanced' section which is its distinctive feature that we shall concern ourselves.

There is one dangerous tendency common to Cubists, Post-Impressionists, and in a measure even to Futurists; and that is a violent reaction in favour of the 'primitive' in treatment. There is nothing really novel in this. In a minor degree we find it in Greek sculpture of the 'archaistic' school. The idea behind it was the same, though the ineradicable sense of balance of the Greek kept him from reverting, say, to the Branchidæ type of sculpture. The latter-day reactionary, within an incredibly short space of time, is finding even the art of ancient Egypt too sophisticated for his ideal, and seeking technical inspiration in the crude fetishes of the South Pacific. It is not the mere germ of an idea, a suggestion of decorative possibilities, that he claims to borrow and develope independently. He holds up the objects themselves as examples of the truest and most complete art. The truth is that the present generation is overstocked with 'clever' young men. The number of technically capable and facile artists is out of all proportion to the percentage of true vocations. Hence the man who is obeying an irresistible internal mandate is apt to be confounded with the common run of mediocrities, whose deliberate eccentricity merely obscures such real ability as they possess. The claim of these would-be primitives is apparently to clear their vision of all sophistication and *parti-paris*, and to envisage the world with the free and unspoilt eyes of a child. There is a great deal to be said at first glance in defence of these aspirations. We all recognize the fascination of childhood, but it is a vision not to be realized a second time. There may be – nay, one is certain there are – many men whose natural outlook is that of the child or the barbarian, but the majority of these would-be neo-primitives are the very antithesis of the type to which they would revert. For the most part their attempts in this direction tend to place them on no higher level than that of the professed mimic. Their aspiration seems to be less towards reformation or creation than towards change for the sake of change.

Akin to this is the affectation of a sort of esoteric symbolism among the 'advanced' section. Their work may possess a latent significance which the uninitiate are unable at present to apprehend; but so long as the keys of the mysteries are retained by the artist and his fellowship, he must be content to forgo the recognition he presumably desires and possibly merits.

In any discussion of the modernists there are, however, two points which deserve the attention of all serious observers. First, this revolution, if it is a real revolution, cannot be checked. Secondly, it may be the immediate herald of a new Renaissance. Vaguely we seem to see here and there tokens that the consummation is close upon us. Many of these paintings and sculptures arouse not indifference, but a curious disquiet which is difficult to analyze.

97. Roger Fry,
'The Allied Artists'

Nation, 2 August 1913, 676–7

The Allied Artists are celebrating their sixth exhibition at the Albert Hall. The early struggles are over, and general recognition seems to be at hand. Their chief danger was not antagonism but indifference, but even that, formidable though it becomes in July, seems to be yielding to the persistence and energy of the promoters. Antagonism there could hardly be, since, by their very nature, the Allied Artists stand for no one artistic creed more than another. They stand merely for equal treatment of all or any. It is, therefore, as impossible for them as a body to be revolutionary as it is to be academic. Their one and only *raison d'être* is to put each artist into direct touch with the public without submitting his work to the censorship of brother artists. And few artists, however doubtful they may feel of their competence to criticise their own work, are convinced of the qualifications of others who may be hostile, incompetent, or else unfairly prepossessed in their favor. Indeed, when once this principle is fairly brought home to the artist, it would seem that the Allied Artists is the only general exhibition where he can exhibit without some loss of dignity, the only place where no one pretends to dispense to him reward or blame – no one at least but that vague, inarticulate abstraction, the public, and the contradictory voices of contemporary criticism.

Perhaps it is nothing but our familiarity with the examination system which has kept alive so long the mischievous or futile activity of the jury. Artists covet the honor of being accepted where thousands are refused, of being hung on the line when most are skyed; it gives them a position; they are fortified in addressing the public by these certificates from fellow-artists; they can say with some confidence, 'You don't like my picture? But at least you know you ought to do so when you consider its excellent testimonials and references.' The Allied Artists, while they appeal to the artist's just pride in his profession, to his responsibility and conscience as a worker, do not therefore appeal to his vanity as the diploma-giving juries do. They merely put the artist on his honor.

If he shows bad work, he can shelter behind no verdict of his *confrères*.

Their attitude to the public is no less courageous. The Allied Artists call on the spectator for an effort which he is little accustomed to in contemporary exhibitions. He is forced to leave at the door the crutches of snobbism and social prestige with which he has hobbled round the Academy and the Salons. No friendly hand is held out to lead him up to this or that 'picture of the year,' no prominent finger points the way to facile enthusiasm over this or that centre piece in the weltering mass of art. He knows that the luck of the ballot may have placed the best work in the corner of the nth cubicle, indistinguishable by any mark but its intrinsic merit from all the *croûtes* which jostle it. It is a hard test, and the novice suffers like the faith-healer's patients; but in the end he may learn to walk alone.

Anyhow, a visit to the Albert Hall, alarming as it is, is profoundly interesting, even to those whose æsthetic sense is vague and dim. I can imagine the social philosopher finding here his proper field for investigation. Here, more than anywhere else, he may realise for himself the incredible jumble of conflicting ideals and aspirations of the modern world. Side by side with Kandinsky, pushing forward his fascinating experiments into a new world of expressive form, he will find some dear old friends of his childhood, sailors in sou'-westers and oilskins with the carefully painted grey reflections on their wet surfaces, helping in the rescue of a wrecked schooner, 'Mother's darling!' 'Love me, love my dog!' and all the other *clichés* of a sentimentality which by now has the faint and pleasant perfume of long-forgotten, old-world things. Side by side with Brancuzi's and Epstein's audacious simplifications he will find the last gasps of the enthusiasm for Roman *pastiches* of Greek art, or the tried and laborious photographic realism of the nineteenth century. Here he can trace the expiring efforts of lost causes, and forecast the menace of creeds that are not yet formulated.

But, naturally, the spectator who is more immediately concerned with art will turn with most interest to those works which would not pass the censorship of any British jury, and which, but for the Allied Artists, would never come to his notice. I have mentioned already the two names that interested me most. Constantin Brancuzi's sculptures have not, I think, been seen before in England. Hiss three heads are the most remarkable works of sculpture at the Albert Hall. Two are in brass and one in stone. They show a technical skill which is almost disquieting, a skill which might lead him, in default of any overpowering imaginative purpose, to become a brilliant *pasticheur*. But it seemed to me that there was evidence of passionate conviction; that the simplification of forms was no mere exercise in plastic design, but a real interpretation of the rhythm of life. These abstract ovoid forms into which he compresses his heads give a vivid presentment of character; they are not empty abstractions, but filled with a content which has been clearly and passionately apprehended. I wish I could say the same of Epstein's exquisitely executed and brilliant deformation, the carving in Flenite. I can only say that my admiration of his skill and of his alert critical sense is unaccompanied, in my own case, with any warmer feeling. I should have to give him the highest marks if I were an examiner, but I should withold my sympathy.

Of the paintings in the arena, Wyndham Lewis's *Group*, No. 998, is remarkable. It is more completely realised than anything he has shown yet. His power of selecting those lines of movement and those sequences of mass which express his personal feeling, is increasing visibly. In this work the mood is Michælangelesque in its sombre and tragic intensity. Mr. Lewis is no primitive. Thérèse Lessore's *Market Day at Amiens* is a praiseworthy attempt at composition on a monumental scale and by means of large masses, but it lacks accent in the color oppositions. Her *Village Shop* on a small scale in the Balcony has real charm.

As usual, Mr. Walter Sickert's pupils make a brave show in this part of the exhibition. They nearly all attain to a certain high level of solid accomplishment, and the women attain at least as high a point as the men. Mr. Sickert has reduced teaching to a fine art. He can push his pupils through all the preliminary stages of their art with unfailing accuracy and dispatch. And that, perhaps, is all a teacher can do. The rest he must leave to the gods, and I am not sure that they have yet sent him the material he deserves. Mr. Gore is already distinguished; Miss Sands in her *Venetian Interior* has gone as far as a refined and discriminating taste can go; but for the most part these artists seem too much preoccupied with the *métier* of painting to become definite creators. Still, the influence of their conscientious and honest craftsmanship makes itself felt, and makes the general average of competence higher at the Albert Hall than at any other annual exhibition. This group of artists alone suffices to prove the contention of the Allied Artists. But by far the best pictures there seemed to me to be the three works by Kandinsky. They are of peculiar interest, because one is a landscape in which the disposition of the forms is clearly prompted by a thing seen, while the other two are pure improvisations. In these the forms and colors have no possible justification, except the rightness of their relations. This, of course, is really true of all art, but where representation of natural form comes in, the senses are apt to be tricked into acquiescence by the intelligence. In these improvisations, therefore, the form has to stand the test without any adventitious aids. It seemed to me that they did this, and established their right to be what they were. In fact, these seemed to me the most complete pictures in the exhibition, to be those which had the most definite and coherent expressive power. Undoubtedly representation, besides the evocative power which it has through association of ideas, has also a value in assisting us to co-ordinate forms, and, until Picasso and Kandinsky tried to do without it, this function at least was always regarded as a necessity. That is why, of the three pictures by Kandinsky, the landscape strikes one most at first. Even if one does not recognise it as a landscape, it is easier to find one's way about in it, because the forms have the same sort of relations as the forms of nature, whereas in the two others there is no reminiscence of the general structure of the visible world. The landscape is easier, but that is all. As one contemplates the three, one finds that after a time the improvisations become more definite, more logical, and closely knit in structure, more surprisingly beautiful in their color oppositions, more exact in their equilibrium. They are pure visual music, but I cannot any longer doubt the possibility of emotional expression by such abstract visual signs.

There are a good many examples of cubism in the gallery. Among the best of these are Per Krohg's *Carnaval* and Wadsworth's *La Route*, but there are also a number of pictures in which a cubist formula has been imposed upon a very ordinary photographic vision. I do not know that this process spoils the work, but it certainly does little to help it, and the pictures remain to all intents and purposes the same as if they had been made up for the Royal Academy instead of the Albert Hall.

Phelan Gibb is an artist whose work has never appealed to me hitherto, but I felt that his *Etude Gothique* had a definite personal character which I could not trace in his more realistic *Huit nus*. I have never seen anything before by Miss Estelle Rice that seemed to me nearly so complete and personal as her *Early Rising* in this exhibition, though the other works she shows are on her more usual level of stylistic accomplishment.

---

98. Frank Rutter, Foreword to the catalogue of the Post-Impressionist and Futurist Exhibition

Doré Galleries, October 1913

This exhibition served to sum up graphically the change that had taken place in the style of English art and the sensibility of the British public since Roger Fry's first Post-Impressionist exhibition in the three years between 1910 and 1913. The old-masters of Post-Impressionism were represented by the paintings of Cézanne, Van Gogh and Gauguin; Fauvism in both its Continental and its English forms in the paintings of Friesz, Derain, Segonzac, J.D. Fergusson and Wolmark. Work by the two most prominent French living artists – Picasso and Matisse – was seen alongside the more conservative 'Neo-realism' of Gore, Ginner and Gilman and there was Futurist painting by Severini together with recent Vorticist work by Wyndham Lewis, Wadsworth and Nevinson. Bloomsbury painting – work by Duncan Grant, Vanessa Bell and Fry was notably absent after the rift between Fry and Lewis over the Omega project.

This Exhibition is an attempt to set forth in a coherent and so far as possible in a chronological order examples of various schools of painting which have made some noise in the world during the last quarter of a century. The loose way in which the term 'post-impressionist' has been used to cover a number of varying, and in some respects contradictory movements, has naturally confused a public seldom inclined to push very far its analysis of modern painting.

Now the French Impressionist movement of 1870 was based on two great principles: –

1. The instantaneous vision of a whole scene as opposed to the consecutive vision that sees nature piece by piece;

2. The substitution of a natural chiaroscura of *colour* based on the solar spectrum for a conventional chiaroscura of *tone* based on black and white.

The first was not a new discovery, it was a principle more or less recognised by most of the old masters, a principle which urged Titian, Rembrandt and others to give more breadth and less detail in their painting as they grew older, a principle which has intuitively prompted art lovers all the whole world over frequently to prefer a sketch to a 'finished picture.'

The second was essentially a new discovery, though earlier painters, and especially Delacroix, had been moving in that direction. The change, amounting to a revolution, brought about by the adoption of this principle showed itself markedly in the painting of shadows. Whereas formerly painters were apt to ask of a gray whether it was merely light or dark, these impressionist painters went further and asked if it was a purplish gray, a blue gray, a greenish gray, and so on. This research into the colour of light and especially into the colour of shadows, begun by Claude Monet, Camille Pissarro and Sisley, was further developed by Seurat and Paul Signac, the leaders of the neo impressionist movement. The school which sought to give the brilliance of light by the juxtaposition of touches of pure, undegraded pigment, had offspring in the groups known as the 'Divisionists,' 'Pointillists,' and 'Intimists.' The two former were dogmatic in their methods, the last adopted the divisionist principles only so far as these suited their subjects and their own temperaments. But all these schools aimed at truth to nature, and many of the leading members have painted the colour of light and shadow with a truth and brilliance never before attained in the history of art.

Camille Pissarro has been appropriately chosen as the starting point for this exhibition. Both in his landscapes and in his figure subjects the two great principles of French Impressionism came to their full fruition. While as the master of Gauguin and Van Gogh he unconsciously brought into being the painters who, with Cézanne, were to be the parents of most that is known as 'post impressionism.'

The 'Divisionists' are represented in this exhibition by Signac, Cros, and the Belgian artist M. Theo van Rysselberg. It is interesting to note that the last, in his recent work has abandoned his former dogmatic pointillism touch.

MM. Bonnard and Vuillard are accepted masters of 'Intimist' painting in France, but in England another turn has been given to this movement. M. Lucien Pissarro, eldest son of Camille Pissarro, brought to England the principles of French Impressionism. Mr. Walter Sickert, once a pupil of Whistler, had already acquainted himself in Paris with the art of the earlier impressionists, and these two quietly working in London and attracting to themselves congenial spirits, have produced the highly promising group of painters associated with the name of Camden Town. Allowing for individual and racial temperamental differences. Messrs. Spencer Gore, Gilman, Ginner and others of this group, might with least confusion be regarded as the 'intimists' of England.

It must always be remembered that classification at best is but a clumsy and inexact contrivance, and art is so much the affair of the individual that any attempt to classify artists is courting disaster. Nevertheless there are closer affinities between some artists than others, though strict classification is impossible when we are dealing with all the ramifications of modern painting. Mention must now be made of three artists exceedingly hard to classify.

Paul Cézanne (1839–1906) was numbered in his lifetime among the Impressionists, with whom he exhibited; but to-day he is commonly regarded as the Father of Post-Impressionism. The unique position he holds between the older and younger schools is best explained by his own words: – 'I wish to make of impressionism something solid and durable like the art of the old masters.' Whereas most of his comrades were pre-occupied with the rendering of transitory, fugitive effects of light, Cézanne seemed concerned with expressing in terms of colour the eternal verities of things themselves. His art was more simple, less complex than that of Monet and Pissarro; his analysis of colour was more summary, his expression more vigorous and forcible. Cézanne has had and still has an immense influence on modern painting. Weaker disciples are apt to imitate his superficial mannerisms and his defects rather than his qualities, but others have learnt from him to have the courage of their own vision and give that 'plain, forcible statement of things seen' which distinguishes much of the best contemporary French work, including notably the paintings of Albert Marquet. Attention may here be called to the remarkable group of young aquarellists who have been influenced by Cézanne as well as by Pissarro. Hitherto, France has had individual artists in water colours of great distinction, like Jacquemart, but never a great water colour school. The growth of this school in the present century is foreshadowed by the brilliant drawings of MM. Asselin, Ludovic Rodo, Doucet, Picart Le Doux, Paul Emile Pissarro, and other young artists represented in this exhibition.

Vincent Van Gogh (1853–90), in turn a shop-assistant, a school-master and a missionary, only began painting in 1885, and was at first influenced by Pissarro and Seurat whom he met in Paris soon after he left Holland in 1886. His comparatively early river scene (No. 15) in this exhibition is a typical neo-impressionist painting. He rapidly developed a very distinctive style of his own, remarkable for its vehemence of attack, fierce strokes of paint being rained almost like blows on the canvas. He was the most passionate of painters, and the extraordinary intensity of his vivid impressions has been likened by a sympathetic connoisseur to our impression of 'things seen momentarily in the duration of a lightning flash.' His work has had an influence on modern art hardly inferior to that of Cézanne. His colour is of a high order and pitch, showing a fine sensibility for the splendour of pigment.

Paul Gauguin (1848–1903) learnt painting from Camille Pissarro, whose style he emulated closely in his earlier works. His mother was a Creole, and he was born with a passion for the tropics, and in 1891 he sailed for Tahiti where he sought to paint primitive folk in a primitive style. Gauguin was not a realist, but an idealist and he found his ideal among the unspoilt barbarians of the Pacific. When a literary friend quarrelled with his ideal, Gauguin replied, 'Your civilization is your disease, my barbarism is my restoration to health.' This sentence gives us the clue to the movement known as 'Fauvism,' which to a great extent was influenced by the example of Gauguin.

Philosophers can easily make out a case for regarding the whole of modern European civilization as a disease. 'If our life is diseased,' said some young painters of Paris, 'our art must be diseased also, and we can only restore art to health by starting it afresh like children or savages.' The old masters were swept aside by these young revolutionaries who sought inspiration from the rudimentary art of savage and barbaric nations. Forcible, child-like scrawls began to appear in the Paris exhibitions, and these earned for their authors the nickname of *fauves* (wild beasts), a term which was accepted by the painters as fitly embodying their hatred of the tame and conventional. M. Henri Matisse has good claims to be regarded the leader of this *fauviste* movement, which has a salient merit in so far as it aims at simplicity and at securing a maximum of expression with the most rigid economy of means.

The weakness of the *fauvist* group is that too many of its members have derived their impetus from art rather than from nature. The figures of Cézanne and Gauguin are the parents of many fauvist pictures, and apart from M. Matisse, the most original of all, the fauves may be sub-divided according to the masters by whom they have most been influenced. Serusier is perhaps the most gifted of those influenced by Gauguin; Doucet and Friesz of those influenced by Cézanne.

Loosely attached to this group are English and American artists like J.D. Fergusson and Miss Rice who have been encouraged by the freedom of the fauves to follow their own instinctive love for bright colour and pattern. Generally the *fauviste* movement may be regarded as an extreme emotional reaction against the too coldly intellectual tendencies of other painters. Art is always swinging between emotion and intellectuality, and painters who keep the balance between the two are rare.

How cubism grew out of fauvism may be studied by the curious in three photographs of works by M. Pablo Piccasso. This young Spanish artist accomplished in early youth a series of masterly drawings and etchings in orthodox styles. Later he developed the angular style seen in his *Lady with a Fan*. The transition from this to *Head of a Lady with a Mantilla*, (No. 74) is easy to follow, and here we have the beginning of cubism. But after this comes the *Portrait of M. Kahnweiler*, (No. 73). Although the father of cubism

M. Piccasso disclaims being a cubist himself, he calls himself a realist. According to the artist, his latest works show 'things as they are and not as they appear'; that is to say they do not show *one* aspect of objects but a number of sectional aspects seen from different standpoints and arbitrarily grouped together in one composition. Of many artists whom Piccasso has influenced Herbin is undoubtedly one of the most gifted and original, his work having a charm of bright colour which the almost monotone paintings of M. Piccasso do not possess.

This idea of the sectional statement of divers aspects of different things has been developed, with an accent on the expression of movement, by the group of Italian painters known as the 'Futurists,' and though there is strong evidence that Signor Severini developed his own style independently of Piccasso – as Segantini did of Monet – the majority of the futurists have been more or less influenced by Piccasso. A similar development in Paris has been given the name of '*Orféism*,' and of this movement M. Delaunay is the protagonist. That 'cubism' and 'futurism' have already stirred English artists is shown by the contributions of Mr. Wyndham Lewis, Mr. Wadsworth, Mr. Nevinson and others.

This foreword has seemed necessary to emphasise, even at the risk of boredom, the fact that all 'post-impressionism' is not a development of 'impressionism.' Much is a reaction against the stern intellectual discipline which true impressionism necessitates. Movements so contrary and so numerous cannot be swallowed whole however they are labelled, and the work of each individual artist must be studied and analysed if justice is to be done to the various groups and fusions of groups concealed under the general term of 'Post-Impressionism.'

The months which led up to the outbreak of the First World War in August 1914 marked the final disintegration of any consensus about the idea of modern art as it related to Post-Impressionism. Until the early months of 1913 modern English aesthetics had been dominated by Bloomsbury and most especially by Roger Fry's views of French painting. Fry's authority was challenged by Wyndham Lewis, who reacted against the gentle neo-classicism of Bloomsbury and removed himself from the Omega project, taking with him Frederick Etchells, Edward Wadsworth and others.

They founded the Rebel Arts Centre and the London Group and their artistic principles were significantly different from those of Fry and Bell. The second group which set themselves up against Fry and his interpretation of Post-Impressionism were the so-called Neo-Realists, Harold Gilman and Charles Ginner. They, too, were members of the London group, but their sources of inspiration were radically different from those of Lewis. Finally there were the writers T.E. Hulme and Ezra Pound who established themselves in the columns of the *New Age* at the expense of Anthony Ludovici with whom they disagreed profoundly over the sculpture of Jacob Epstein. In 1914, however, Bloomsbury re-established its position with the publication of Clive Bell's *Art* which, though it was philosophically suspect in its reasoning about significant form and aesthetic emotion, was at the same time a huge popular success.

99. Percy Wyndham Lewis, 'The Cubist Room'

Egoist, 1 January 1914, 8–9

Lewis (1882–1957), painter, novelist and critic, was a student at the Slade School between 1898 and 1901. In his subsequent travels on the Continent he became familiar with developments in French art and in 1911 became, together with Sickert, Gore and others, a founder member of the Camden Town Group and later the London Group. The exhibition of English Post-Impressionists, Cubists and Others at the Brighton art gallery between December 1913 and January 1914 was divided into two sections. The catalogue introduction to the more conservative Camden Town painting was written by J.B. Manson and included painting by Ginner, Bevan, Gore, Gilman, Sickert, Pissarro, Sands, John and Paul Nash, Drummond, Bayes and others. Wyndham Lewis wrote the introduction to the second section (later published in *The Egoist*) which included the work of Epstein, Hamilton, Wadsworth, Etchells, Bomberg, Nevinson and Lewis himself.

Futurism, one of the alternative terms for modern painting, was patented in Milan. It means the Present with the Past rigidly excluded, and flavoured strongly with H.G. Wells' dreams of the dance of monstrous and arrogant Machinery, to the frenzied clapping of men's hands. But futurism will never mean anything else, in painting, than the Art practised by the five or six Italian painters grouped beneath Marinetti's influence. Gino Severini, the foremost of them, has for subject matter the night resorts of Paris. This, as subject matter, is obviously not of the future. For we all foresee in a century or so everybody being put to bed at 7 o'clock in the evening by a State Nurse. Therefore the Pan Pan at the Monaco will be, for Ginos of the Future, an archaistic experience.

Cubism means, chiefly, the art, superbly severe and so far morose, of those who have taken the genius of Cézanne as a starting point, and organised the character of the works he threw up in his indiscriminate and grand labour. It is the reconstruction of a simpler earth, left as choked and muddy fragments by him. Cubism includes much more than this, but the 'cube' is implicit in that master's painting.

To be done with terms and tags, post impressionism is an insipid and pointless name invented by a journalist, which has been naturally ousted by the better word 'Futurism' in public debate on modern art.

This room is chiefly composed of works by a group of painters, consisting of Frederick Etchells, Cuthbert Hamilton, Edward Wadsworth, C.R.W. Levinson, and the writer of this foreword. These painters are not accidentally associated here, but form a vertiginous but not exotic island, in the placid and respectable archipelago of English art. This formation is undeniably of volcanic matter, and even origin; for it appeared suddenly above the waves following certain seismic shakings beneath the surface. It is very closely-knit and admirably adapted to withstand the imperturbable Britannic breakers which roll pleasantly against its sides.

Beneath the Past and the Future the most sanguine would hardly expect a more different skeleton to exist than that respectively of ape and man. Man with an aeroplane is still merely a bad bird. But a man who passes his days amid the rigid lines of houses, a plague of cheap ornamentation, noisy street locomotion, the Bedlam of the press, will evidently possess a different habit of vision to a man living amongst the lines of a landscape. As to turning the back, most wise men, Egyptians, Chinese or what not, have remained where they found themselves, their appetite for life sufficient to reconcile them, and allow them to create significant things. Suicide is the obvious course for the dreamer, who is a man without an anchor of sufficient weight.

The work of this group of artists for the most part underlines such geometric bases and structure of life, and they would spend their energies rather in showing a different skeleton and abstraction than formerly could exist than a different degree of hairiness or dress. All revolutionary painting to-day has in common the rigid reflections of steel and stone in the spirit of the artist; that desire for stability as though a machine were being built to fly or kill with; an alienation from the traditional photographer's trade

and realisation of the value of colour and form as such independently of what recognisable form it covers or encloses. People are
invited, in short, to change entirely their idea of the painter's
mission, and penetrate, deferentially, with him into a transposed
universe as abstract as, though different from, the musicians.

I will not describe individually the works of my colleagues. In
No. 165 of Edward D. Wadsworth; No. 161 of Cuthbert
Hamilton; Nos. 169 and 181 of Etchells; No. 174 of Nevinson,
they are probably best represented.

Hung in this room as well are three drawings by Jacob Epstein,
the only great sculptor at present working in England. He finds in
the machinery of procreation a dynamo to work the deep atavism
of his spirit. Symbolically strident above his work, or in the midst
of it, is, like the Pathe cock, a new-born baby, with a mystic but
puissant crow. His latest work opens up a region of great
possibilities, and new creation – David Bomberg's painting of a
platform, announces a colourist's temperament, something between the cold blond of Severini's earlier paintings and Vallotton.
The form and subject matter are academic but the structure of the
criss-cross pattern new and extremely interesting.

---

100. Charles Ginner, 'Neo-Realism'

New Age, 1 January 1914, 271–2

Ginner (1878–1952) studied at the Académie Vitti between
1904 and 1908. His early work suggests some familiarity with
the painting of Van Gogh.

All great painters by direct intercourse with Nature have extracted
from her facts which others have not observed before, and
interpreted them by methods which are personal and expressive of

themselves – this is the great tradition of Realism. It can be traced in Europe down from Van Eyck and the early French primitives of the Ecole d'Avignon. It is carried through the dark period of the Poussins and Lebruns by Les Frères le Nain; in the eighteenth century by Chardin; in the nineteenth by Courbet and the Impressionists, and unbroken to this day by Cézanne and Van Gogh. Realism has produced the *Pieta* of the Ecole d'Avignon, the *Flemish Merchant and Lady* of Van Eyck, the old man and child of Ghirlandajo at the Louvre, *La Parabole des Aveugles* of Breughel (Le Vieux), the *Repos de Paysans* of Les Frères le Nain. Greco, Rembrandt, Millet, Courbet, Cézanne – all the great painters of the world have known that great art can only be created out of continued intercourse with nature.

It should be our endeavour to maintain this tradition through this present dark period of bad 'Academism' – the result, as ever, of the adoption by weak or commercial painters of the creative artist's personal methods of interpreting nature and the consequent creation of a formula; it is this which constitutes Academism. The further they go the more they see only this formula, and, losing all sight of nature, become Formula-machines. Art goes then from bad to worse; through the history of Art we see this continually. It has resulted in the decadence of every Art movement. We have the downfall of Egypt, the downfall of Greece, and the bad art of Rome. The Italian Renaissance going to Rome and not to nature ended in the quagmire of Giulio Romano, Carracci, etc. Poussin, Lebrun, and others, going to the Italian Renaissance, stultified French Art for hundreds of year until it finally ended in the 'débacle' of Bouguerau, Gérome, of the British Royal Academy, and of those of all the nations.

It is this shrinking from the Life around them, this hunting after a something as remote from life as possible, this race for Formula-Illusions, which destroys Art.

The creative power has always been realist. We can take as examples the early Egyptians, the early Greeks, the early Italians, the early French, the early Flemish.

The Academic painters merely adopt the visions which the creative artists drew from the source of nature itself. They adopt these mannerisms, which is all they are capable of seeing in the work of the creative artist, and make formulas out of them.

They are copyists. They are the poor of mind.

But in this article I wish to deal with our own times, with the Art of to-day.

The old Academic movement which reigned at Burlington House and the Paris Salon counts no more. In these precincts it has been replaced by a Naturalism just as bad, but which I will speak of anon.

There is a new Academic movement full of dangers. Full of dangers because it is disguised under a false cloak. It cries that it is going to save Art, while in reality it will destroy it. What in England is known as Post-Impressionism – Voila l'ennemi! It is all the more dangerous since it is enveloped in a kind of rose-pink halo of interest. Take away the rose-pink and you find the Academic skeleton. There are several forms of painting which I understand to be included under the journalistic term 'Post-Impressionism.' One is the adoption of a formula founded on the special interpretation of nature we find in Cézanne the Realist. He felt nature simply and interpreted it accordingly by dividing the object into separate simplified planes of colour which strengthened the feeling of solidity and depth and gave in certain cases a cubistic appearance to the depicted objects. His words that the forms of nature 'peuvent se ramener au cône, au cylindre et à la sphère,' was simply his mode of expressing his feelings of simplified nature.

The Post-Cézannes adopted this superficial aspect of his work without searching into the depth of his emotions and his mind, and created a formula.

Cubism is a development of Post-Cézannism.

Besides Cézannism and Cubism there is another form of Post-Impressionism of which exactly the same may be said: that of Matisse and his followers. Matisse hunts up formulas in Egypt, in Africa, in the South Seas, like a dog hunting out truffles. The formula once found ready made, the work is easy. The smaller Matisse fry find it even easier, as they have not the trouble of hunting.

The Matisse movement is a misconception of Gauguin as the rest of this Post-Impressionist movement is a misconception of Cézanne. Gauguin, who had a strong romantic touch, went to the South Seas and painted the South Sea islanders. Out of this a Post-Gauguin school arose, of which Matisse would seem to be the most important development. Out of Gauguin's Romantic Realism and his personal interpretation Matisse and Co. created a formula to be worked quietly at home in some snug Paris studio, as far away as possible from the South Seas or any other exotic country.

And so we come to my point that Cézannism and Gauguinism, i.e., Post-Impressionism, are academic movements as preached in England, being Art based on formula.

To Art, Academism means Death.

Every new Post-Cézanne, Post-Gauguin, or Post-Cubist will get worse than his predecessor as he gets further and further away from the light. The brain ceases to act as it ceases to search out its own personal expression of Nature, its only true and healthy source. Lying with ease on a bed of formulas the brain becomes dull and the Art becomes bad.

To this new Academism, which will eventually destroy Art, already so sorely tried by a recent bad Academic movement, we must oppose a young and healthy realistic movement, a New Realism, i.e., 'Neo-Realism.'

But that the conception of Realism, more especially that of Neo-Realism, may not be confounded with the Naturalism of Burlington House, I will say a few words about this dying naturalistic movement.

Naturalism is a kind of poor relation of Realism. It is the production of a Realist with a poor mind. A mind that goes to search out and reveal the secrets of Life and Nature, but has not the power to find. Naturalism is the photography of Nature. The Naturalist, with infinite care, goes out to her and copies the superficial aspect of the object before him. He only sees Nature with a dull and common eye, and has nothing to reveal. He has no personal vision, no individual temperament to express, no power of research. Nature remains a mystery to him in spite of all his work. Plastic Art then ceases, the decorative interpretation and intimate research of Nature, i.e., Life, are no more. It is in the R.A. that the last embers of this short-lived Naturalism are burning out.

Having given a summary of the place Realism holds with regard to Academism and Naturalism, I will try and develop the ideas that must guide Neo-Realism, the New Realism that is to oppose the headlong destructive flow of the New Academism, i.e., Post-Impressionism.

The aim of Neo-Realism is the plastic interpretation of Life through the intimate research into Nature. Life and Nature are the sources of the greatest variety. The artist who, with his personal ideal, his personal vision of nature and attitude towards life, makes an intimate study of what is round him is bound, even if he has not a strong individuality, to reveal an interesting work. Formula, on the other hand, being especially destructive to the smaller minds.

When this method of intimate research has been followed we find that the infinite variety of colour, pattern and line which is to be found in Nature and the arrangements evinced by them under the artist's personality 'create a whole which is a decorative composition.' This resulting decorative composition is an unconscious creation produced by the collaboration of Nature and the Artist Mind. A striking example of this unconscious decoration is to be seen in the works of the most intense of modern Realists, Vincent Van Gogh. A room at Bernheim's private house in Paris hung only by works of this great realist (who confessed to Gauguin that he could not work from imagination) makes one of the finest decorative wall-spaces I have ever seen.

A decorative formula tends to fall into monotony. The individual relying on his imagination and his formula finds himself very limited in comparison with the infinite variety of Life.

It is a common opinion of the day, especially in Paris (even Paris can make mistakes at times), that Decoration is the unique aim of Art. Neo-Realism, based on its tradition of Realism, has another aim of equal importance, a message deeper than the simple decorative Ideal, and on which it relies for its greatest strength. It must interpret that which, to us who are of this earth, ought to lie nearest our hearts, i.e., Life in all its effects, moods and developments. Each age has its landscape, its atmosphere, its cities, its people. Realism, loving Life, loving its Age, interprets its Epoch by extracting from it the very essence of all it contains of great or of weak, of beautiful or of sordid, according to the individual temperament. Realism is thus not only a present intimate revelation of its own time, but becomes a document for future ages. It attaches itself to history.

Neo-Realism must be a deliberate and objective transposition of the object (man, woman, tree, apple, light, shade, movement, etc.) under observation, which has for certain specific reasons appealed to the artist's ideal or mood, for self-expression. When the artist is carried away by an intense desire to interpret an object or an agglomeration of objects, the only sure means at his disposal to find and express that unknown quantity in the object which raised his desire, mood, or ideal, and which united his inner self with the aforesaid unknown quantity, is a deliberate research, concise study and transposition. It is only this intimate relation between the artist and the object which can produce original and great work. Away from this we fall into unoriginal and monotonous Formula.

Now let us consider Neo-Realism from a technical standpoint.

Which is the latest and most important realistic movement? It is unquestionably the impressionist movement in France.

The Impressionists, by their searching study of light, purified the muddy palettes by exchanging colour values for tone values, and thus strangely brought modern painting nearer to the great works of the Primitives; and they further revealed what till then seemed an unknown quantity: Light in Nature. This was an important discovery that no modern painter could afford to neglect, and the Neo-Impressionists pushed their study further and succeeded in relating Impressionist painting to Science. But with their eyes entirely fixed on this scientific study of colour and neglecting to keep themselves in relationship with Nature they began gradually to sink into the Formula Pit. On the other hand, we find Cézanne, Gauguin, Van Gogh, all three children of Impressionism learning from it, as a wholesome source, all that it had to teach, but keeping their minds and ideals open and independent, and with their eyes fixed on the only true spring of Art: Life itself. By this direct intercourse with Nature they brought out of Impressionism a new development by creating a personal Art and self-expression. So much so, that we have had learned, but short-sighted, men in France, Germany, and England demonstrating, amidst much noise, that these three painters were a reaction against impressionistic realism.

Far from being a reaction they were the very outcome, as stated above, the very development of it. They knew and that is what they have taught us, that great Art can be generated by the artist only through continued renewal with Life.

No masters could be further apart from each other than Cézanne, Gauguin, Van Gogh, and yet all their teaching is the same. Neo-Realism must take to heart the lesson so strikingly demonstrated by the comparison between the failure of the Neo-Impressionists, who created a formula out of Impressionism, and the success of the three French modern masters. Let those who are making a formula out of Cézanne or Van Gogh get entangled in the formulas and fall, only he who takes from Cézanne or Van Gogh that which he finds in them relating to Nature and not that which is merely personal to themselves will ever produce an original and great work of Art.

This deliberate research by the artist into Nature, this collaboration, this objective transposition must necessarily bring with it good and sound craftsmanship, a thing sorely lacking in these days. Of whatever interest a sketch may be as expressing a fleeting note, a mood, an 'état d'âme,' it can only be a small part of what the artist has in him to say. A pictorial work of Art must be a complete expression of the artist in relation to Nature, and must result in a strong and solidly built up work to be of any lasting purpose. Good craftsmanship must be the natural result of a strong, forcible, and deliberate self-expression. An artist who cannot go beyond a sketch is but a poor artist. Neo-Realism by its very ideals finds itself opposed to the slap-dash, careless, and slick painting which has been and is still so much in vogue.

The good craftsman loves the medium and the tools he uses. The real painter loves his paint as the sculptor his marble, for it is through these mediums that he reveals himself, is himself, and finds all his joy. In the great artist one must feel revealed, his love and passion for the medium with whose help he works to create the Art resulting from his desire to express those emotions awakened in him by Nature and Life around him.

Furthermore, in this matter of medium, it is only out of a sound and solid pigment that good surface and variety can be got, and durability in the ages to come.

Neo-Realism means intimate study of Nature, deliberate objective transposition, good craftsmanship, and a love of the medium. These, with a continued renewal with Life, i.e., collaboration of the Artist and Nature, must result in a strong, individual, and interesting interpretation of Life.

Neo-Realism must oppose itself to slave-ridden formula and be creative.

101. T.E. Hulme, 'Modern Art – 1: The Grafton Group'

New Age, January 1914, 341–2

Hulme (1883–1917) was sent down from St John's College, Cambridge, and joined Pound, Lewis and Epstein in their defence of what they saw as the weak Liberalism of Bloomsbury.

I am attempting in this series of articles to define the characteristics of a new constructive geometric art which seems to me to be emerging at the present moment. In a later series, to be called the 'Break up of the Renaissance,' I shall attempt to show the relation between this art and a certain general changed outlook.

I am afraid that my use of the word 'new' here will arouse a certain prejudice in the minds of the kind of people that I am anxious to convince. I may say then that I use the word with no enthusiasm. I want to convince those people who regard the feeble romanticism which is always wriggling and vibrating to the stimulus of the word 'new,' with a certain amount of disgust, that the art which they incline to condemn as decadent is in reality the new order for which they are looking. It seems to me to be the genuine expression of abhorrence of slop and romanticism which has quite mistakenly sought refuge in the conception of a classical revival. By temperament I should adopt the classical attitude myself. My assertion then that a 'new' art is being formed is not due to any desire on my part to perceive something 'new,' but is forced on me almost against my inclination by an honest observation of the facts themselves.

In attempting thus to define the characteristics of a new movement a certain clearance of the ground is necessary. A certain work of dissociation and analysis is required, in connection with what is vaguely thought of as 'modern' in art. A writer on art may perform a useful function in pointing out that what is generally thought of as one living movement consists really of many parts, some of which are as a matter of fact quite dead. The words 'modern,' 'Post-Impressionist' and 'Cubist' are used as synonyms, not only in the more simple form of instinctive reaction to an unpleasing phenomena, but also in a more positive way, the psychology of which seems to me to be rather interesting. The Post-Impressionist or Cubist appearances, at first perceived chaotically as 'queer' and rejected as such, became after a mysterious act of conversion, a signal for exhilarated acceptance, irrespective of the quality of the painting itself. They give every picture, good or bad, which possesses them, a sort of cachet. But although this complex of qualities passes from the stage in which it is repulsive to that in which it is attractive, yet for most people it remains unanalysed. It must be pointed out that what has been grouped together as one, really contains within itself several diverse and even contradictory tendencies. One might separate the modern movement into three parts, to be roughly indicated as Post-Impressionism, analytical Cubism and a new constructive geometrical art. The first of these, and to a certain extent the second, seem to me to be necessary but entirely transitional stages leading up to the third, which is the only one containing possibilities of development.

This show at the Alpine Club provides a convenient illustration of these points. Mr. Fry organised the first Post-Impressionist exhibition in London and was thought to have established a corner in the movement. He probably regards himself, and is certainly regarded by many others, as the representative of the new direction in art. The earlier shows of the Grafton group were sufficiently comprehensive and varied to make this opinion seem plausible ... There was a mixture of a sort of æsthetic archaism and a more vigorous cubism which corresponded very well to the loose use of the words 'modern art' which I have just mentioned, and helped to maintain the illusion that the whole formed in reality one movement. But the departure of Mr. Wyndham Lewis, Mr. Etchells, Mr. Nevinson and several others has left concentrated in a purer form all the worked-out and dead elements in the movement. It has become increasingly obvious that Mr. Fry and his group are nothing but a kind of backwater, and it seems to me to be here worth while pointing out the character of this backwater. As you enter the room you almost know what to expect, from the effect of the general colour. It consists almost uniformly of pallid chalky blues, yellows and strawberry colours, with a strong family resemblance between all the pictures; in every case a kind of anæmic effect showing no personal or constructive use of colour. The subjects also are significant. One may recognise the whole familiar bag of tricks – the usual Cézanne landscapes, the still lifes, the Eves in their gardens, and the botched Byzantine. As the Frenchmen exhibited here have really no connection with the Grafton group, I will omit them and confine myself to the English painters. In Mr. Fry's landscape you can see his inability to follow a method to its proper conclusion. The colour is always rather sentimental and pretty. He thus accomplishes the extraordinary feat of adapting the austere Cézanne into something quite fitted for chocolate boxes. It is too tedious to go on mentioning mediocre stuff, so I should like to point out the two things which are worth seeing, No. 29, a very interesting pattern by Mr. Roberts, and M. Gaudier-Brzeska's sculpture.

However, I find it more interesting to escape from this show for a minute, by discussing a general subject which is to a certain extent suggested by it – the exact place of archaism in the new movement. I want to maintain (i) that a certain archaism was a natural stage in the preparation of a new method of expression, and (ii) that the persistence of a feeble imitation of archaism, such as one gets in this show, is an absolutely unnecessary survival when this stage has been passed through.

In the first place then, how does it come about that a movement towards a new method of expression should contain so many archaic elements? How can a movement whose essence is the exact opposite of romanticism and nostalgie, which is striving towards a hard and definite structure in art, take the form of archaism? How can a sensibility so opposed to that which generally finds satisfaction in the archaic, make such use of it? What happens, I take it, is something of this kind: a certain change of direction takes place which begins negatively with a feeling of dissatisfaction with and reaction against existing art. But the new tendency, admitting that it exists, cannot at once find its own appropriate expression. But although the artist feels that he must have done with contemporary means of expression, yet a new and more fitting method is not easily created. Expression is by no means a natural thing. It is an unnatural, artificial and, as it were, external thing which a man has to install himself in before he can manipulate it. The way from intention to expression does not come naturally as it were from in outwards. It in no way resembles the birth of Minerva. A gap between the intention and its actual expression in material exists, which cannot be bridged directly. A man has first to obtain a foothold in this, so to speak, alien and external world of material expression, at a point near to the one he is making for. He has to utilise some already existing method of expression, and work from that to the one that expresses his own personal conception more accurately and naturally. At the present moment this leads to archaism because the particular change of direction in the new movement is a striving towards a certain intensity which is already expressed in archaic form. This perhaps supplements what I said about the archaism of Mr. Epstein's 'carvings in flenite.' It perhaps enables me to state more clearly the relation between those works and the more recent work represented by the drawings. You get a breaking away from contemporary methods of expression, a new direction, an intenser perception of things striving towards expression. And as this intensity is fundamentally the same kind of intensity as that expressed in certain archaic arts, it quite naturally and legitimately finds a foothold in these archaic yet permanent formulæ. But as this intensity is at the same time no romantic revival, but part of a real change of sensibility occurring now in the modern mind, and is coloured by a particular and original quality due to this fact, it quite as naturally develops from the original formula one which is for it, a purer and more accurate medium of expression. [That the great change in outlook is coming about naturally at the present moment, I shall attempt to demonstrate later by a consideration which has nothing whatever to do with art. I shall then be able to explain what I meant by the 'dregs of the Renaissance.']

To return then to the discussion from which I started. A certain archaism it seems is at the beginning a help to an artist. Although it may afterwards be repudiated, it is an assistance in the construction of a new method of expression. Most of the artists who prepared the new movement passed through this stage. Picasso, for example, used many forms taken from archaic art, and other examples will occur to everyone. It might be objected that a direct line of development could be traced through Cézanne showing no archaic influence. But I think it would be true to say of Cézanne, even in much of his later work, that he seeks expression through forms that are to a certain extent archaic. So much then for the function of archaism. Apply this to what you find in the Grafton group. If it were only a matter of serious experimentation in archaic forms, after the necessity for that experimentation had passed by, the thing would be regrettable but not a matter for any violent condemnation. But you do not find anything of that kind, but merely a cultured and anæmic imitation of it. What in the original was a sincere effort towards a certain kind of intensity, becomes in its English dress a mere utilisation of the archaic in the spirit of the æsthetic. It is used as a plaything to a certain quaintness. In Mr. Duncan Grant's *Adam and Eve*, for example, elements taken out of the extremely intense and serious Byzantine art are used in an entirely meaningless and pointless way. There is no solidity about any of the things; all of them are quite flimsy. One delightful review of the show described Mr. Fry's landscapes as having 'the fascination of reality seen through a cultured mind.' The word 'cultured' here explains a good deal. I feel about the whole show a typically Cambridge sort of atmosphere. I have a very vivid impression of what I mean here by Cambridge, as I have recently had the opportunity of observing the phenomenon at close quarters. I know the kind of dons who buy these pictures, the character of the dilettante appreciation they feel for them. It is so interesting and clever of the artist to use the archaic in this paradoxical way, so amusing to make Adam stand on his head, and the donkey's ear continue into the hills – gentle little Cambridge jokes.

It is all amusing enough in its way, a sort of æsthetic playing about. It can best be described in fact as a new disguise of æstheticism. It is not a new art, there is nothing new and creative about it. At first appearance the pictures seem to have no resemblance to pre-Raphaelitism. But when the spectator has overcome his first mild shock and is familiarised with them, he will perceive the fundamental likeness. Their 'queerness,' such as it is, is not the same serious queerness of the pre-Raphaelites, it is perhaps only quaint and playful; but essentially the same English æsthetic is behind both, and essentially the same cultured reminiscent pleasure is given to the spectator. This being the basic constituent of both arts, just as the one ultimately declined into Liberty's, so there is no reason why the other should not find its grave in some emporium which will provide the wives of young and advanced dons with suitable house decoration.

What is living and important in new art must be looked for elsewhere.

---

102. Clive Bell, 'Aesthetics and Post-Impressionism',

Art (1914), ed. J.B. Bullen (1987), pp. 38–48

In the first chapter of *Art* Bell had put forward the idea that all works of art appeal primarily to what he called the 'aesthetic emotion' which is dependent upon 'significant form' within those works of art. This second chapter develops these ideas specifically in the context of Post-Impressionist painting.

By the light of my aesthetic hypothesis I can read more clearly than before the history of art; also I can see in that history the place of the contemporary movement. As I shall have a great deal to say about the contemporary movement, perhaps I shall do well to seize this moment, when the aesthetic hypothesis is fresh in my mind and, I hope, in the minds of my readers, for an examination of the movement in relation to the hypothesis. For anyone of my generation to write a book about art that said nothing of the movement dubbed in this country Post-Impressionist would be a piece of pure affectation. I shall have a great deal to say about it, and therefore I wish to see at the earliest possible opportunity how Post-Impressionism stands with regard to my theory of aesthetics. The survey will give me occasion for stating some of the things that Post-Impressionism is and some that it is not. I shall have to raise points that will be dealt with at greater length elsewhere. Here I shall have a chance of raising them, and at least suggesting a solution.

Primitives produce art because they must; they have no other motive than a passionate desire to express their sense of form. Untempted, or incompetent, to create illusions, to the creation of form they devote themselves entirely. Presently, however, the artist is joined by a patron and a public, and soon there grows up a demand for 'speaking likenesses.' While the gross herd still clamours for likeness, the choicer spirits begin to affect an admiration for cleverness and skill. The end is in sight. In Europe we watch art sinking, by slow degrees, from the thrilling design of Ravenna to the tedious portraiture of Holland, while the grand proportion of Romanesque and Norman architecture becomes Gothic juggling in stone and glass. Before the late noon of the Renaissance art was almost extinct. Only nice illusionists and masters of craft abounded. That was the moment for a Post-Impressionist revival.

For various reasons there was no revolution. The tradition of art remained comatose. Here and there a genius appeared and wrestled with the coils of convention and created significant form. For instance, the art of Nicolas Poussin, Claude, El Greco, Chardin, Ingres, and Renoir, to name a few, moves us as that of Giotto and Cézanne moves. The bulk, however, of those who flourished between the high Renaissance and the contemporary movement may be divided into two classes, virtuosi and dunces. The clever fellows, the minor masters, who might have been artists if painting had not absorbed all their energies, were throughout that period for ever setting themselves technical acrostics and solving them. The dunces continued to elaborate chromophotographs, and continue.

The fact that significant form was the only common quality in the works that moved me and that in the works that moved me most and seemed most to move the most sensitive people – in primitive art, that is to say – it was almost the only quality, had led me to my hypothesis before ever I became familiar with the works of Cézanne and his followers. Cézanne carried me off my feet before ever I noticed that his strongest characteristic was an insistence on the supremacy of significant form. When I noticed this, my admiration for Cézanne and some of his followers confirmed me in my aesthetic theories. Naturally, I had found no difficulty in liking them since I found in them exactly what I liked in everything else that moved me.

There is no mystery about Post-Impressionism; a good Post-

Impressionist picture is good for precisely the same reasons as any other picture is good. The essential quality in art is permanent. Post-Impressionism, therefore, implies no violent break with the past. It is merely a deliberate rejection of certain hampering traditions of modern growth. It does deny that art need ever take orders from the past; but that is not a badge of Post-Impressionism, it is the commonest mark of vitality. Even to speak of Post-Impressionism as a movement may lead to misconceptions; the habit of speaking of movements at all is rather misleading. The stream of art has never run utterly dry: it flows through the ages, now broad now narrow, now deep now shallow, now rapid now sluggish: its colour is changing always. But who can set a mark against the exact point of change? In the earlier nineteenth century the stream ran very low. In the days of the Impressionists, against whom the contemporary movement is in some ways a reaction, it had already become copious. Any attempt to dam and imprison this river, to choose out a particular school or movement and say: 'Here art begins and there it ends,' is a pernicious absurdity. That way Academization lies. At this moment there are not above half a dozen good painters alive who do not derive, to some extent, from Cézanne, and belong, in some sense, to the Post-Impressionist movement; but tomorrow a great painter may arise who will create significant form by means superficially opposed to those of Cézanne. Superficially, I say, because, essentially, all good art is of the same movement: there are only two kinds of art, good and bad. Nevertheless, the division of the stream into reaches, distinguished by differences of manner, is intelligible and, to historians at any rate, useful. The reaches also differ from each other in volume; one period of art is distinguished from another by its fertility. For a few fortunate years or decades the output of considerable art is great. Suddenly it ceases; or slowly it dwindles: a movement has exhausted itself. How far a movement is made by the fortuitous synchronisation of a number of good artists, and how far the artists are helped to the creation of significant form by the pervasion of some underlying spirit of the age, is a question that can never be decided beyond cavil. But however the credit is to be apportioned – and I suspect it should be divided about equally – we are justified, I think, looking at the history of art as a whole, in regarding such periods of fertility as distinct parts of that whole.

Primarily, it is as a period of fertility in good art and artists that I admire the Post-Impressionist movement. Also, I believe that the principles which underlie and inspire that movement are more likely to encourage artists to give of their best, and to foster a good tradition, than any of which modern history bears record. But my interest in this movement, and my admiration for much of the art it has produced, does not blind me to the greatness of the products of other movements; neither, I hope, will it blind me to the greatness of any new creation of form even though that novelty may seem to imply a reaction against the tradition of Cézanne.

Like all sound revolutions, Post-Impressionism is nothing more than a return to first principles. Into a world where the painter was expected to be either a photographer or an acrobat burst the Post-Impressionist, claiming that, above all things, he should be an artist. Never mind, said he, about representation or accomplishment – mind about creating significant form, mind about art. Creating a work of art is so tremendous a business that it leaves no leisure for catching a likeness or displaying address. Every sacrifice made to representation is something stolen from art. Far from being the insolent kind of revolution it is vulgarly supposed to be, Post-Impressionism is, in fact, a return, not indeed to any particular tradition of painting, but to the great tradition of visual art. It sets before every artist the ideal set before themselves by the primitives, an ideal which, since the twelfth century, has been cherished only by exceptional men of genius. Post-Impressionism is nothing but the reassertion of the first commandment of art – Thou shalt create form. By this assertion it shakes hands across the ages with the Byzantine primitives and with every vital movement that has struggled into existence since the arts began.

Post-Impressionism is not a matter of technique. Certainly Cézanne invented a technique, admirably suited to his purpose, which has been adopted and elaborated, more or less, by the majority of his followers. The important thing about a picture, however, is not how it is painted, but whether it provokes aesthetic emotion. As I have said, essentially, a good Post-Impressionist picture resembles all other good works of art, and only differs from some, superficially, by a conscious and deliberate rejection of those technical and sentimental irrelevancies that have been imposed on painting by a bad tradition. This becomes obvious when one visits an exhibition such as the *Salon d'Automne* or *Les Indépendants*, where there are hundreds of pictures in the Post-Impressionist manner, many of which are quite worthless.[1] These, one realises, are bad in precisely the same way as any other picture is bad; their forms are insignificant, and compel no aesthetic reaction. In truth, it was an unfortunate necessity that obliged us to speak of 'Post-Impressionist pictures,' and now, I think, the moment is at hand when we shall be able to return to the older and more adequate nomenclature, and speak of good pictures and bad. Only we must not forget that the movement of which Cézanne is the earliest manifestation, and which has borne so amazing a crop of good art, owes something, though not everything, to the liberating and revolutionary doctrines of Post-Impressionism.

The silliest things said about Post-Impressionist pictures are said by people who regard Post-Impressionism as an isolated movement, whereas, in fact, it takes its place as part of one of those huge slopes into which we can divide the history of art and the spiritual history of mankind. In my enthusiastic moments I am tempted to hope that it is the first stage in a new slope to which it will stand in the same relation as sixth-century Byzantine art stands to the old. In that case we shall compare Post-Impressionism with that vital spirit which, towards the end of the fifth century, flickered into life amidst the ruins of Graeco-Roman realism. Post-Impressionsim, or, let us say the Contemporary Movement, has a future; but when that future is present Cézanne and Matisse will no longer be called Post-Impressionists. They will certainly be called great artists, just as Giotto and Masaccio are called great artists; they will be called the masters of a movement; but whether that movement is destined to be more than a movement, to be something as vast as the slope that lies between Cézanne and the masters of S. Vitale, is a matter of much less certainty than enthusiasts care to suppose.

Post-Impressionism is accused of being a negative and destructive creed. In art no creed is healthy that is anything else. You cannot give men genius; you can only give them freedom – freedom from superstition. Post-Impressionism can no more make good artists than good laws can make good men. Doubtless, with its increasing popularity, an annually increasing horde of nincompoops will employ the so-called 'Post-Impressionist technique' for presenting insignificant patterns and recounting foolish anecdote. Their pictures will be dubbed 'Post-Impressionist,' but only by gross injustice will they be excluded from Burlington House. Post-Impressionism is no specific against human folly and incompetence. All it can do for painters is to bring before them the claims of art. To the man of genius and to the student of talent it can say: 'Don't waste your time and energy on things that don't matter: concentrate on what does: concentrate on the creation of significant form.' Only thus can either give the best that is in him. Formerly because both felt bound to strike a compromise between art and what the public had been taught to expect, the work of one was grievously disfigured, that of the other ruined. Tradition ordered the painter to be photographer, acrobat, archaeologist and littérateur: Post-Impressionism invites him to become an artist.

NOTE

[1] Anyone who has visited the very latest French exhibitions will have seen scores of what are called 'Cubist' pictures. These afford an excellent illustration of my thesis. Of a hundred cubist pictures three or four will have artistic value. Thirty years ago the same might have been said of 'Impressionist' pictures; forty years before that of romantic pictures in the manner of Delacroix. The explanation is simple, – the vast majority of those who paint pictures have neither originality nor any considerable talent. Left to themselves they would probably produce the kind of painful absurdity which in England is known as an 'Academy picture.' But a student who has no original gift may yet be anything but a fool, and many students understand that the ordinary cultivated picture-goer knows an 'Academy picture' at a glance and knows that it is bad. Is it fair to condemn severely a young painter for trying to give his picture a factitious interest, or even for trying to conceal beneath striking wrappers the essential mediocrity of his wares? If not heroically sincere he is surely not inhumanly base. Besides, he has to imitate someone, and he likes to be in the fashion. And, after all, a bad cubist picture is no worse than any other bad picture. If anyone is to be blamed, it should be the spectator who cannot distinguish between good cubist pictures and bad. Blame alike the fools who think that because a picture is cubist it must be worthless, and their idiotic enemies who think it must be marvellous. People of sensibility can see that there is as much difference between Picasso and a Montmartre sensationalist as there is between Ingres and the President of the Royal Academy. [Bell's note]

103. Roger Fry,
'A New Theory of Art'

Nation, 7 March 1914, 937–8

Though many of Bell's theories are closely related to those of
Roger Fry, Fry's review is neither adulatory nor uncritical
and envisages numerous objections in the future to Bell's
arguments about significant form and aesthetic emotion.

Those questions of aesthetics theory which at any time in these last
four years have ruffled the pages of the *Nation* are treated in a new
book with such intensity of feeling and in such a vivid style as to
claim general attention more insistently than ever. Mr. Clive Bell's
book is as simple and suggestive as its title. He sets out to state a
complete theory of visual art. He says in his preface that he differs
profoundly from me. I feel bound, therefore, to do my best to
return the compliment. But I can do it but half-heartedly, for
although I have never stated a complete theory of art, my various
essays towards that end have by very slow steps been approaching
more and more in the direction which Mr. Bell has here indicated
with an assurance denied to me. It needs some courage to state a
complete theory of art, and leave it there for all the æsthetes to have
a shy at. And perhaps it is the high courage, the good-natured
pugnacity and outspokenness of its author, that make this book the
most readable of abstract treatises. Mr. Bell hits out freely all
round, and trails his coat as he goes along, but in so gay a spirit that
one thinks that none of his victims, Academicians, artists, critics,
experts, men of science, or even the general public, will bear him a
grudge even if they take up the challenge. And he has a gift of terse
and lucid explanation which has enabled him in this short book not
only to state his complete theory of visual art, but to show its
implications, metaphysical, ethical, social, religious, and even
historical. This last is perhaps the most thrilling and exhilarating
part of Mr. Bell's æsthetical joy-ride, for he runs us through from
Sumerian sculpture to Cézanne, from 3000 B.C. to 1900 A.D., at
breathless speed, and with some sharp turns and shattering jolts for

the cultured. But it would be a mistake to suppose that because the book is so pleasant and exhilarating to read, because the author keeps us entertained and either delighted or irritated from beginning to end, that it is flippant or superficial. Whether we agree with it or not we must take the argument seriously, for it is meant in all sincerity, and stated with a cogency that claims close attention.

He begins by inquiring what quality is common to all works of visual art and peculiar to them. He finds it to be the possession of 'significant form.' How do we recognize significant form? By its power to arouse æsthetic emotion. The reader will probably at this point ask: What is æsthetic emotion? And Mr. Bell will reply, the emotion aroused by significant form. Which seems to bring us full circle. But we have really got somewhere. For consider the concrete cases Mr. Bell gives. What is the quality common to Sta Sophia, a Persian bowl, a Chinese carpet, Giotto's frescoes at Padua? Now, if thinking of Giotto's frescoes, we begin talking of the emotions of pity, love, tenderness, and what not, we are pulled up over the Persian bowl and Sta Sophia, where certainly this class of emotion does not occur; so that we are driven to suppose that, however valuable or desirable the expression of these emotions may be in a particular case, it is not the fundamental and universal quality of works of visual art. We have separated out the emotions aroused by certain formal relations from the emotions aroused by the events of life, or by their echoes in imaginative creations. This is one of Mr. Bell's most important contributions to the question. But it is at this point that I wish Mr. Bell had been more ambitious and more comprehensive. I wish he had extended his theory, and taken literature (in so far as it is an art) into fuller consideration, for I feel confident that great poetry arouses æsthetic emotions of a similar kind to painting and architecture. And to make his theory complete, it would have been Mr. Bell's task to show that the human emotions of *King Lear* and *The Wild Duck* were also accessory, and not the fundamental and essential qualities of these works.

Perhaps Mr. Bell rightly felt that this would demand another book to itself. I hope he will undertake it in order to round off his theory, and if he should succeed in making this good, as I believe may be done, it would open up the possibility of a true art of illustration, which Mr. Bell at present refuses to contemplate.

Since, if in words images may be evoked in such an order, and having such a rhythmic relation, as to arouse æsthetic emotion, there is no apparent reason why images may not be similarly evoked by painting having a similar formal relation. This would be, as in literature, not a visible, but an ideal form. If, on the other hand, it cannot be shown that the essence of poetry is also one of formal relations, but is due to an admixture of form with content, then there would be nothing surprising in discovering that the art of painting was of a similar composite nature.

Song, again, would appear to be in exactly the same position as painting as ordinarily understood. For here we have the formal relations of music, and the content of ideas conveyed by words. Is there, then, a true art of song? Can we get a pure æsthetic emotion from this mixture? Probably, in proportion as people feel keenly the æsthetic quality of the music, they are content to let the words pass unheeded. I suspect, however, that most people feel æsthetic emotions so slightly that they treat them merely as stimulants to those echoes of the emotions of life which the words of a song arouse, and here, too, we should find the explanation of why many people are moved by pictures without having any keen æsthetic perception of form.

Mr. Bell expresses rather a pious belief than a reasoned conviction that the æsthetic emotion is indeed an emotion about ultimate reality, that it has, therefore, a claim as absolute as the religious emotion has upon those who feel it. The real gist of his book is a plea that to those who feel the æsthetic emotion, it becomes of such importance, so intimately and conclusively satisfying to their spiritual nature, that for them to have it interfered with by any other considerations, by the intrusion of any human emotion, however intense and valid it may be, is to miss the greatest value of art. His book is in praise of contemplative as against practical virtue.

Let us turn to another objection to Mr. Bell's theory, which is likely to arise. He himself admits that the artist has very rarely set out to create significant form; that the early Christian masters set out to express dogmatic theology, the fifteenth-century Florentine and the Impressionist to state laws of optics, Giotto and Rembrandt to express human emotion. Indeed, he considers that few things are more disastrous to the artist than the desire to create significant form in the abstract. It leads to naked and empty æstheticism. The artist, Mr. Bell rightly says, must have something to get into a passion about. But surely this significant form, with all its possible implications of ultimate reality, this form about which Mr. Bell himself rises to genuine artistic passion, is a thing of passionate import. Why, then, should it not suffice better than anything else for the artist? Mr. Bell has a theory about the artistic problem which I will not state, because I do not fully understand it, but it seems to me to be in the nature of a buttress run up to support a weak patch in the wall of his argument. It is just here, indeed, that I feel that the ultimate nature of aesthetic experience still eludes Mr. Bell and all of us, as indeed it may well do for some centuries to come.

Why must the potter who is to make a superbly beautiful pot not think only of its significant form, but think first and most passionately about its functions as a pot? Why must the architect get excited about engineering, as all great architects have, and why must the painter begin by abandoning himself to the love of God or man or Nature unless it is that in all art there is a fusion of something with form in order that form may become significant. And is it not just the fusion of this something with form that makes the difference between the finest pattern-making and real design? For the most ingenious and perfect pattern – a pattern which we judge to be absolutely impeccable – has not significance, while some quite faulty and stumbling efforts possessing this other thing in them move us profoundly. We should have to admit that this something, this x in the equation, was quite inconstant, and might be of almost any conceivable nature, but I believe it would be possible, applying Mr. Bell's logical methods of deduction, to restate his answer to the inquiry what is common and peculiar to all works of art in some such way as this: The common quality is significant form, that is to say, forms related to one another in a particular manner, which is always the outcome of their relation to x (where x is anything that is not of itself form). I throw out this horrible mathematical formula as a possible suggestion for future investigation, to those at least, whom Mr. Bell's brilliant logic and persuasive eloquence still leave incompletely satisfied.

It would need a separate article or perhaps a separate book to deal with our author's historical survey of art. It is certainly one of the most brilliant, provocative, suggestive things that have ever been written on the subject. It is in a way a complete vindication of

Ruskin's muddle-headed but prophetic intimations of the truth. For here at last the whole history of art, as it has been taught by extensive lecturers and text books for ages, is just turned upside down. Barbarian becomes *Blüthezeit*, and *Blüthezeit* complete decadence. Dark Ages become points of shining light, and Renaissance merely prolonged putrefaction.

This revaluation is not, of course, entirely new; the process has been gradually going on for many years; even the despised collectors and archæologists have done something towards it. Did not the late Mr. Morgan pay more for some Byzantine enamels than for an Alma Tadema? But the implications of this turn of taste have never been so freely accepted, and certainly never stated with such passionate conviction. No one yet has ventured to push over the ancient idols of culture with so light and unconcerned a touch. It is this that dates the work; we of an older generation have felt qualms about the worship of culture, have carped at this or that reputation, have pushed in a query here and a caution there, but we have always had a sense of the awful responsibility of profaning the temples; we have been apologetic and deferential even while we were undermining the foundations.

But Mr. Bell walks into the holy of holies of culture in knickerbockers with a big walking-stick in his hand, and just knocks one head off after another with a dexterous back-hander that leaves us gasping. Many a bounder has been in before and had a cock-shy at the reverend figures, but precisely because he was a bounder his missiles have had no effect. But Mr. Bell knows the ritual of culture better than the pious hierophants. It is that that makes him so deadly in his aim.

No doubt, if I were to go through his history in detail, I should try to set up a good many of his victims on their pedestals once more, try to repair with pious hands some of what seems to me unnecessary damage. But what a breath of fresh air this iconoclast brings in with him, what masses of mouldy snobbism he sweeps into the dust heap, how salutary even for the idols themselves – those at least that survive at all – is such a thorough turning out! It will be seen that this is a book that all who care for art must read; the surprising good fortune that has befallen them is that it is so eminently readable.

Chronology

1905

GREAT BRITAIN

Derain and Vlaminck painting in London
Walter Sickert returned to England from the Continent

January–February	*Grafton Galleries*: Boudin, Cézanne, Degas, Manet, Monet, Morisot, Pissarro, Renoir, Sisley
March–April	*New Gallery*: Whistler Memorial exhibition. Transferred to Paris in May
May–June	*Baillie Gallery*: Fergusson
Autumn	*Friday Club* formed by Vanessa Stephen (later Bell)

FRANCE

March–April	*Indépendants*: O'Conor, Denis, Derain, Van Gogh (retrospective), Matisse, Seurat (retrospective), Vallotton, Vuillard
October–November	*Salon d'Automne*: Cézanne, Derain, Matisse, Kandinsky, Lucian Pissarro, Rouault, Sickert, Vallotton, Vlaminck, Vuillard. So-called *Cage des Fauves*

1906

GREAT BRITAIN

George Moore, *Reminiscences of the Impressionist Painters*

January	*New Gallery*: International Society: Cézanne, Manet, Pissarro, Degas, Morisot, Fergusson, Renoir
February	*Lafayette Gallery*: Pictures from the 1905 Salon d'Automne
February–March	*Agnew*: Some Examples of the Independent Art of Today: Brown, Conder, Dewhurst, Fry, Holmes, Lavery, MacColl, Nicholson, Orpen, Ricketts, Rothenstein, Shannon, Sickert, Steer, Strang, Tonks

FRANCE

March–April	*Indépendants*, inc.: Braque, O'Conor, Cross, Denis, Derain, Van Dongen, Marquet, Matisse, Rouault, Vallotton, Vlaminck, Vuillard
October	Death of Cézanne
October–November	*Salon d'Automne*, inc.: Cézanne, O'Conor, Derain, Van Dongen, Friesz, Gauguin (retrospective), Kandinsky, Matisse, Metzinger, Sickert, Vlaminck, Vuillard, Weber

1907

GREAT BRITAIN

Spring	*Fitzroy Street Group* founded by Sickert, Gore, Gilman and others

FRANCE

Picasso paints *Les Demoiselles d'Avignon*
Manet's *Olympia* and *Déjeuner sur l'herbe* enter the Louvre

January	*Bernheim Jeune*: Sickert

January–February	*Bernheim Jeune*: Signac
February	*Druet*: Marquet
March–April	*Indépendants*: Braque, O'Conor, Cross, Delaunay, Derain, Gilman, Gore, Kandinsky, Matisse, Signac, Vallotton, Vuillard
June	*Bernheim Jeune*: Cézanne watercolours
October–November	*Salon d'Automne*: Braque, Cézanne (retrospective), Van Dongen, Gauguin, Matisse, Rouault
December	*E. Blot*: Cézanne, Van Gogh, Matisse, Picasso
December–January 1908	*Bernheim Jeune*: Portraits d'hommes: Cézanne, Van Dongen, Gauguin, Van Gogh, Matisse, Seurat, Signac, Sickert, Vallotton, Valtat, Vuillard

1908

GREAT BRITAIN

Foundation of the Allied Artists' Association
Meier-Graefe, *Art*, English translation
Wyndham Lewis returns to England from the Continent

January–February	*International Society*: Bonnard, Cézanne, Cross, Degas, Denis, Gauguin, Van Gogh, Matisse, Monet, Signac, Vallotton, Vuillard
June–July	*Baillie Gallery*: Friday Club: Bell, Innes, Leech, Pissarro, Renoir
July	*Albert Hall*: 1st Allied Artists' Exhibition: Bell, Bevan, O'Conor, Fergusson, Gilman, Ginner, Gore, L. Pissarro, Sickert, Steer

FRANCE

Braque's style christened as 'cubism' in November
Matisse, 'Notes d'un peintre', *La Grande Revue*, December

January	*Bernheim Jeune*: Van Gogh
	Druet: Van Gogh
March–May	*Indépendants*: Braque, O'Conor, Cross, Derain, Gilman, Gore, Kandinsky, Luce, Severini, Sickert, Signac, Vallotton, Vlaminck.
October–November	*Salon d'Automne*: Bonnard, Bussy, O'Conor, Denis, Derain, Fergusson, Friesz, Matisse, Metzinger, Rice, Rouault, Sands, Sickert, Vallotton, Valtat, Vlaminck, Weber
December–January 1909	*Druet*: Bonnard, Cézanne, Cross, Denis, Gauguin, Van Gogh, Seurat, Signac, Vallotton
December–January 1909	*Galerie Notre-dame-des-champs*: Braque, Derain, Picasso

1909

GREAT BRITAIN

April	*Carfax Gallery*: Fry
July	*AAA*: Bevan, Fergusson, Gilman, Gore, John, Kandinsky, Sickert
	International Society: Denis, Van Gogh

FRANCE

Duncan Grant meets Matisse and Picasso

April	*Indépendants*: Braque, Cross, Denis, Derain, Van Dongen, Friesz, Lhote, Luce, Matisse, Metzinger, Rouault, Signac, Vallotton, Valtat, Vlaminck
October	*Salon d'Automne*: O'Conor, Desvallières, Van Dongen, Fergusson, Laprade, Manguin, Marquet, Matisse, Puy, Rice, Sickert
November	*Druet*: Van Gogh

1910

GREAT BRITAIN

T. Duret, *Manet and the French Impressionists*, English translation
James Huneker, *Promenades of an Impressionist*
Futurist Manifesto published in the *Tramp*, August
Frank Rutter, *Revolution in Art*
C.J. Holmes, *Notes on the Post-Impressionist Painters*

April	Marinetti's first visit to England
April–May	*International Society*: Denis, Peploe, Shannon, Vallotton, Vuillard
June	*Alpine Gallery*: Friday Club: Bell, Grant
June–August	*Brighton Public Art Galleries*: Modern French Artists: Cézanne, Cross, Denis, Derain, Gauguin, Luce, Martin, Matisse, Signac, Vallotton, Valtat, Vlaminck, Vuillard
July	*AAA*: Bevan, Fergusson, Gilman, Ginner, Gore, Kandinsky, Innes, Steer
November– January 1911	*Grafton Galleries*: Manet and the Post Impressionists: Cézanne, Denis, Derain, Gauguin, Van Gogh, Manguin, Maillol, Manet, Marquet, Matisse, Picasso, Redon, Rouault, Sérusier, Seurat, Signac, Vallotton, Valtat, Vlaminck

FRANCE

May	*Vollard*: Gauguin
April	*Indépendants*: Van Dongen, Cross, Friesz, Luce, Marquet, Laprade, Pissarro, Van Rysselberghe, Signac
October– November	*Salon d'Automne*: Bonnard, Denis, Derain, Van Dongen, Fergusson, Kandinsky, Matisse, Manguin, Metzinger, Picasso, Rice, Vallotton, Valtat, Vlaminck, Vuillard

GREAT BRITAIN

C.L. Hind, *The Post Impressionists*
Roger Fry offered directorship of the Tate Gallery

January	*Grafton Galleries*: Epstein
January– February	*Chenil Gallery*: Gill, Innes
February	*Alpine Gallery*: Friday Club: Bell, Fry, Gertler, Grant, Leech
April–May	*International Society*: Bonnard, Denis, Bussy
March	*Women's International Art Club*: Rice, Wright
May–June	*Carfax Gallery*: Camden Town Group, Ist exh.: Bevan, Gilman, Ginner, Gore, L. Pissarro, Sickert
July	*AAA*: Bevan, Fergusson, Gilman, Ginner, Gore, Kandinsky, Nevinson, Sickert, Wright
	Esperantist Vagabond Club: Bevan, Fergusson, Ginner, Gilman, Gore
Summer	First issue of *Rhythm*, ed. John Middleton Murry and Katherine Mansfield. Art editor: J.D. Fergusson
Autumn	*Borough Polytechnic Murals*: Adeney, Fry, Etchells, Gill, Grant, Rutherston
November	*Stafford Gallery*: Cézanne and Gauguin
	Goupil Gallery Salon: Fergusson, John, Peploe, Wolmark
November	Reproductions of cubist pictures by Picasso and Herbin in the *New Age*
December	*Carfax Gallery*: Camden Town Group: Bevan, Gilman, Ginner, Gore, Grant, Innes, Lamb, L. Pissarro, Sickert

FRANCE

June	*Indépendants*: Van Dongen, Denis, Friesz, Herbin, Lhote, Matisse
October	*Salon d'Automne*: Van Dongen, Doucet, Friesz, Fergusson, Gleizes, Léger, Manguin, Matisse, Marquet, Metzinger, Picabia, Puy, Rouault, Sands, De Segonzac, Valtat, Vlaminck, Wolmark

1912

GREAT BRITAIN

The Letters of a Post-Impressionist: being the Familiar Correspondence of Vincent Van Gogh, ed. A.M. Ludovici

January	*Stafford Gallery*: Fergusson
	Goupil Gallery: Denis, Desvallières, Sérusier
February	*Alpine Gallery*: Friday Club: Bell, Gertler, Grant, Etchells, Wadsworth
March	*Sackville Gallery*: Italian Futurist Painters: Boccioni, Carrà, Russolo, Severini
April	*Stafford Gallery*: Picasso
	Goupil Gallery: Wolmark
April–May	*International Society*: Bonnard, Denis, Gauguin, Van Gogh, Vuillard
July	*AAA*: Adeney, Bell, Bevan, Dismorr, Epstein, Fergusson, Fry, Gilman, Ginner, Gore, Kandinsky, Leech, Lewis
Summer	Opening of Madame Strinberg's *Cave of the Golden Calf*
October	*Stafford Gallery*: Dismorr, Fergusson
October–January 1913	*Grafton Galleries*: 2nd Post-Impressionist Exhibition: Bell, Bonnard, Braque, Cézanne, Derain, Etchells, Fry, Gill, Gore, Grant, Lamb, Lewis, Matisse, Picasso, Spencer, Vlaminck

December   *Carfax Gallery*: Camden Town Group:
Bevan, Gilman, Gore, Lamb, Lewis, L.
Pissarro, Sickert

FRANCE

Maurice Denis, *Théories*

**February–
March**   *Bernheim Jeune*: Peintres futuristes italiens:
Balla, Boccioni, Carrà, Russolo, Severini
April   *Indépendants*: Delaunay, Gleizes, Laprade,
Lhote, Luce, Manguin, Marquet, Metzin-
ger, Rouault, De Segonzac, Signac,
Vlaminck
May   *Barbazagnes*: Quelques artistes indépendants
anglais: Bell, Etchells, Fry, Ginner, Gore,
Grant, Lewis
October   *Salon d'Automne*: Cézanne, Fergusson, Friesz,
Gleizes, Laprade, Matisse, Metzinger, Puy,
Vallotton, Vlaminck

1913

GREAT BRITAIN

Gleizes and Metzinger, *Cubism*

January   *Alpine Club*: The Grafton Group: Bell, Et-
chells, Fry, Grant, Kandinsky, Lewis,
Weber
Carfax Gallery: Gore, Gilman
March   *Friday Club*: Bomberg, Gertler, Hamilton,
Rutherston
April   *Marborough Gallery*: Severini
Opening of *Omega Workshops*
April–May   *Goupil Gallery*: Martin
Baillie Gallery: Rice

499

July	*Doré Gallery*: Post-Impressionist Posters
	AAA: Kandinsky, Lewis, Rice, Sands, Wadsworth
October	Lewis, Etchells, Hamilton, and Wadsworth secede from the *Omega Workshops*
October–November	*Chenil Gallery*: Von Anrep
November	*Goupil Gallery Salon*: Bevan, Gilman, Ginner, Gore, Innes, Steer
	Doré Gallery: Post-Impressionists and Futurist Exhibition: Lecture by Marinetti (Dec.): Asselin, Bonnard, Bevan, Cézanne, Cross, Delaunay, Derain, Doucet, Epstein, Etchells, Fergusson, Freisz, Gauguin, Gore, Gilman, Ginner, Van Gogh, Herbin, Lewis, Manson, Marquet, Matisse, Nevinson, Picasso, C. Pissarro, L. Pissarro, Rice, Signac, De Segonzac, Severini, Sickert, Vuillard, Wadsworth, Wolmark
December–January 1914	*Brighton Public Art Galleries*: English Post-Impressionists, Cubists and Others: Etchells, Bomberg, Epstein, Hamilton, Wadsworth, Lewis, Nevinson

FRANCE

April	*Indépendants*: Van Dongen, De la Fresnaye, Gleizes, Lhote, Metzinger, Signac
December	*Salon d'Automne*: Van Dongen, Gleizes, De Segonzac

GREAT BRITAIN

Clive Bell, *Art*

March   Formation of *Rebel Arts Centre*
Goupil Gallery: Ist exhibition of the London
Group: Adeney, Bayes, Bevan, Bomberg,
Epstein, Fergusson, Gilman, Ginner, Gore,
Hudson, Lewis, Manson, Nevinson, Pep-
loe, Ratcliffe, Roberts, Sands, Wadsworth

Suggestions for Further Reading

Baron, Wendy, *The Camden Town Group*, London: Scolar Press, 1979.

Bell, Clive, *Art* (1914), ed. J.B. Bullen, Oxford University Press, 1987.

Bell, Clive, 'How England Met Modern Art', *Art News*, October 1950, 24–7.

Cooper, Douglas, *The Courtauld Collection*, London: Athlone Press, 1954.

Cork, Richard, *Vorticism and Abstract Art in the First Machine Age*, 2 vols, London: Gordon Fraser, 1976.

Falkenheim, Jacqueline V., *Roger Fry and the Beginnings of Formalist Art Criticism*, Ann Arbor: University of Michigan Press, 1980.

Flint, Kate, ed., *Impressionists in England: The Critical Reception*, London, Boston, Melbourne and Henley: Routledge & Kegan Paul, 1984.

Fry, Roger, *Vision and Design*, ed. J.B. Bullen, Oxford University Press, 1981.

Harrison, Charles, *English Art and English Modernism 1900–1939*, London and Bloomington: Penguin Books and Indiana University Press, 1981.

Ingamells, John, 'Cézanne in England 1910–1930', *British Journal of Æsthetics*, 1955, viii, 341–50.

MacCarthy, Desmond, 'The Art Quake of 1910', *Listener*, 1 February 1945, 123–4 and 129.

Nicolson, Benedict, 'Roger Fry and Post-Impressionism', *Burlington Magazine*, January 1951, xciii, 11–15.

Post-Impressionism: Crosscurrents in European Painting, catalogue, Royal Academy of Arts, London, Weidenfeld & Nicolson, 1979.

Rewald, John, *Post-Impressionism from Van Gogh to Gauguin*, New York: Museum of Modern Art, 3rd edn, 1976.

Rothenstein, Sir John K.M., *Modern English Painters*, 2 vols, London: Eyre & Spottiswoode, 1952, 1956.

Rutter, Frank, *Art in My Time*, London: Rich & Cowan, 1933.

Shone, Richard, *Bloomsbury Portraits*, London: Phaidon Press, 1976.

Shone, Richard, *The Post-Impressionists*, London: Octopus Books, 1979.

Spalding, Frances, *Roger Fry: Art and Life*, London, Toronto, Sydney and New York: Granada Publishing, 1980.

Watney, Simon, *English Post-Impressionism*, London: Studio Vista, 1980.

Index